Digital Transformation and Modernization with IBM API Connect

A practical guide to developing, deploying, and managing high-performance and secure hybrid-cloud APIs

Bryon Kataoka

James Brennan

Ashish Aggarwal

BIRMINGHAM—MUMBAI

Digital Transformation and Modernization with IBM API Connect

Copyright © 2022 Packt Publishing

Associate Group Product Manager: Alok Dhuri

Senior Editor: Ruvika Rao

Content Development Editor: Tiksha Lad

Technical Editor: Maran Fernandes

Copy Editor: Safis Editing

Project Coordinator: Deeksha Thakkar

Proofreader: Safis Editing

Indexer: Tejal Daruwale Soni

Production Designer: Vijay Kamble

First published: December 2021

Production reference: 1071221

Published by Packt Publishing Ltd.
Livery Place
35 Livery Street
Birmingham
B3 2PB, UK.

ISBN 978-1-80107-079-9

www.packt.com

To my wife, Cheryl Bertini, who has been my life partner for 36 years. The last 2 years have been difficult, but together we have thrived. To my mentors over my career, Wil Kimura, Lynn Bailey, and Peter Ling, for their career/business coaching and friendship. And finally to Dale Nilsson, who introduced me to his passion for writing books. Now you only play golf and that, my friend, will be my next passion.

– Bryon Kataoka

To my wife, Jennifer Brennan, and my three beautiful children, Emily, Patrick, and Madison, who give me the drive and support to do my best to succeed every day. They have stood by my side and encouraged me even though the demands on my time are significant at times. To my parents, who instilled a sense of pride in me, and to God, who has provided me with the ability to learn, grow, and flourish.

– James Brennan

To the heavens, for providing me with infinite growth opportunities. To my parents, for their blessings. To my wife, Nidhi, and my children, Dhruv and Devika, for their endless faith, love, and support.

– Ashish Aggarwal

Foreword

Throughout my 20+ years at IBM, I've had the pleasure of working in a wide range of roles in product development, as well as spending several years working onsite with customers as a Software Services Consultant. A formative part of that development timeline was my role as a Technical Lead through the first five releases of IBM API Connect from 2012 until I moved to another role in the Integration space at the beginning of 2017. I then returned to the API Connect space in 2020 as the Senior Technical Staff Member for the Cloud Pak for Integration, of which API Connect is a key part.

There is a consistent theme across all the product releases and client engagements over that time. This theme is the appetite and enthusiasm our customers have for content that goes beyond the typical task-based description we provide in the formal product documentation to illustrate not just the "what" but also the "why" of a product. By incorporating practical experience and lessons learned from real-world project scenarios we significantly enhance the value of the information we provide. I saw this personally in 2016 when I wrote the first API Connect Deployment Whitepaper, which presented practical guidance on high availability production deployment scenarios, based on my experiences with customers and leadership of the API Connect SaaS service. The emphatically positive reaction to that publication cemented in my mind the value of practitioner-driven product content.

Fast-forward five years and that's why I'm excited to introduce you to this major work from Bryon, Jim, and Ashish. Their combined track record of successfully delivering API Connect and DataPower gateway solutions to customers makes them the ideal candidates to codify their knowledge so that everyone can benefit from it! Between them, they have decades of experience and have led implementation projects across the spectrum of customers and industries, and you will see the fruits of that labor permeate through these pages.

As a reader, you'll find this book provides a single reference that covers the entire breadth of the API Connect product. It begins with the business context in which the product should be used, before progressing through the basic introductory concepts. Having established these foundations, the majority of the book then provides detailed and easily reusable examples of the powerful features that will enable you to be successful in your API journey.

My current role as CTO for the IBM Integration portfolio gives me broad visibility across the Integration space, and it is examples like these where our partners and practitioners share their experience that makes me truly excited to be a part of the IBM ecosystem. It's my sincere hope that you find this book a highly-valuable companion to your use of API Connect and that it enables you to accelerate the successful delivery of your digital transformation projects!

Matt Roberts

Chief Technology Officer, IBM Integration

Winchester, UK

Contributors

About the authors

As Trusted Advisor and CTO of the iSOA Group, **Bryon Kataoka** has established a reputation as a leader in architecture and technology, focused around IBM API Connect, Red Hat OpenShift, and DataPower gateways. Bryon has developed his style of professionalism, where education and knowledge is the driving innovative force to success. Bryon participates in IBM certification exams, has presented at conferences, and has authored *The WebSphere Application Server Bible*. He is a father of four, a home winemaker with his own vineyard, a decent golfer, and practices blues guitar.

As an owner and lead consultant of McIndi Solutions, **James Brennan** is a recognized leader in the DataPower and IBM API Connect space. Jim has participated in developing IBM certification exams, has spoken at several IBM internal and customer-facing conferences, and is a coauthor of the *IBM WebSphere DataPower SOA Appliance Handbook*, as well as the individual IBM DataPower volumes available. He is an accomplished musician with a prior career in musical performance, a father of three, and a fitness enthusiast.

As the CEO and Technology Advisor of ANDD Techservices (USA and India), **Ashish Aggarwal** leads complex application modernization, integration, and API/SOA enablement projects. He has helped numerous industries, including telecoms, E&U, e-tailers, BFSI, and animal disease management implement API/SOA-oriented solutions. He has extensive architectural, operational, and development experience in multiple IBM integration products; for example, APIC, DataPower, ACE/IIB, and MQ. Ashish is a graduate of Harvard Business School and Delhi University.

About the reviewers

Sudhir Mohith is a Technology Partner Architect at IBM with 25 years of experience in the computer industry, working with business partners and clients all over the world on complex enterprise-scale solutions. Outside of work, he devotes his time to his two children and enjoys DIY, hiking, and traveling.

Ricky Moorhouse is a Cloud Architect for the IBM API Connect cloud service with 16 years of experience working with APIs and automation. He has worked at IBM for over 20 years and has been part of the API Connect team since the product's inception, primarily leading the team that operates and manages the hosted API Connect services in IBM Cloud.

Table of Contents

Preface

Section 1: Digital Transformation and API Connect

1

Digital Transformation and Modernization with API Connect

API-led digital transformation	4
A journey back in time	4
Fast forward to the 21st century	6
Business transformation to a digital transformation framework	9
Digital framework considerations	11
Hybrid reference architecture	14

| API-led architectural approach | 15 |
| API flavors | 15 |

Your responsibilities for APIs	18
Digital modernization and APIs	18
Approaches to modernization	19
API Connect enabling digital transformation	20
API Connect aligns with the goals of digital transformation	21
Summary	26

2

Introducing API Connect

Technical requirements	28
API Connect	28
The components of API Connect	31
Deployment models	48
The on-premises implementation	49
Cloud implementations	50

Hybrid cloud	53
HA	54
Introduction to the CLI commands	56
Having fun playing with FHIR	58
Summary	66

3

Setting Up and Getting Organized

Technical requirements	68	Utilizing Spaces	80
The big picture	68	Configuring your spaces	81
Provider Organizations	70	Developer Portal	84
Configuring your Provider Organization	70	Configuring the Developer Portal	85
API Catalogs	74	Summary	90
Configuring your API Catalog	75		

Section 2: Agility in Development

4

API Creation

Technical requirements	94	What is an OpenAPI design?	104
Development tools	95	Creating an API Proxy	107
Installing Designer	96	Testing APIs	112
Installing the LTE	97	Using variables	115
Connecting Designer to the LTE	102	Adding policies	119
Creating APIs	104	Summary	151

5

Modernizing SOAP Services

Technical requirements	154	Create a REST proxy that invokes a SOAP service	166
SOAP capabilities in APIC	156		
Creating a SOAP proxy that invokes a SOAP service	159	Creating a REST proxy	166
		Review the REST proxy configuration	168
Creating a SOAP proxy	160	Testing the REST Proxy	171
Review SOAP Proxy configuration	161	Summary	174
Testing the SOAP Proxy	164		

6

Supporting FHIR REST Services

Technical requirements	176	Playing with FHIR	182
Introducing FHIR	176	**Applying logic policies**	
FHIR resources	177	**to your FHIR API**	**193**
FHIR server	179	The If and Switch logic policy	194
Government-mandated		The operation switch logic policy	200
FHIR interfaces	180		
Creating a RESTful FHIR API	181	**Summary**	**219**

7

Securing APIs

Technical requirements	222	Creating a client	242
Out-of-the-box security		Testing OAuth flow	245
capabilities of APIC	**223**	**Implementing OpenId**	
Preparing for the APIC security		**Connect (OIDC)**	**251**
implementation	224	OAuth provider changes	252
Protecting APIs with Basic		OAuth flow changes	255
authentication and Client ID		**Using JWT policies**	**258**
(API key)	**234**	JWT generation	259
Implementing Basic authentication		JWT verification	264
and ClientID in API security	234	**Adding additional**	
Applying OAuth 2.0	**239**	**security measures**	**269**
Enabling API with an OAuth		**Summary**	**270**
security definition	239		

8

Message Transformations

Technical requirements	272	Using a Map policy	273
Introduction to API		Redacting fields	283
Connect pre-built		Applying JSON to XML	
transformation policies	272	or XML to JSON policies	287

Implementing advanced transformations with XSLT and GatewayScript   292

XSLT   292
GatewayScript   297

Summary   301

9
Building a GraphQL API

Technical requirements   304
Why GraphQL?   304
GraphQL anatomy   306
Installing a GraphQL Express server   309

Creating a GraphQL API   311
Adding the GraphQL proxy   312
Addressing warnings in GraphQL   319

Addressing the warnings   319

Setting weights, costs, and rate limits   321
Considerations on performance   322

Removing fields from GraphQL   327
Summary   330

10
Publishing Options

Technical requirements   332
Working with Products and Plans   332
Configuring Products and Plans   334

Creating Rate Limits   339
Defining Rate Limits in a Plan
(consumer Rate Limits)   340

Defining Rate Limits in an assembly   346

Publishing and configuring Catalogs   356
Publishing to your Catalogs   359

Consumer interaction   362
Summary   371

11
API Management and Governance

Technical requirements   374
Understanding the API and product lifecycle   374
The "what" of the lifecycle   375
The "how" of the lifecycle   376

Roles   386

The "who" of the lifecycle   386
Cloud Manager roles   387
API Manager roles   389
Developer portal roles   392

Managing versions   395
Version numbering scheme   395

Segregating environments   400
Logical segregation   400
Physical segregation   404

Applying standards to your API

environment   406
Discoverability   406
Naming standards   407

Summary   409

12
User-Defined Policies

Technical requirements   412

Understanding user-defined
policy types   412

Creating catalog scoped
user-defined policies   413
Building your UDP .yaml file   414
Completed .yaml   418
Installing your UDP   419

Implementing your UDP   421

Global scoped
user-defined policy   422
Planning your global UDP   422
Creating a UDP configuration file 426
Packaging and publishing
your global UDP   430

Summary   434

Section 3:
DevOps Pipelines and What's Next

13
Using Test and Monitor for Unit Tests

Technical requirements   438
Configuring unit tests   438
Working with environments   451

Monitoring the test cases   454

Summary   457

14
Building Pipelines for API Connect

Technical requirements   460
Introducing pipelines   461
Choosing between CLIs or
Platform API Interface use   462

Using the CLIs   462
Using the Platform APIs   463
Calling the publish Platform API 466
Understanding Ansible automation 467

Testing and monitoring hooks   470
Using Git as your SCM   477
Constructing the
Jenkins pipeline   479
Jenkins in a nutshell   480

Exploring the scripted and
declarative methods   481

Working on an API
Connect sample pipeline   483
Jenkins housekeeping   483
Building a Pipeline item   485

Summary   491

15
API Analytics and the Developer Portal

Customizing the
Developer Portal   494
Reviewing Drupal 9 capabilities   495
Customizing the Portal
as the administrator   498

Introducing Analytics   504
Understanding the analytics
initial setup   505
Viewing analytics   508

Understanding dashboards   510

Creating visualizations
and dashboards   516
Creating a new dashboard   517
Creating a new visualization   519
Deploying analytics
to new environments   527

Summary   528

16
What's Next in Digital Transformation Post-COVID?

Technical requirements   533
Understanding API Connect
late-breaking changes   533
Understanding API Connect
and Hybrid Cloud   535
OpenShift   535
Cloud Pak for Integration (CP4I)   539

Understanding Artificial
Intelligence for
IT Operation (AIOps)   542
The role of AIOps   543

Exploring 5G Edge computing   544
Advantage of 5G Edge computing   545

Summary   547

Other Books You May Enjoy

Index

Preface

We, as authors, all agree that the reason for this book is to share our API Connect experience, as consultants with multiple clients, to guide new clients with the benefits of our knowledge. All of us agree that when confronted with new digital transformation challenges, we all try our best to transfer knowledge to those who have not had the experience we have had.

We embarked on this journey with mutual friendship and comradery in helping our users and customers, making our best effort to ensure they are successful. Your success is our success. Obviously, today's rapid expansion in delivering new features and functions instills a tremendous demand for rapid performance. It is our hope that this book provides the energy to deliver digital transformation in your organization.

Who this book is for

This book is for the architect or developer who has embarked on a digital transformation journey. Your background in web services and API development reinforces the need for API management. You know that building APIs quickly and making them discoverable is critical to achieving your corporate initiatives. This book will allow you to understand how IBM API Connect will enable your API management vision.

What this book covers

Chapter 1, *Digital Transformation and Modernization with API Connect*, looks at digital transformation and modernization – two adaptations that organizations are moving towards to help them be more effective. API integration plays an important role in the success of these initiatives.

Chapter 2, *Introducing API Connect*, introduces API Connect and its associated components. You'll learn about the various deployment models and take a deeper dive into the associated components, such as the Developer Portal, the Gateway Service, and Analytics. You will also learn about the management capabilities provided by the CLI commands.

Chapter 3, Setting Up and Getting Organized, looks at how, before starting to build APIs there is some housekeeping work that should be accomplished to ensure your digital transformation follows a framework for success. This means you address various factors, such as who is providing the APIs, how you will address consumer expectations, and setting up for operational efficiency. This chapter will help you get started on the correct footing.

Chapter 4, API Creation, helps you learn how to use API Connect to develop APIs. You will discover how easy it is to navigate through and immediately publish new APIs.

Chapter 5, Modernizing SOAP Services, addresses the fact that many organizations still support many SOAP web services. These valuable assets still perform necessary business functions and there are often external consumers who need access to them. API Connect has the ability to incorporate these services so they can be managed. In this chapter, you will learn how to add SOAP services as APIs.

Chapter 6, Supporting FHIR REST Services, looks at **FHIR** (short for **Fast Healthcare Interoperability Resources**), an HL7 standard that supports providing resources using REST APIs and JSON. This chapter introduces how to develop REST APIs that support the FHIR specification.

Chapter 7, Securing APIs, addresses the fact that APIs need to be secured. In this chapter, you will learn how to secure your APIs with basic authentication, OAuth2/OpenID, and JWT. You also will learn how to configure LDAP and authentication URLs as user registries.

Chapter 8, Message Transformations, introduces methods of transforming payloads, headers, and variables of APIs. We will discuss XSLT and GatewayScript transformations as well as JSON to XML, XML to JSON, Redaction, Parse, and Map policies.

Chapter 9, Building a GraphQL API, looks at how managing multiple versions of APIs can complicate your API strategy. GraphQL is one option to address the concern. In this chapter, we introduce GraphQL and how API Connect provides safeguards to GraphQL APIs.

Chapter 10, Publishing Options, explores how to customize your deployment to an environment where users can access your APIs. Since it's all about the consumer experience, this chapter will show you how to publish APIs that are easy to discover, well documented, and mindful of overages.

Chapter 11, API Management and Governance, delves into how you go about managing your APIs. You will learn about how consumers access your APIs and how versioning has an impact on prior releases. You will also learn how other roles can be used to limit what organizations can perform.

Chapter 12, User-Defined Policies, covers how API Connect provides the ability to create your own custom policies. These become reusable components that you can drag onto the Gateway Policy editor. Often, this is created to enforce a security requirement but it can be used to create additional helper functionality that doesn't exist out of the box.

Chapter 13, Using Test and Monitor for Unit Tests, API Connect Test and Monitor is an add-on feature of API Connect. You can use this facility to generate on-the-fly unit tests for your API deployments that can be rerun anytime.

Chapter 14, Building Pipelines for API Connect, customizes the API Connect Developer Portal with new themes as well as applying additional functionality provided by the Drupal content manager. As you know, digital transformation puts an emphasis on user experience and social interaction. Learning how to customize the Developer Portal is one of the keys to success. Also, in this chapter, you will learn how to customize the analytics of your APIs. You can create new dashboards and visualizations and even provide analytics to your consumers' applications.

Chapter 15, API Analytics and the Developer Portal, looks at how managing multiple versions of APIs can complicate your API strategy. GraphQL is one option to address this concern. In this chapter, we introduce GraphQL and how API Connect provides safeguards to GraphQL APIs.

Chapter 16, What's Next in Digital Transformation Post-COVID?, informs you of any new changes coming to API Connect or any late-breaking news on additional fix packs and/or updates. As you know, digital transformation doesn't stop with APIs. As new capabilities become available, some of these capabilities may fit well within your organization.

To get the most out of this book

You must understand that API Connect is not open source. You will need to have API Connect available to you. You can procure it in IBM Cloud or perhaps your employer has purchased a license of API Connect Enterprise. Regardless, you will need to access API Connect on the cloud or within your organization. While the book will guide you without a licensed API Connect implementation system, not having one will hamper your experience. The book is based on v10.0.1.5.

Software/hardware covered in the book	Operating system requirements
IBM API Connect version 10	Windows, macOS, or Linux

If you are using the digital version of this book, we advise you to type the code yourself or access the code from the book's GitHub repository (a link is available in the next section). Doing so will help you avoid any potential errors related to the copying and pasting of code.

Download the example code files

You can download the example code files for this book from GitHub at `https://github.com/PacktPublishing/Digital-Transformation-and-Modernization-with-IBM-API-Connect`. If there's an update to the code, it will be updated in the GitHub repository.

Download the color images

We also provide a PDF file that has color images of the screenshots and diagrams used in this book. You can download it here:

`https://static.packt-cdn.com/downloads/9781801070799_ColorImages.pdf`

Conventions used

There are a number of text conventions used throughout this book.

`Code in text`: Indicates code words in text, host names, userids, catalogs, and organizations. Here is an example: "To run the actual tests, you will use the same URL, `Key`, and `Secret` and just change the path from `/test` to `/test/run`."

A parameter block is set as follows:

```
<api manager host> = api-manager-ui.apicisoa.com
<userid> = isoadeveloper
<catalog> = sandbox
<organization> = middleware
```

Any command-line input or output is written as follows:

```
curl -v -k -X POST
  -F "product=@/home/[user]/jenkins/workspace/basic-product_1.0
.0.yaml;type=application/yaml"
  -F "openapi=@/home/[user]/jenkins/workspace/basic-
test_1.0.0.yaml;type=application/yaml"
```

```
-H "Authorization: Bearer [GENERATED TOKEN]"
-H 'Accept: application/json'
 https://[apimserver]/api/catalogs/[organization name]/
[catalog]/publish
```

Bold: Indicates a new term, an important word, or words that you see onscreen. For instance, words in menus or dialog boxes appear in **bold**. Here is an example: "Select **System info** from the **Administration** panel."

Tips or important notes
Appear like this.

Get in touch

Feedback from our readers is always welcome.

General feedback: If you have questions about any aspect of this book, email us at customercare@packtpub.com and mention the book title in the subject of your message.

Errata: Although we have taken every care to ensure the accuracy of our content, mistakes do happen. If you have found a mistake in this book, we would be grateful if you would report this to us. Please visit www.packtpub.com/support/errata and fill in the form.

Piracy: If you come across any illegal copies of our works in any form on the internet, we would be grateful if you would provide us with the location address or website name. Please contact us at copyright@packt.com with a link to the material.

If you are interested in becoming an author: If there is a topic that you have expertise in and you are interested in either writing or contributing to a book, please visit authors.packtpub.com.

Share Your Thoughts

Once you've read *Digital Transformation and Modernization with IBM API Connect*, we'd love to hear your thoughts! Scan the QR code below to go straight to the Amazon review page for this book and share your feedback.

https://packt.link/r/1801070792

Your review is important to us and the tech community and will help us make sure we're delivering excellent quality content.

Section 1: Digital Transformation and API Connect

Organizations are adopting digital transformation to transform their business with APIs. This section will show you how to use API Connect's capabilities to achieve those goals.

This section comprises the following chapters:

- *Chapter 1, Digital Transformation and Modernization with API Connect*
- *Chapter 2, Introducing API Connect*
- *Chapter 3, Setting Up and Getting Organized*

1
Digital Transformation and Modernization with API Connect

Digital transformation and modernization are two adaptations that organizations are moving toward to help their organizations be more effective. API integration plays an important role in the success of these initiatives.

In this chapter, you will be introduced to digital transformation and how your role, as a solution architect or developer, plays an active part in achieving the benefits of digital transformation. You will understand the motivation and history behind what is driving this transformation and modernization. You'll also be introduced to the benefits of API-led architectures and strategies. With this understanding, you will be able to apply IBM API Connect capabilities that map to the transformation journey.

In this chapter, we're going to cover the following main topics:

- API-led digital transformation
- Digital modernization and APIs
- API Connect enabling digital transformation

By the end of this chapter, you will have a good conceptional view of the reasons why your company is adopting digital transformation and how your efforts in using API Connect will help.

API-led digital transformation

Digital transformation and digital modernization are both common in business discussions and literature and are often being morphed into a marketing drumbeat. What does this mean? Is it just the latest buzzword? How does it fit into the project you are just starting or have been asked to architect a solution for? As an architect, project/product manager, or developer, having a good idea of what your end goal is and how you are participating in the eventual vision should be important.

API-led digital transformation is how you achieve connectivity to support your digital transformation, but before we dive deeper into the *hows* and *whys*, it's important to look at, historically, the business reasons that make this endeavor important.

A journey back in time

To gain a better grasp of the benefits of digital transformation, let's take a journey back in time. Imagine that you owned a family business where you make blankets. Like any business, you required the necessary raw materials (cotton or wool) to perform the painstaking weaving to produce your product. You had to reflect that in the price. Too high and it wouldn't sell, too low and you ran out of goods and it took you considerable time to create them. You hung your shingle outside to advertise your craft so townspeople would walk to your place of business and inquire about your goods.

Being a crafty business person, you also tried to create as many blankets as possible and have them transported to other towns or ships to sell on street markets or to other stores. Every evening you would sit down and review your books to count your sales, determine how many blankets you could complete, ensure you had enough raw materials to create the blankets, and write letters to other business owners to see if they would like to buy some of your blankets. You often wondered that if you had a storefront in the town square, whether that would increase your sales. In the 18th century, digits referred to fingers and toes, which you perhaps used to count your sales!

Luckily for you, you were living in the 18th century and it was the beginning of the Industrial Revolution. The Industrial Revolution was a period of innovation that transformed largely rural, agricultural societies in Europe and America into industrialized, more urban societies:

Figure 1.1 – Textiles of the Industrial Revolution, by Illustrator T. Allom – History of the cotton manufacture in Great Britain by Sir Edward Baines. Public domain: https://commons.wikimedia.org/w/index.php?curid=9430141

It was a time where inventions/technologies that were adopted by businesses provided improved business efficiencies/values and changed people's expectations on the availability of goods.

You found that by purchasing a Spinning Jenny, which made weaving faster, you saw work shift from family-led home production to factory production. Later, you were able to buy faster weaving equipment that ran on steam power and your output increased drastically.

Of course, this changed how you thought about your business. You had greater productivity and you needed to transport your product more efficiently. Horse and cart wouldn't be as effective as a locomotive, so you started shipping by train. Being the only blanket maker in town, you increased your storage space and people from all over flocked to your store. Given you were more efficient in making your blankets, you were able to garner better deals for increased raw materials and passed that on to consumers at lower costs. You even hired an accountant to keep track of your books (because you ran out of digits). You were a success and ahead of your time.

Fast forward to the 21ˢᵗ century

As we fast forward to the 21ˢᵗ century, we can see how the Industrial Revolution changed businesses (even our fictional ones). These factories could employ hundreds and even thousands of workers who produced mass batches of blankets more cheaply than they could be produced in homes.

So, how does that all relate to digital transformation, and where does API-led digital transformation come into play? We'll answer this question with some interesting correlations in the following sections.

Marketing mix – 4Ps

With the Industrial Revolution, it was inventions that drove increased productivity. Depending on the factory's location, their products were limited to transporting the product and finding a place to sell it. Getting the word out about their product was done through storefront advertising so that growth was steady.

As time passed on, the Industrial Revolution continued to bring improvement. Transportation improved, telephones became prevalent, and even typewriters came to light. Eventually, along came the internet, and the possibilities exploded. So, what did these revolutions have in common and why is digital transformation key? It all revolves around marketing. Digital transformation has its roots in marketing.

In marketing, there is the concept of 4Ps:

- **Product**: This refers to goods or services that a company offers to customers. A product should be something that satisfies a customer's need, want, or desire. A product could also be a new invention that generates a demand to have one. It's important to note that a product has a shelf life. It may go through various cycles of reinvention to maintain demand. In subsequent chapters, you will learn how the product in API Connect is directly related.

- **Price**: Price is the cost consumers will pay for your product. This will be based on several factors, including the cost of the materials, how quickly the product can be produced, and how to transport the product. Product managers must determine the price of the product's monetized value, but they may also consider various plans to support valued customers or try and buy scenarios. When you learn more about API Connect, you will learn how to monetize your API products.

- **Place**: How you market your product is critical when determining where to place your products. In the traditional sense, you often place your products where customers can see them. In a brick-and-mortar establishment, this placement can be near a checkout display or with highly visible advertisements.

  More often than not, television shows, phones, kiosks, and web pages are the best way to attract attention to the product. API Connect allows you to showcase your APIs on the Consumer Developer Portal. You'll learn more about that later.

- **Promotion**: Promotion is your chosen method of publicly advertising your product. This promotional strategy can be accomplished in various ways. The goal of promoting your product is to show the value of your product to consumers or business entities.

> **Note**
>
> Place and promotion are somewhat interconnected/dependent. In today's environment, most promotion is online. You are probably not surprised that most promotion is done with social media (digital word-of-mouth).

In our fictional blanket entrepreneur example, you probably recall where the 4Ps were important. For instance, the product was the blankets. You saw that the price of the blankets was contingent on the cost of raw materials and the cost of creating the blankets. The store location was important as you wanted as many potential buyers to purchase your blankets. Other than breadboards and your shingle hanging outside your home, you didn't have customers from outside of your town, but you were successful because you were the only store in town.

As you review the company today, there have been considerable improvements over the centuries that have been propelled by the second and third Industrial Revolutions. Inventions such as the telephone, engines, automobiles/trucks, and machine tools from the second Industrial Revolution greatly improved business ability. The advent of computers, telecommunications, and electronics in the third Industrial Revolution further advanced business processing. Factories got bigger, stores (brick and mortar) were replicated across the country, accountants were replaced with accounting software, and marketing continued to follow the 4Ps to grow business.

Best laid plans go astray

An interesting phenomenon occurred with the age of the internet, **Service Oriented Architect (SOA)**, and Web 2.0+. As systems evolved, we have seen a great many implementations in silos. Enterprises went from centralization to decentralization and back again. SOA introduced the concept of **Enterprise Service Buses (ESBs)** and high governance (or assumed governance). Web Services APIs were introduced and Centers of Excellence and integration teams were formulated. The culmination of these factors and the siloed demarcations eventually led to hot projects oozing like lava. The 4Ps, while still applicable, were being hindered by a lack of agility and changing demographics and attitudes.

Adding more Ps to the 4Ps

As we progress toward API-led digital transformation in this chapter, you are probably aware the technology today has brought to light how people, processes, and passion (the new Ps) about products have changed the way marketing is strategized:

- **Passion**: Consumers today are passionate. They demand innovations, speed, and personalization. You may have been considered a "Best of Breed" before but in today's markets, you can go from "Hero to Zero" in a heartbeat. Just consider the passionate responses on Yelp and other social media. As a part of your digital transformation, you have to take into account how passionate your consumers are. Doing so will gain the respect and loyalty of your customers.

- **People**: Being customer-focused is imperative. The more you know about your consumer and how they view your products or services, the better. You learned about passion and how it is paramount to manage expectations. When you consider adding people to our mix, you should also consider your internal people. They are the ones responsible for achieving your marketing goals. It will be these same people who will buy into your digital transformation and participate in making the transformation holistic and broad.

- **Process**: Process is the final consideration. It will be the process that ensures the delivery of your product to the customer. There will be policies and compliance aspects that will need to be considered. Failure to execute could be catastrophic to your goals. When you learn about the features of API Connect, you will be able to incorporate these processes into the life cycle of your products. Some of these processes will enable you to be more agile and deliver more in less time. You'll learn more about this in detail in *Chapter 11, API Management and Governance.*

> **Additional Resources**
>
> Here are some good resources about Agile integration:
>
> ```
> https://www.youtube.com/watch?v=IB001-
> j8bOg&list=PL_4RxtD-BL5tYINDy6tntfYx4t6N7ulNe&ind
> ex=11
> ```
>
> ```
> https://community.ibm.com/community/user/
> integration/viewdocument/integration-
> modernization-the-journ?CommunityKey=77544459-
> 9fda-40da-ae0b-fc8c76f0ce18&tab=librarydocu
> ments&LibraryFolderKey=028cc27f-9de4-478d-a8-
> 18-f0ba67621cb2&DefaultView=folder
> ```

The concepts you've learned about in this chapter provide a basis on the key aspects that drive business. Relating these concepts to strategic and tactical digital planning is what you'll learn next.

Business transformation to a digital transformation framework

At this point, you've learned about some of the history and marketing tactics that establish reasons for the business's transformation. As you have probably recognized by now, being successful in your business transformation involves taking advantage of the digital capabilities that are a part of today's everyday life.

Let's summarize some of the key business changes that you should be aware of:

- New startups are going digital at inception.
- Customers/consumers are changing and so are their expectations. Customer loyalty is waning.
- How information and customer feedback affects your products.
- Innovations are changing rapidly. You need to surpass or be left behind.
- Delivery of goods has drastically improved, which affects pricing.
- Similar products have made innovation and pricing differences indifferent. Best of Breed is no longer a consideration.

So, how do you get started and how can you ensure you are successful? All too often, digital transformation initiatives fail.

Avoiding failure

So, how can you move thoughtfully toward transformation? Prior business models and processes may cloud your direction. If you recall the key business changes in the previous section, you should be aware of how old methods could be a roadblock to your digital goals. As an architect, it will be contingent on you to thoughtfully set goals and guidelines to ensure continuity and compliance with your strategic endeavor. It should be a holistic view that considers all the departments or lines of businesses and how each contributes to the overall digital transformation.

What you should attempt to avoid is allowing a single silo to be implemented digitally, and then another, without first establishing a digital integration plan between the two and recognizing the benefits this brings to the company. Always review alternatives before launching each new effort. However, you need to coordinate these efforts; the incremental steps should be appropriate and should follow a logical path.

Getting off to the right start should begin with establishing a digital startup checklist:

- Are all the executives on board and supportive? An executive demanding change and expediency may easily destroy the framework. It must be a sponsored endeavor.

- Consider a design thinking session where multiple parties participate and contribute their insight and experience. This should include parties from compliance/governance, marketing, content delivery, analytics, architecture and development teams, and executive input. The team should be open to change and radical ideas.

- Identify stumbling blocks and technical challenges.

- Determine how you measure success. Not all implementations by department or line of business will be measured the same.

- Take stock of your existing resources. Are they sufficiently trained? Are there business cultural barriers? Are you organized by silos? Is there a DevOps team?

- Look at yourself. Are you engaged and excited? Do you feel supported by executives?

- Are you considering frameworks to assist in your digital vision and provide guidance as you begin your journey?

You should now have a good idea of how the business ties in with digital transformation and have been given some simple guidelines on how not to fail. If failures do occur, revisit the checklist and see what you could have done better or if there was something you didn't consider.

The last item in the checklist was considering frameworks. Having an architectural picture will provide your company with some guidelines and potential solution additions to help you achieve your goals.

Digital framework considerations

So far, you have been provided a business background on why to consider various factors when improving your business. You have also been introduced to a startup checklist to begin to formalize change and improve the process so that you can measure success. You know that your customer has changed and demands more, changes often, and seeks better experiences. With all this information, establishing or adopting a digital transformation framework will help you ensure you thoughtfully consider all the aspects covering the facets of customer experience, operational improvement, and the modernization strategy.

The following table helps organize everything we have discussed into categories. These categories will transition to digital practices and implementations:

Digital Transformation Guidelines	Customers	Process and Performance	Business Case
	Social Awareness	Performance Improvements	Enterprise integration
	Customer Service	Operational Improvements such as DevOps	Digital Products
	Self Service	Work from anywhere	Blurring Silos
	Market Analytics	Sharing among organization	Transformation to Digital
	Customer Passions	Data Driven	Decision sharing
	Social Feedback	Operational transparencies	Design Thinking
	Digital Selling		Sharing digital services
	Streamline Processes		
	Multi Channel deliveries		
API-Led	UX APIs	Process/Interaction APIs	System APIs

Figure 1.2 – Categories of digital transformation

The three categories shown in the preceding table highlight tactical directions within the digital transformation adoption. We'll have a look at each in the following subsections.

Customers

One constant theme in digital transformation frameworks is the *focus on customers*. The Customers category lists items that are important in getting to know your customer's needs, desires, and ability to provide feedback (good or bad).

As you learned previously, people's passions can provide positive or negative results. So, for example, if customers would like to sign up for your application, make sure the self-service capabilities are implemented. Ensure that the approach is streamlined and proper customer service capabilities (such as reset password) are easy to perform. Perhaps consider chat sessions and/or alternative ways to access your application (mobile, iPad, kiosk, and so on) so that the customer feels important.

As you look into building digitally, utilize this category to verify you have considered and addressed those items and have communicated to the other teams how they will be accomplished.

Processes and performance

Having a focus on the customer also leads to considerations to improve processes and performance. Delivering a new digital product should include communications throughout the organization so that everyone is on board. One critical aspect of digital transformation is ensuring that changes and improvements are delivered quickly. Your customers do not want to wait for extended periods for new features or improvements. Investing in an agile delivery mechanism, as well as considering how application scaling can improve performance at peak times for seasonal events, will help you out here.

You should also consider where you would like to implement your digital solutions. Considerations should be taken into account for hybrid implementations so that you can take advantage of cloud capabilities/efficiencies such as IaaS, PaaS, and SaaS, as well as capitalize on your on-premises assets and solutions.

While digital transformation is generally about applications and solutions being executed effectively, what it really takes is people. Many of the implementations could include new technologies and platforms. Ensuring your teams are trained on new technologies and communication flows between organizations is critical.

Business cases

One potential stumbling block when adopting digital transformation is the misunderstanding that one single implementation of a digital project doesn't make your entire organization digitally transformed. A digital island is just that – an island.

To be an effective digitally transformed enterprise, you will need to be constantly focused on the holistic view. You should be thinking about the enterprise integration of digital implementations working cohesively together with common products, shared decision making, and creative design thinking. Your implemented digital services should be shared. Don't build digital islands. Focus on blurring silos and folding in and adopting existing digital successes.

As a reminder of all of the common considerations, the following word cloud diagram can be referenced at your leisure:

Figure 1.3 – Digital framework word cloud

It's time to get technical and dive into API-led frameworks.

Hybrid reference architecture

Working within a framework will give you some themes to follow when you are just starting. Taking those themes, you can now map them to an architecture that will provide you with the flexibility to implement them.

The following hybrid reference architectural diagram provides you with a holistic look at the components you can use to bring your digital solution together:

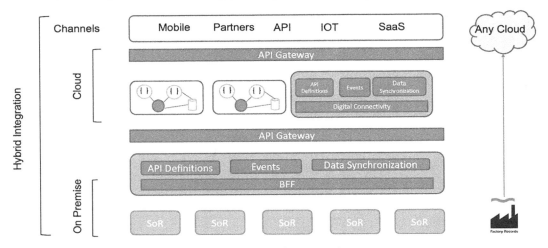

Figure 1.4 – Hybrid reference architecture

This reference diagram depicts the intersection of your on-premises applications and services and new or migrated functionality to cloud infrastructures.

The on-premises functionality (at the lowest level) would be your existing applications utilizing SOA, Java, messaging, and web services, all of which will be coordinating and managing your system of records.

As you begin extending this functionality with external partners, you begin integrating with another implementation layer that provides APIs, events, and various ways of exchanging data. This layer will operate with general-purpose APIs, or what is called **backend for frontend (BFF)** APIs. Introduced by Sam Newman, this design strategy reduces bloated services with too many responsibilities into a tightly coupled specific user experience API that is maintained by the same team as the UI.

The benefit of BFF is this API layer, which can now be implemented by your digital teams as a new service offering in the public or private clouds.

With API management, these new APIs can be exposed to various channels with proper monitoring, governance, and security. These system APIs are the workhorses within this architecture.

With the reference architecture in place, you can now explore the capabilities of **Infrastructure as a Service (IaaS)** on cloud service providers, build new applications using **Platform as a Service (PaaS)**, and incorporate and integrate with other **Software as a Service (SaaS)** solutions such as Workday, ServiceNow, and Salesforce.

API-led architectural approach

API-led, API-first, and the API economy are all mantras that bring to the forefront that APIs are key implementation strategies for digital services. These are all architectural approaches that center around APIs as the mechanism to communicate between applications and business services that result in revenue. APIs provide an invisible delivery functionality that allows applications to run across all digital channels (as noted in the *Hybrid reference architecture* section).

An API is an API is an API. Yes, in general, that is correct, but the placement and responsibilities of an API may change. Let's explore these differences.

API flavors

APIs are developed in different flavors, with each for a different target audience. When you learn about API Connect, it will be beneficial to understand the different types of APIs you may need to manage and/or interact with. We will use the hybrid reference architecture to help label the different types of APIs:

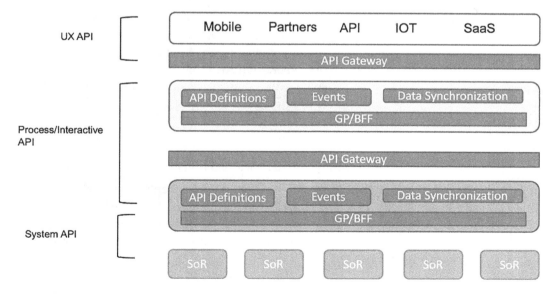

Figure 1.5 – API-led flavors

As you can see, there are four types of APIs. The following is a brief description of each:

API Type	Description
UX APIs	These are used by developers to build upon the user experience. These are single-use APIs because they are so customizable.
Process/Interactive APIs	These APIs perform actions such as securing APIs from unauthorized access, throttling/rate limiting, providing the security necessary for backend authorization, and routing. These APIs can reside on-premises or in cloud infrastructure. In API terminology, these would be created as API definitions.
System APIs	System APIs define backend service API products. They are often flows that are deployed on internal messaging systems and/or part of the SOA ESB connecting backend systems and SOA services. Other uses could be the Java WS service exposing a data access layer on top of a **system of records (SoRs)**.
General-Purpose (GP)/Backend for Frontend (BFF) APIs	These APIs are multi-use APIs. General-purpose APIs are predominately God-like APIs that support many different combinations in support of a general-purpose API. These types of APIs come with a lot of baggage. The suggested way is to use BFF APIs, which are specialized APIs that support the **user interface (UI)** and are generally coded together with the UI.

Table 1.1 – API types based on where they're utilized

Our primary focus will be developing process/interactive APIs when we are working with API Connect. Although API Connect has Node.js capabilities to create backend APIs, we will not be showing you how to code with Node.js. There are many other resources available to help you with that. You can start learning about Node.js by going to `https://nodejs.dev/learn`.

If we return to our digital framework categories, the following diagram highlights the specific points that can be accomplished using API Connect:

Customers	Process and Performance	Business Case
Social Awareness	Performance Improvements	Enterprise integration
Customer Service	Operational Improvements such as DevOps	Digital Products
Self Service	Work from anywhere	Blurring Silos
Market Analytics	Sharing among organization	Transformation to Digital
Customer Passions	Data Driven	Decision sharing
Social Feedback	Operational transparencies	Design Thinking
Digital Selling		Sharing digital services
Streamline Processes		
Multi Channel deliveries		

Figure 1.6 – API Connect supported technologies

As you can see, the table has highlighted the majority of the goals. When you look at the Customers column, you can see that you can provide consumers with the ability to subscribe to and create UX APIs/applications for multi-channel devices. API Connect provides you with the ability to engage with consumers using the Developer Portal. The Developer Portal has built-in capabilities to interact with social media applications. It lets you start forums and allows the company to add customer support (FAQ and Contact Us). As added benefits, analytics is provided to the consumers to show how well their apps are performing.

As we mentioned earlier, you can include operational capabilities (such as automatic deployment and testing), improved performance within the gateway runtimes, and the ability to share APIs between consumers and internal teams.

Your responsibilities for APIs

Now that you understand what an API-led architecture is and the types of APIs, you will have to consider the following responsibilities and address them:

- Securing APIs from unauthorized access (OAuth, JWT, and others).
- Defining security authentication/authorization for backend systems.
- Ensuring that consuming applications are routed to the appropriate API endpoint.
- Setting a rate and burst limit to limit the number of calls that are made to an API.
- Error handling with catch blocks and preventing error propagation to the backend.
- Begin working on using the API Connect capabilities for life cycle management, CLI interaction for future DevOps integration, and generating unit tests for deployment.
- Documenting your API so that consumers can quickly adopt your services.

So, you might be wondering where digital modernization fits in. You will learn about this next.

Digital modernization and APIs

You have already learned a lot about what digital transformation is. As you know already, digital transformation is the adoption of new, advanced, digital technologies, processes, people, and culture to transform your business. You have looked at ways to approach transformation and understand business benefits such as improved customer experience.

What you might be confused about is what digital modernization is. Is it the same as digital transformation? While many of the concepts are similar, the adoptions are looked at slightly differently.

Digital modernization is the practice of upgrading and/or implementing new technology, platforms (cloud or hybrid), and software innovations to meet the needs of today's organizations. Digital modernization has a focus on making infrastructure and operations improved to a point where the maintenance of systems is reduced – both financially and operationally. Legacy systems were expensive to provision and maintain and as time passed, finding resources to maintain them was difficult. IT functions are constantly under pressure to support new capabilities such as data analytics, continuous integration, and integration with SaaS applications and other state-of-the-art vendors.

The easiest way to do this is through a platform that has a cloud-based digital infrastructure. IBM's Cloud Paks provides such a capability.

Modernization can be implemented in a variety of ways. You should consider different approaches. Let's identify some of them.

Approaches to modernization

Just like what you learned about digital transformation, there are many ways to approach modernization. Depending on your current digital makeup, consider the following approaches:

- **Migrate your existing systems to the cloud**: Move your legacy systems to the cloud. If you run application servers, you can now move your deployments to containerized application servers such as Liberty. This would reduce costs in your data center and simplify operations.

- **Utilize APIs to modernize**: There may be some legacy systems that are just too difficult to containerize. In those cases, you can extend their capabilities by exposing APIs to new customers.

- **Chip away at your existing monolithic systems**: As a sculptor, you can take a massive rock and begin chipping away to make it more manageable. These newer deployable components can then be applied either on-premises or in the cloud.

- **Begin microservices projects**: As we mentioned chipping away, one way to begin is to start developing microservices that you can build incrementally. Since this type of development may be new to your organization, it will prove to be invaluable as this will reduce the complexity, introduce deployment experience, and provide valuable experience for future microservice projects. Orchestrating containers can be managed by Kubernetes. Kubernetes is a piece of software that automates container orchestration, thereby replacing – with automation – operational tasks such as automatic scaling, deployment, and self-healing. Kubernetes can run on-premises as well as on cloud environments, giving you the flexibility to digitally transform and modernize. Nothing builds confidence more than successful outcomes.

- **Leave nothing behind**: To achieve modernization and capitalize on the digital transformation, you need to begin moving away from older technologies and focus on the new. When the goal is to modernize, there is nothing worse than having multiple silos with bits and pieces of legacy systems that never were sunset. This is one of the causes of digital transformation failures.

There are many benefits to digital modernization. By modernizing your applications, you help accelerate your digital transformation goal. And by adopting cloud-native architecture and containerization, your applications can be deployed faster. Your ability to utilize a hybrid cloud approach gives you greater flexibility in deployments and improved delivery. With modernized APIs and DevOps, your transition will be more agile, more consistent, and you'll deliver solutions that meet your digital transformation for your organization.

API Connect enabling digital transformation

At this point, you have learned about digital transformation and how digital modernization helps implement that transformation. You have learned about the various tactics you can use to begin the modernization process and understand that you can utilize APIs to modernize and achieve transformations.

Since this book is titled *Digital Transformation and Modernization with IBM API Connect*, you will now learn how to begin developing, securing, deploying, and managing agile and multi-cloud APIs. Your efforts will enable your transformation and let you achieve many of the key benefits of digital transformation. Your motivation will be to do the following:

- Develop new capabilities while extending existing applications by exposing APIs.

- Leverage these new APIs with other applications while maintaining the methodology of incorporating digital solutions holistically.

- Create and deploy your APIs on-premises or in the cloud, depending on your modernization effort. Implement a hybrid cloud to combine modernization without compromising business assets.

- Expose business capabilities as RESTful services and incorporate event processing to support the many channels your customer uses.

- Not only will you create APIs, but you will ensure the management of those APIs using API Connect's capabilities to support improved security and provide optimal performance to business partners and consumers.

- Continue to build with agility and promote the utilization of DevOps pipelines and automated testing and deployment.

Let's learn how to take these motivating factors and map them to the capabilities of API Connect.

API Connect aligns with the goals of digital transformation

API Connect is not only about creating APIs – it's also about API management. API Connect is configured with a program called Cloud Manager and an API Manager console where you can configure your environment to fit your needs. Cloud Manager is shown here:

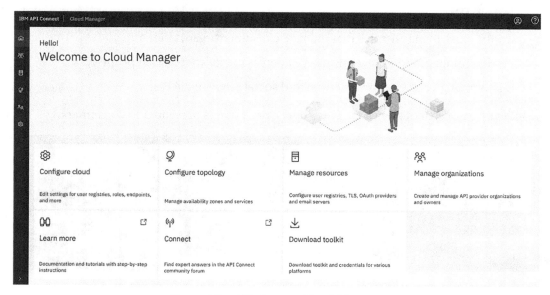

Figure 1.7 – API Connect Cloud Manager

As you can see, Cloud Manager lets you configure your cloud, specify a topology that can include multiple availability zones, manage resources such as user registries, create provider organizations to establish a development team, and download a toolkit that provides offline development. It also provides a command-line interface so that you can begin building DevOps pipelines to drive agility in your deployments.

API Connect lets you create APIs, develop process/interactive APIs, and integrate with general-purpose or BFF APIs that work with system APIs. All of these APIs need to be managed. The consumers of the APIs need to be able to discover them, quickly understand how to interface with them, and review the documentation on how to apply security and quickly test the APIs. Here are some of the things that need to be managed

- Versions of your API Products and Open API source code.

- Creating and managing provider organizations. Provider organizations create the APIs, so you will need to add/maintain users to your provider organization.

- Creating Catalogs for provider organization users that they can deploy to testing and production. You must also maintain users and roles for the respective Catalogs.

- Creating and managing user registries for authentication/authorization.

- Creating and managing security for your APIs with TLS certificates and OAuth provider resources.

- Managing the life cycles of API products as you publish them to Catalogs and portals.

- Community manager roles that manage consumer organizations.

- Product manager roles that ensure API deployments are released properly.

- Customizing your Consumer Portal and socializing your APIs.

The following screenshot shows the Developer Portal that is presented to consumers:

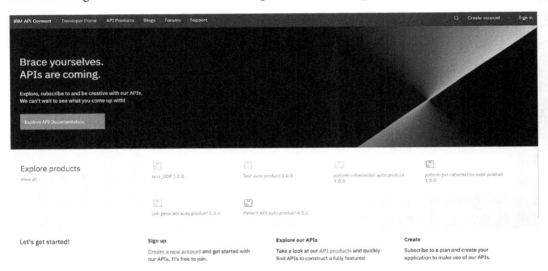

Figure 1.8 – API Connect Developer Portal

This un-customized portal shows features that you have learned are important in satisfying digital user experiences such as self-service, social interaction, search, and much more.

You'll learn more about these activities in *Chapter 3, Setting Up and Getting Organized.*

Modernizing implementation choices with API Connect

When you are modernizing, having the flexibility to choose where you deploy is important. If you are modernizing your existing on-premises systems with APIs, you can deploy API Connect on-premises using Kubernetes, OpenShift, or VMware.

The following is an OpenShift screenshot showing pods in a user-friendly interface:

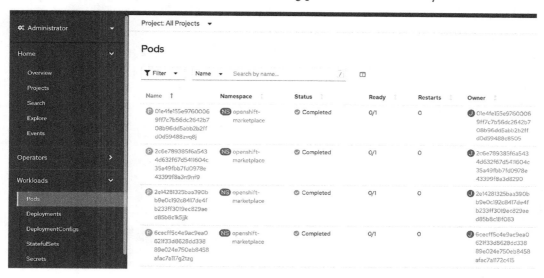

Figure 1.9 – RedHat OpenShift platform

The **OpenShift Container Platform (OCP)** provides the container orchestration layer for API Connect. API Connect is built on microservices and running on top of OCP allows your organization to choose which cloud provider to use. You can run OpenShift on-premises or on any of the major cloud providers (AWS, GCP, Azure, or IBM Cloud). OpenShift includes an enterprise-grade Linux operating system, container runtime, networking, monitoring, a registry, and authentication and authorization solutions.

The following is a screenshot of the OpenShift Cluster management screen. You can see that it provides critical information for an administrator to understand the health of the cluster and what activities are running:

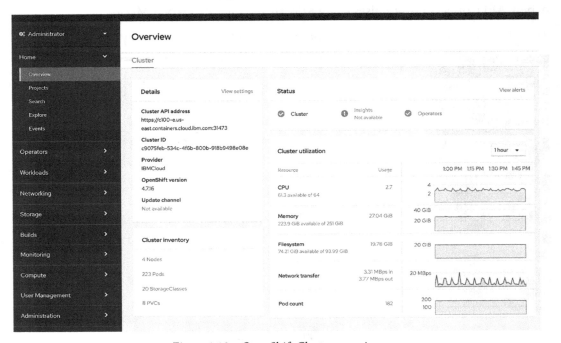

Figure 1.10 – OpenShift Cluster overview

If your company is leaning toward containerization in your digital journey, then OpenShift or Kubernetes are excellent options. If you're taking advantage of IaaS or PaaS, you can install API Connect on the IBM Cloud or as a reserved instance where your instance of API Connect is managed separately from a shared instance.

One of the PaaS options is Cloud Pak for Integration, which is part of IBM Automation. **Cloud Pak for Integration (CP4I)** is a powerful integrated platform for deploying containerized integration capabilities as part of an OpenShift deployment environment. By making use of CP4I, an organization can connect applications, systems, and services quickly and simply as part of a managed, controlled, scalable, and secure environment. This includes the following:

- IBM App Connect Enterprise
- IBM API Connect
- IBM DataPower Gateway Virtual Edition

- IBM MQ and IBM MQ Advanced

- IBM Event Streams and Confluent OEM (add-on)

- IBM Aspera High-Speed Transfer Server

The following is a screenshot of CP4I:

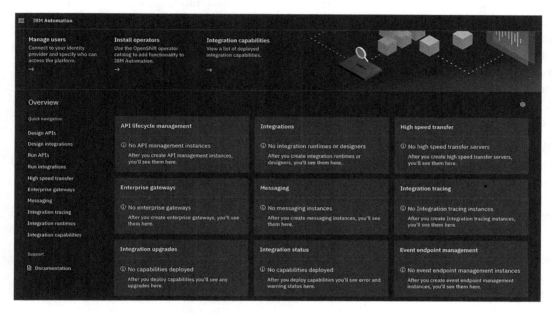

Figure 1.11 – IBM Cloud Pak for Integration

These capabilities of CP4I (excluding API Connect) are not covered in this book because they support other capabilities and functionality that are not part of API Connect. If you would like to learn more, then there is plenty of information on the web regarding CP4I. You can start by going to the official website: `https://www.ibm.com/cloud/cloud-pak-for-integration`.

Summary

For some time now, digital transformation and digital modernization have been discussed in countless articles and references. There has been a success in these transformations, as well as many failures. At the beginning of this chapter, you were provided with background knowledge of the driving influences of businesses and how improvements in technology have driven new business practices. By understanding how the 7Ps of marketing (price, product, place, promotion, passion, people, and process) has led digital transformation and modernization into a holistic consolidation where all the parts of the organization work toward a common goal. By understanding your customers and taking advantage of new ways of doing business through digital technology, you learned that organizations can be more agile, run more efficiently, and produce products at a high level of efficiency and creativity so that consumers will react passionately about your product and do the promotion for you.

You learned that, of the many ways to modernize, APIs become the flexible glue that ties the solutions together. APIs are apparent at many levels of your architecture. Of the three major layers of APIs (UX APIs, process/interactive APIs, and system APIs), you learned that API Connect is dominant in the latter two by providing important management capabilities that follow the digital transformation's holistic practice.

In the latter half of this chapter, you were provided with more details on how API Connect supports your digital transformation. You were introduced to how API Connect can be installed on various platforms that follow your modernization efforts, whether that be on-premises, on cloud platforms using Kubernetes, or with OpenShift/Cloud Pak for Integration. By learning about all the available options, you now have choices that you can map to your strategic goals.

You are now at a point where having a deeper insight into API Connect will give you the tools to begin building, deploying, securing, and managing your digital transformation with APIs. The next chapter will provide that introduction.

2
Introducing API Connect

In the previous chapter, you were briefly introduced to API Connect and how it can support your adoption of digital transformation and modernization. Now is a good time to learn, in greater detail, the full capabilities of API Connect along with specific highlights in terms of how it can impact digital transformation and/or your modernization effort.

In this chapter, we are going to cover the following main topics:

- The components of API Connect
- Deployment models
- Introduction to the **Command-Line Interface (CLI)** commands

Some of the new skills you will learn in this chapter include the following:

- Properly describe the components of API Connect and how they provide you with all the tools you need to successfully create, manage, secure, and govern your APIs.
- Understand the different deployment options of API Connect to fit your digital strategy.
- Learn how to run CLI commands that perform the API configuration and deployments of your DevOps pipeline.

As you might have gathered, API Connect has several valuable components, and having a better understanding of each of them will be helpful in your understanding of how best to take advantage of their capabilities.

Technical requirements

In this chapter, you will be referencing a number of Swagger definition files to assist you with your learning experience. You can find these files in the Git repository at `https://github.com/PacktPublishing/Digital-Transformation-and-Modernization-with-IBM-API-Connect`.

You should copy the files for `Chapter02` to your development environment.

In addition, you should download the API Connect toolkit by visiting IBM Fix Central and searching on API Connect: `https://www.ibm.com/support/fixcentral/`.

API Connect

In the previous chapter, you learned about several API-led responsibilities. In order to accomplish API activities (such as applying security, establishing rate limits, building and deploying using DevOps pipelines, and socializing your APIs), you need to be familiar with how API Connect makes those tasks easy to accomplish.

Currently, IBM supports three versions of API Connect (that is, version 5, version 2018, and version 10). At the time of writing, support for version 5 will be removed in April 2022. Version 2018 was released in April 2018. At that time, IBM was using release numbers based on the year of release. In 2020, they changed it back again to use a version number of 10, so **v10.0.1.5** is the current latest version. IBM uses a version number strategy (**VRMF - Version, Release, Modification, Fix pack**) based on the following:

- **Version**: Major feature changes
- **Release**: The release of minor feature changes
- **Modification**: Manufacturing refresh also known as Mod Pak – full release with functionality updates
- **Fix Pack**: Releases with cumulative fix packs

There is also a special release called an **iFix**. These are special updates that do not modify the VRMF versioning. They have the `_iFix<number>` text appended to the version. An example of a previous iFix is v10.0.1.2_iFix002.

You can learn more about this terminology at `https://www.ibm.com/support/pages/vrmf-maintenance-stream-delivery-vehicle-terminology-explanation`.

> **Note**
>
> IBM refers to supported versions of API Connect as **Long-Term Support (LTS)**. As new features are developed, IBM releases a **Continuous Deployment (CD)** version so that customers can try out the new features. These CD releases are not supported but are a way for customers to test out new functionality before their official release. The CD versions of the v10.0.2.0 and v10.0.3.0 enhancements were packaged into v10.0.1.5 as a new LTS version.

Since this book is focused primarily on developing APIs, you will not be learning much about the installation and cloud administration of API Connect. However, there are a few things you should be aware of to help with your overall understanding of API Connect and its components.

The most important change you should be aware of is that beginning with version 2018, all of the implementations of API Connect are now built on microservices. This means all servers (including VMware implementations of OVA files) are running on Kubernetes. As discussed in *Chapter 1, Digital Transformation and Modernization with API Connect*, containerization might be one of your company's goals in modernization. With API Connect running containerized components, you are already on your way. You have multiple options in which to deploy API Connect running Kubernetes, such as the following:

- Kubernetes on bare metal (Ubuntu and RHEL)
- VMware
- OpenShift or as part of IBM's **Cloud Pak for Integration (CP4I)**

To check the requirements for installing API Connect, please refer to the IBM Software Product Compatibility Report at `https://www.ibm.com/software/reports/compatibility/clarity/softwareReqsForProduct.html`.

As shown in the following screenshot, you can choose various tabs to obtain additional details about the system requirements:

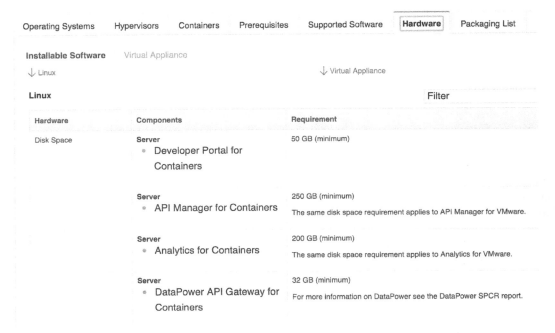

Figure 2.1 – Software compatibility report

Note

One excellent resource is the deployment white paper on API Connect. You can locate it here:

```
https://community.ibm.com/community/user/
integration/viewdocument/ibm-api-connect-v10x-
deployment-wh?CommunityKey=2106cca0-a9f9-45c6-
9b28-01a28f4ce947
```

Now that you have been introduced to the various deployment implementations, we'll move on to the components of API Connect.

The components of API Connect

There are four major components of API Connect and two additional add-on components that integrate and provide added value to developers. Those components are as follows:

- Cloud Manager
- API Manager Server
- The Developer Portal service
- Gateway
- The Analytics subsystem
- **Local Testing Environment** (LTE; an add-on)
- API Connect Test and Monitoring (an add-on)

The two add-on components are discussed, in greater detail, in *Chapter 4, API Creation*, and *Chapter 13, Using Test and Monitor for Unit Tests*.

> **Note**
> Cloud Manager and API Manager can be considered separate components, but when you deploy the API Connect cloud, these two components are part of the same service. These two components comprise a management cluster. We'll discuss these separately in the coming sections because each has different capabilities and responsibilities that require different roles from users.

The following diagram gives you a high-level picture of the components and their relationships:

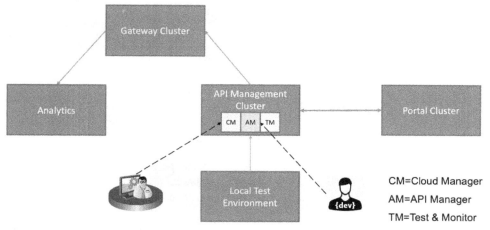

Figure 2.2 – The API components

Here, the solid lines represent the integration between the components. The dashed lines represent users interacting with the Cloud Manager user interface and the API Manager user interface. You will learn, in the *Introduction to the CLI commands* section, all about the CLI interface, which also allows administrators and developers access to the Cloud Manager and API Manager components.

You are probably wondering how you will be able to take all that you learn here and put it into practice. There are several places where you can get access to API Connect:

- Your company's implementation of API Connect, whether it's on-premises or a part of CP4I.
- If you are an IBM business partner, then you can download the full API Connect product and/or the LTE.
- Utilize the IBM Cloud platform and subscribe to an API Connect service.

You should be aware that the IBM Cloud version might be some levels behind and not on the same version as this book.

Understanding the integration between the individual components will be very helpful. You'll learn about this next.

Cloud Manager

Cloud Manager is where the administrator will construct the cloud topology. The administrator is responsible for configuring the cluster, which comprises the Gateway, Portal, and Analytics services:

Figure 2.3 – Cloud Manager

Setting up **High Availability (HA)** and **Availability Zones (AZs)** is also a part of the responsibilities of the administrator. These can be very simple or very complex depending on the desired implementation. For instance, gateways can be placed in different zones (such as west, east, and other designations), and these zones can be in different clouds, providing the capability to put host runtimes closer to user communities or provide compliance standards for situations such as **Payment Card Industry (PCI)** security compliance.

The cloud topology is where you set up the AZs for API Connect. Additionally, this is where you establish which services are within the AZ. *Figure 2.4* shows the default AZ with three components registered (the Analytics service, DPGatewayV5, and Developer Portal):

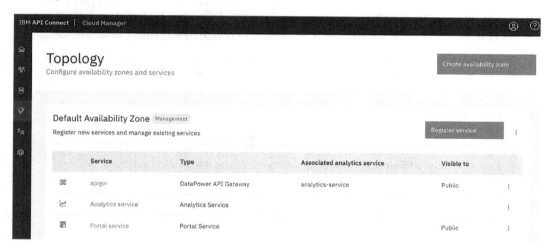

Figure 2.4 – Cloud Topology for API Connect

Additional responsibilities include creating Provider organizations. Provider organizations are the creators of APIs, and they manage the deployment and life cycles of the APIs that are created.

In addition to this, the administrators set roles for other activities within the cloud along with the setup of alternative administrators. All administrators are responsible for the health of the API Connect cloud.

Other activities that the administrators might configure include the following:

- The **Simple Mail Transport Protocol (SMTP)** server for notifications (along with default email templates)

- User registries such as local registries, the **Lightweight Directory Access Protocol (LDAP)**, **Uniform Resource Locator (URL)**, common services, and more

- Roles

- Catalog gateway defaults

- Audit settings

- Notification and timeout settings

From a developer's perspective, you won't be interacting with Cloud Manager, but there are areas you should be aware of. For instance, you should have a solid understanding of the following:

- Provider organizations enable your team to work on API Connect.

- If you have special requirements for gateway implementations (for example, if you are building APIs on AWS), you can have a gateway service deployed in AWS that is associated with the API Connect Management service to improve performance.

- If you have the need for external registries, you can have facilities such as Okta or Ping registries associated by your administrator.

> **Important Note**
>
> The minimum configuration requires a topology comprising a gateway service, a portal service, and an analytics service. In order to utilize that configuration, first, you must add an email server and define it in the notification settings.

Once your Provider organization has been established, the owner of that organization can invite users to API Manager to start developing and managing APIs.

API Manager

API Manager is where you manage the packaging and deployment of APIs and their life cycles. The following screenshot is of a logged-in user in API Manager:

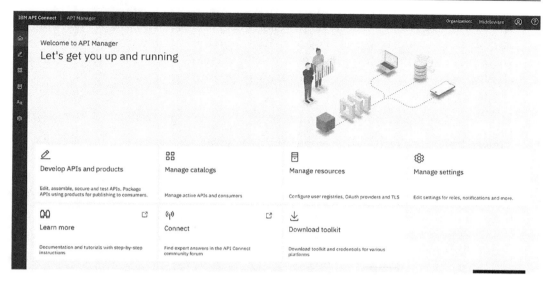

Figure 2.5 – API Manager

When you access API Manager, you represent the user of a Provider organization. As a member of the organization, you have the ability to develop APIs and products, manage Catalogs, manage resources, and manage other supplemental settings depending on your role.

Roles

Roles can vary and are initially assigned by the owner of the Provider organization. You have choices that enable or limit permissions for the organization. These permissions allow various actions to be performed. The following predefined roles are available in API Connect. A short description is provided for each of these roles to enhance your learning:

- **Organization owner**: They have full permissions, by default, for all of the API Connect manager functions.

- **Administrator**: They have the same functions as the owner but can be updated.

- **API administrator**: They manage API life cycles and the publishing of APIs to the Developer Portal.

- **Community manager**: They manage the relationship between the Provider organization and the app developers, provide API analytics, and provide support to the app developer teams within the community.

- **Developer**: They design and create APIs and products with the ability to stage and publish those products to a Catalog or space within the Provider organization. These permissions do not apply when a developer is assigned to a specific Catalog or Space. Those users can only manage products within the Catalog or Space assigned.

- **Member**: They are a read-only member of the Provider organization.

- **Viewer**: They are a read-only member of the Provider organization.

> **Tip**
>
> For more details regarding these roles, please refer to the v10.0.1.5 API Connect Knowledge Center at `https://www.ibm.com/support/knowledgecenter/SSMNED_v10/com.ibm.apic.overview.doc/overview_apimgmt_users.html#overview_apimgmt_users__apim_manager_roles`.

Knowing who is assigned to what role is something that should be decided from the outset. You can create a custom role by using the Manage Settings option and then adding a new role. Generally, one out-of-the-box role should suffice, but if you desire more granular roles, you have that capability through a custom role.

> **Important Note**
>
> API Manager holds a database that keeps track of the APIs, products, life cycles, subscriptions, and other relevant information. When a backup of API Connect is conducted, it is this database that maintains the state of the environment.

Developing APIs and products

Of course, the reason you have API Connect is to create APIs and manage them. You will be creating various types of APIs and publishing them through a packaging mechanism called a product. Similar to the product that you learned about in the 4 Ps of Marketing, in the *Chapter 1, Digital Transformation and Modernization with API Connect*, this product is what consumers subscribe to. A product is the packaging of the services your consumers will use within their user experience APIs. How you create an API definition is shown in Figure 2.6:

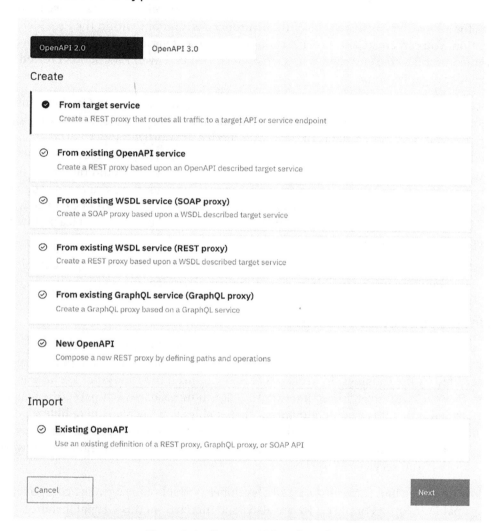

Figure 2.6 – Creating APIs and products

As you might have gathered, there are various ways and types of APIs that you can create. You will learn how to create these APIs in a three steps, in *Chapter 4*, *API Creation*.

Catalogs

A Catalog is a target for deploying APIs where it has a gateway and (if desired) a Developer Portal service. A Catalog is a logical partition; therefore, you can have many Catalogs for a Provider organization. With the proper permissions within the Provider organization, you can create additional Catalogs to represent various environments. It's not uncommon to see environments such as Test, QA, and even Production created for a Provider organization.

As mentioned earlier, Developer Portal services can be created for a given Catalog. This is not a requirement. So, why would you not want a portal for every Catalog? Well, Catalog Developer Portals are really tenants of the portal service. If your Developer Portal service is limited in capacity, having multiple portal tenants might stretch your resources. In addition, Catalog Developer Portals require configuration, so if you would rather not have to configure a Developer Portal service for each of your Catalogs, then you can decide not to have a Developer Portal. You will learn about Catalogs in greater detail in *Chapter 3, Setting Up and Getting Organized*. Additionally, in the same chapter, you will learn how to create a portal tenant for a Catalog.

That was a quick introduction to the Cloud Manager and API Manager components along with their capabilities. Next, you'll learn about the portal service.

The Developer Portal service

In this section, you will learn about the portal service. As mentioned earlier, the Developer Portal service is a multi-tenant portal server that holds the Developer Portal tenants created by the Catalogs. In version 5 of API Connect, the Developer Portal was based on **Drupal 7**; however, in version v10.0.1.5, it is now based on Drupal 9. The benefit of using Drupal is that it is a content manager that enables you to create incredible digital experiences. What you should know is that this is an enterprise-enhanced version of Drupal that includes plugin components of API Connect. These allow integration between the Developer Portal service and API Manager.

The following screenshot shows the default Developer Portal home page. The components of the Drupal Developer Portal service are also containerized, so this Drupal instance is running under Kubernetes:

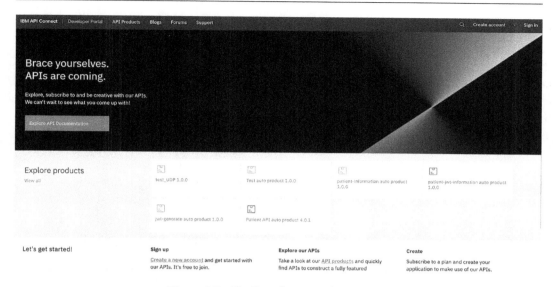

Figure 2.7 – The Portal instance home page

As mentioned earlier, Drupal is packed with additional functionality for API Connect. So not only is it an awesome content manager, but it's also a full-fledged API socialization platform for APIs. From a digital transformation viewpoint, it fits right within the plan.

You'll learn more about the customization capabilities of the Developer Portal service in *Chapter 15, API Analytics and the Developer Portal*.

In the meantime, you should attempt to understand the API Connect capabilities that are incorporated in the Developer Portal.

Digital transformation portal API Advertising

The subheading is a fancy way of introducing the capabilities of API Connect, but in a nutshell, you can simply refer to it as the Developer Portal. The Developer Portal can be used to discover APIs and develop new applications. This utilization is not just for external consumers and business partners, but it is also a perfect way to integrate internal applications. In fact, many API Connect enterprises utilize it internally as part of their initial digital transformation effort. Making their internal applications successful provides them with the confidence to extend their API effort beyond the firewall.

So, what are the capabilities of the Developer Portal? First and foremost, it has the ability to showcase your APIs. This is accomplished with the content manager by providing a Product page. The Product is the packaging for your APIs. A consumer will visit the portal and peruse the Product page, review the documentation for the product and its APIs, and, finally, determine whether the product is something they would like to try out.

With the Developer Portal, you can configure it to only allow logged-in users the ability to view these products. That will depend upon your business requirements. Can anyone see them? Or do you want to implement a mechanism that requests the user to log in? If that is the desired behavior, you can set up self-service registrations where a new user registers and creates a consumer organization. A consumer organization is an external organization or developer. One marketing strategy is to capture the consumer organization information as a means of doing more business. There is a built-in workflow that can be initialized to verify self-registering consumers to your Developer Portal. This can be accomplished via an email to a community manager who can then authorize the new registration request. That gatekeeper can accept or reject that person and/or use the information provided to reach out to the person to sell your APIs as the best in the world – this is a powerful capability.

Regardless of which method is used to get the consumer on board, once the consumer has access, they can review the Product and APIs and choose to subscribe to the Product. When subscribing, the user selects a plan to subscribe to. Plans set limits regarding the utilization of your APIs. Similar to a healthcare plan, the user chooses options that give them the capabilities within the multiple plans your Provider organization has agreed to. Once subscribed, the consumer can begin testing the APIs. This is similar to a test drive of your API. Try it out. See how it behaves; see what type of security is required and what return codes are provided – everything you need as a developer to build a robust application.

It is important to note that we have mentioned a few capabilities that are provided by the portal. You have heard about Consumer organizations, consumers, Products, APIs, subscriptions, and plans. All of this information is part of the API Connect additions to the Drupal portal software, and it's all persisted. Beneath the covers, the portal is communicating with API Manager. It's the API Manager component that contains the database that holds all of API Connect's critical assets.

> **Important Note**
>
> When you do a backup on API Manager, you are backing up the configuration of API Connect and all APIs along with the portal subscriptions and consumers.

As you learned in the previous chapter, your customers love to express their passion for your products. The Developer Portal provides you with that option along with out-of-the-box capabilities for blogs and forums. If you are interested in hearing from your customers, leave the capabilities enabled. If you don't desire them, you have the option to disable them.

The world is at your fingertips with the Developer Portal. With the help of a Drupal developer, you can customize the look and feel; and with API Connect's built-in capabilities, you can build a fantastic portal. Make sure you take full advantage of it:

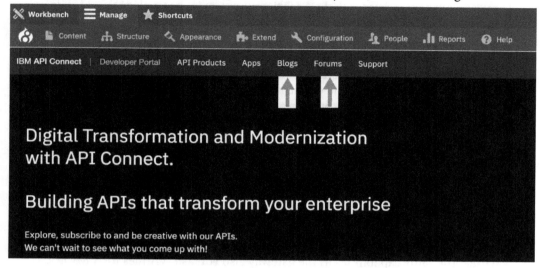

Figure 2.8 – Portal admin customization

You have just learned, in depth, about the Developer Portal component of API Connect. Next, you will learn where the APIs are actually executed. The API runtime of API Connect is all handled in the gateway. You'll learn about that next.

Gateway

The gateway within API Connect is critical for ensuring your APIs are securely executed, highly available, measured, and performant. The gateway in API Connect is called **DataPower**. DataPower has been an IBM product since 2005 and is considered one of the best gateways on the market. For years, customers have been using DataPower in SOA environments within the DMZ or inside the firewall providing security, routing, transformation, supporting multiple protocols and capturing metrics for analytics. These are all the capabilities that are important to the success of your digital transformation.

> **Important Note**
>
> The gateway runtime is where the rubber meets the road. When defining your API Connect topology, the placement of your API gateway is critical. The API gateway can be associated with the Analytics subsystem in the topology to provide valuable statistics to you and your consumers.
>
> If your APIs are deployed in the cloud, having a gateway within the same cloud will improve performance. You should note that with API Connect, you can place multiple gateways in different clouds, on-premises, or within a hybrid architecture (both the cloud and on-premises). In addition to this, you can configure HA with the gateway and spread them between data centers if you wish.

As you might have gathered, the gateway is very important. Because it can be deployed in heterogeneous environments, it must also support many form factors. The DataPower gateway is available in four different form factors:

- **Physical**: For example, a hardware appliance with multiple network interfaces and significant CPU/memory.

- **Virtual**: The same capabilities in the gateway also run as OVA files for VMware.

- **Linux**: You can install the Linux form factor of DataPower on-premises and/or in cloud environments.

- **Docker**: DataPower is also containerized and can be part of your API Connect Kubernetes installation. The container version is also useful for testing your APIs when you are using the LTE:

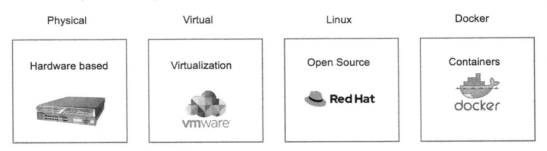

Figure 2.9 – DataPower has four form factors

You can mix and match multiple form factors within your hybrid cloud environment to adhere to your infrastructure needs and corporate standards. Note that cost might also play a role in your decision.

> **Important Note**
>
> Bear in mind that these form factors can be deployed on most common cloud vendors' cloud infrastructures such as AWS, Azure, IBM Cloud, and GCP. The appropriate form factor you pick will depend on the capabilities of the vendor.

So, now that you are aware of the various form factors, you should have a good understanding of how the DataPower gateway can be clustered to provide scale, self-healing, and autonomous gateway management.

DataPower gateway clustering

Gateway clustering is essential to support HA and resiliency. When you add a gateway to the Cloud Manager topology, it is actually referring to a gateway cluster. You can always have a single DataPower gateway within your cluster in non-HA environments, but in general, for all HA environments, you must have a minimum of three DataPower appliances.

> **Important Note**
>
> Having three DataPower gateways for HA is mandatory to establish a **quorum**. In version 5 of API Connect, the only component that required a quorum was the Developer Portal. Now, with versions v2018 and v10, all components (such as Cloud/API Manager, Gateway, Developer Portal, and Analytics) require a quorum for HA.

You can configure gateway clustering by grouping gateways together in an on-premises implementation, or you can spread them out across data centers, AZs, or within cloud implementations. The loss of a gateway in one data center will be picked up by another member of the cluster without any impact. This is all possible because of the quorum.

Briefly, we will discuss why gateway clustering is important. However, for additional details about how to set it up, please refer to the IBM Knowledge Center on this subject. It can be located at `https://www.ibm.com/support/knowledgecenter/SSMNED_v10/com.ibm.apic.install.doc/tapic_install_datapower_gateway.html`.

A properly setup gateway cluster provides you with the ability to scale your gateways to handle spikes in requests and meet predefined **Service-Level Agreements** (**SLAs**). When your gateways establish a quorum, a primary gateway is elected and the other two become secondary gateways. As requests are made (including OAuth token requests), the gateways synchronize. So, if the primary gateway goes down, the secondary gateway kicks in. When the primary gateway becomes available again, resynchronization will occur so that all gateways are available and up to date.

Figure 2.10 depicts a three-instance quorum where there is a primary gateway and two secondary gateways. The disks represent the synchronization of the runtime data:

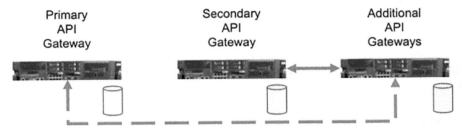

Figure 2.10 – API gateways are scalable

> **Important Note**
>
> Remember that the gateway is the runtime. In the event that other cluster members experience an outage due to a lost quorum, as long as the gateway is still running, your API calls will be processed. You will have to encounter a loss in capability between the interaction of gateways to managers and analytics, but business processing will continue.

Before we move on to the other components of API Connect, there is one important aspect of the gateway that solution architects and developers should be aware of. In version 5 of API Connect, the gateway service was built using the existing DataPower multi-protocol gateway service With v10.0.1.5, an additional API gateway service was created that is built as an out-of-the-box native service. This new service is dramatically faster.

> **Important Note**
>
> The new gateway service is called the **API Connect Gateway Service (apigw)**. It can increase performance by 10 times, and possibly even more, depending on your APIs. Version 5 customers should be aware that during the migration to version 10, you have the choice to convert to this new API Connect Gateway Service.

With API Connect version 10, there are two options to add gateways to your topology. You just learned about the API Connect Gateway Service, which is one of those options. The other option is called the **Version 5 Compatibility (v5c)** gateway. This gateway was also slightly modified to help improve performance, but it is primarily available to support existing version 5 APIs. If your company is migrating from version 5 and is not ready to move to the new API Connect Gateway Service, then you would utilize the v5c gateway instead.

However, there might be legitimate reasons not to move to the latest and faster gateway. There were changes in the API Connect Gateway Service that could hinder the way you code your APIs. While the version 5 to version 10 migration would convert as much as possible, certain custom code might not migrate successfully. So, for migration efforts from version 5 to version 10, you should clearly evaluate the ramifications of converting all of your APIs into the gateway service type. Coding changes of previously running APIs could lead to negative results, rendering your APIs useless.

> **Important Note**
>
> Another important consideration to bear in mind when migrating to version 10 is if you have created user-defined policies based on version 5. There are new methods of supporting user-defined policies, and you might have to update those in order for your APIs to continue using them.

You have just learned about the two types of gateways that are available. If you are new to API Connect version 10 and are not migrating to version 5, you should default to using the API Connect Gateway Service.

As mentioned earlier, the gateway will capture transactional information to route to the Analytics subsystem. The Analytics subsystem is the final of the four components that comprise API Connect. You will learn about it next.

Analytics

The Analytics subsystem is what receives the transaction details from the gateway cluster. After you install the Analytics service, you can then associate the Analytics subsystem with a defined gateway cluster. Once associated, the gateway will begin sending transaction data to the Analytics subsystem:

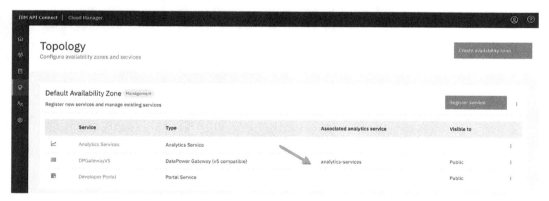

Figure 2.11 – Associating the Analytics service to a gateway

Following this, the analytics data can be reviewed by the Analytics user interface that is displayed in API Manager and Developer Portal. The Analytics capabilities that come with API Connect are based on the open source products from the Elastic Stack (Elasticsearch, Logstash, and Kibana or ELK). Here, the user interface is using Kibana.

As an API developer, users can view analytics within the Provider organization Catalogs they are associated with. This includes spaces that they have been added to.

You can only customize analytics from the Provider organizations. Whenever you click on the **Analytics** tab, you are presented with the choice of preconfigured dashboards:

Figure 2.12 – The Analytics tab within a Catalog

Within each dashboard, there is a preconfigured visualization of the metric data. The screenshot that follows shows you the visualization for the default dashboard. You have the capabilities to modify the visualizations and/or create your own:

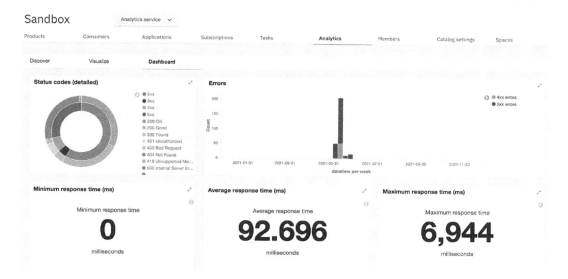

Figure 2.13 – Visualizations for the dashboard

With Analytics, you can also search for analytics produced from a different period of time. On the right-hand side, you can find the default time period of **Last 15 Minutes**. As shown in the following screenshot, if you click on that link, you will be given the option to select a different time period:

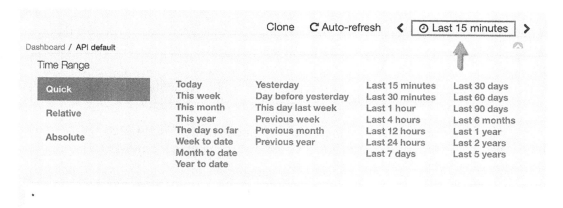

Figure 2.14 – Changing the time range

You might have noticed a dashboard called **Portal default**. This is where you can view and modify the analytics that are available for the Developer Portal.

Within the Developer Portal, API consumers can view analytics under two categories (consumer apps and consumer organizations). They can view analytics under the **Apps** tab for the consumer apps that have been created. The apps are the implementations of the subscribed product and APIs. The analytics that can be viewed include API stats, total calls, and total errors.

The other place in which to view them is based on the consumer organization. The consumer organization is the organization of the developers subscribing to your Product APIs. The information categories are the same (that is, API stats, total calls, and total errors) but rolled up just for that consumer organization.

There is a lot more you can learn about the Analytics subsystem. You can learn how to configure the Analytics dashboards and visualizations by referring to the documentation on the web about the Elastic Stack.

> **Important Note**
>
> While not covered in this section, you should be aware that the analytics data can be exported from the Analytics subsystem. Additionally, you can configure API Connect to send the analytics to another server for analysis and reporting, such as Splunk. If you would like to do so, please refer to the IBM Documentation web page.

So, you have learned about the four major components of API Connect. You learned that Cloud Manager is where you configure the API Cloud topology and link the components. Additionally, you discovered that API Manager is where you create products and APIs. The Developer Portal is the second component where you can publicize your products and APIs as well as allow subscriptions to consumers to develop applications. We introduced the DataPower gateway where your APIs actually run, and we described how analytics are sent from the gateway to the Analytics subsystem for analysis and reporting.

By now, you should be familiar with the various components that comprise an API Connect cloud. Next, you will learn about the deployment models for an API Connect hybrid cloud.

Deployment models

When it comes to the hybrid cloud implementation for API Connect, you will find it difficult to find another API Management product that deploys to as many cloud platforms as IBM API Connect. In this section, you will be provided with information regarding how API Connect is packaged to run on-premises and on cloud platforms using various methods of deployment.

So, how is it that API Connect can be deployed to so many different destinations? Well, it all has to do with how the product was developed. Since version v2018, API Connect has to be built from the ground up as microservices running in containers. To manage these containers, IBM utilizes Kubernetes and, now, Red Hat OpenShift.

You'll start learning about the models, beginning with the most prominent one – the on-premises model.

The on-premises implementation

You have three choices when you want to implement API Connect on-premises within your data center:

- VMware ESX
- Bare-metal Kubernetes
- **OpenShift Container Platform (OCP)**

While the implementation using VMware utilizes an OVA file to install each of the components, beneath the covers, it is still Kubernetes managing the API Connect microservices. This model of implementation is suited for customers who are in the initial stages of modernization and would rather use platforms their teams are familiar with. Since prior versions of API Connect were based on VMware installations, many customers are already familiar with the VMWare and OVA installations. Learning a new API platform along with Kubernetes can be a challenge for many companies, and the VMware option is preferred in such cases.

Installing your own master and worker nodes on bare metal or VMware and implementing Kubernetes requires more skill for the administrator. This requires a good understanding of how Kubernetes is configured and all of its dependencies, just to get started. The installation has been made easier with the addition of Kubernetes operators. In this scenario, operators install the API Connect components. However, administrators still need to have experience of Kubernetes commands, load balancers, ingress controllers, and how to troubleshoot Kubernetes issues.

The third option is to use OCP. So, what is the benefit of OCP? First, OCP has a browser-based interface that makes managing Kubernetes easier. It enables DevOps and corporate-wide collaborations, making deployments easier. Additionally, OCP provides a CLI, providing multiple ways to interface with the platform.

Setting up an OCP platform has similar requirements to getting the cluster configured. You still have master and worker nodes that need to be configured. You will install OCP using the **User Provisioned Infrastructure (UPI)**. To learn about OpenShift and UPI, please visit the Red Hat website at `https://docs.openshift.com/container-platform/4.1/installing/installing_bare_metal/installing-bare-metal.html`.

As you have learned, you have multiple options in which to deploy API Connect on-premises. Each has its pros and cons based on your skillset and budget. Cloud deployments are another deployment model. Let's learn about those options next.

Cloud implementations

Choices, choices, and more choices. The draw toward cloud computing with multiple vendors providing **Infrastructure as a Service (IaaS)** and **Platform as a Service (PaaS)** is compelling.

The discussion regarding installing on-premises is very similar to when installing using IaaS. The differences will be in each cloud vendor's implementation of setting up the environment. For instance, each of the major cloud vendors has its own implementation and setup for networking, LDAP, DNS, and storage. Of course, cost is also important, as all vendors have different cost models. Another slight difference between the vendors is that each might have slightly different versions of Kubernetes.

Despite all the differences, API Connect can be installed on major cloud providers such as Azure, GCP, AWS, and IBM Cloud. All of these implementations allow you to install API Connect; however, only IBM Cloud offers a SaaS implementation of API Connect. The API Connect for IBM Cloud provides an instance of API Connect to allow you to build products and APIs. This implementation provides you with access to API Manager to carry out development and other Provider organization activities. What you do not have access to is Cloud Manager:

Figure 2.15 – API Connect on IBM Cloud

> **Important Note**
>
> API Connect for IBM Cloud is expected to be migrated to API Connect version 10. At present, API Connect may not be at the latest fix pack.

Earlier, you learned that each cloud vendor has slight differences when it comes to installing and supporting Kubernetes. One option is to mitigate, that is, to implement API Connect on OCP. OpenShift is supported on Azure, GCP, and AWS. Azure and AWS each provide self-service managed implementations of OpenShift. Now you can merely choose to stand up OpenShift on one of these platforms and deploy API Connect using operators in order to have an API Connect environment up and running in no time.

IBM CP4I

Another cloud offering from IBM is CP4I. CP4I is a platform that runs on OpenShift and offers IBM integration products that have been containerized to run on top of OCP. One of those products is API Connect.

CP4I has the following capabilities:

- **API management**: This capability is the API Connect implementation.

- **Application integration**: This is App Connect Enterprise, and it allows you to integrate a system of record, SaaS, B2B, and more.

- **Event streaming**: This enables you to build responsive applications using Kafka and integrate with the other capabilities listed.

- **Enterprise messaging**: This allows asynchronous messaging for the enterprise that extends into hybrid architecture.

- **End-to-end security**: The containerized DataPower form factor has been implemented to support your security needs.

- **High-speed data transfer**: This capability allows extremely fast secure file transfer using a product called Aspera.

Referencing the hybrid reference architecture you learned about earlier in *Chapter 1, Digital Transformation and Modernization with API Connect*, the features in CP4I provide all of the necessary toolings to address your digital modernization.

> **Important Note**
>
> The AWS marketplace has now added CP4I as one of its offerings: *Quick Start for IBM Cloud Pak for Integration*. This is a great addition for organizations that want to use IBM integration products on AWS. To learn more, please visit
> `https://aws.amazon.com/quickstart/architecture/`
> `ibm-cloud-pak-for-integration/`.

IBM Cloud (reserved instance)

You already know that you can quickly spin up an instance of API Connect on IBM Cloud. If you recall, this instance provides you with the ability to create and manage APIs using API Manager. Since it's a managed service instance, what is transparent to you is that you are part of a shared deployment. For customers who would prefer to have their own segregated instance, there is also a reserved instance for API Connect.

The reserved instance also runs version 10 of API Connect. The value customers get with the reserved instance is that it is managed and monitored by the IBM team, so your operations teams need not be involved. Additionally, since it's an isolated environment, you do not need to worry about users who are using the same IBM Cloud public service.

When you visit IBM Cloud and review the offers, you will see the reserved instance option, as shown in *Figure 2.16*:

<figure>Figure 2.16 – A Reserved instance of API Connect</figure>

As you can see, there are many cloud deployment models. You now have the ability to choose how and where you want to implement API Connect based on cost, skillset, digital modernization directives, and tooling. The one model that we haven't discussed yet is hybrid cloud. Let's discuss that next.

Hybrid cloud

The last deployment model is **Hybrid cloud**. As you review the following diagram, you should be able to view where the separation between on-premises and the cloud is demarcated. While the diagram specifies any cloud, an example might be helpful for you:

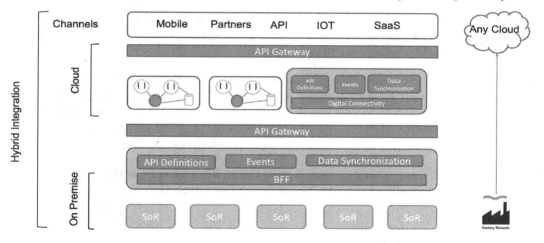

Figure 2.17 – The Hybrid cloud reference architecture

As an example, let's use a company that is building new APIs in support of **Fast Healthcare Interoperability Resources (FHIR)**. The company has decided to use AWS to develop the APIs, but the data is still within the on-premises data center. This group of providers is *autonomous*. Perhaps it is a newly acquired software company. The challenge is how to make this work seamlessly.

If you have deployed API Connect in AWS, you can manage all of the FHIR APIs to the FHIR server. Your backend to the data and system of records is on-premises, but you also have API Connect implemented on-premises to accept calls from the AWS APIs and support API integration between various departments.

Of course, having multiple deployments of API Connect is not necessary. You can implement your API Connect on-premises and still have the APIs developed on AWS. With the configuration options within API Connect, you can place your gateways in AWS to improve performance while still having gateways in your DMZ and inside the firewall to support other APIs.

The benefit of this is that you are no longer hindered by preexisting infrastructure and antiquated legacy applications. You can mix and match your deployment to match your digital modernization as you wish.

We can't wrap up this section without a brief discussion on HA in API Connect. So, let's review that next.

HA

HA is an advanced topic and deserves a lot of planning. Here, you'll learn just the basics, as there is much to consider when setting up HA, and there are various options you would need to bear in mind based on cost, SLA, and effort.

To achieve HA in API Connect, the API Manager, Developer Portal, Analytics, and gateway components all require a quorum. A quorum is a minimum of three instances or nodes in a cluster. Both Kubernetes and API Connect components require a quorum to support HA. So, whether you are using a deployment model of API Connect on VMware or API Connect using Kubernetes, the requirement is still the same.

The condition the quorum is trying to prevent is referred to as a **split brain**. In this scenario, your cluster believes there are two primaries within the HA cluster quorum. Normally, there is one primary synchronizing with secondaries. When the primary goes down, the secondary usually takes over as the primary. However, what if the primary is functioning but the network link goes down?

Well, a secondary could believe the primary has gone down, so it makes itself the primary so that it can continue processing. Sound good? Not really. The original primary believes it is still functioning. So, now we have two primaries, which means database synchronization is running independently. When the network link becomes active, we have two primaries. So, who's the boss? When you have two primaries, this is called a split-brain scenario. It can lead to data inconsistencies.

Quorums aren't unique to API Connect. You can find them with databases (such as MySQL and MariaDB) or with Kafka, where ZooKeeper requires a quorum.

In API Connect, the way to calculate the node failure tolerance is *N/2+1*. So, in our minimal case of three nodes, we require two instances to form a quorum. Some examples are as follows:

- *4 nodes = 4/2+1 = 3 nodes for a quorum*
- *5 nodes = 5/2+1 = 3 nodes for a quorum*
- *7 nodes = 7/2+1 = 4 nodes for a quorum*

So, what split-brain scenario happens to API when a quorum is not achieved? We'll review by each component:

- **Gateway**: APIs will continue to run, but API configurations are constrained. Additionally, you will be unable to store or share revoked OAuth refresh tokens and rate limiting in the **API Connect Gateway** (**apigw**) service.

> **Important Note**
> Rate limiting will continue if you are using the v5c gateway service.

- **Manager**: API Manager will continue to work but only in read-only mode. You will not be able to publish APIs nor create applications.

- **Analytics**: You can view existing analytics, but new data from the gateway is stopped.

- **Developer Portal**: The Developer Portal cannot be accessed. You will not be able to register new consumers nor register new applications.

You now have a good understanding of the various deployment models and how API Connect handles HA. Having this information puts you in the best position to enable you to architect your hybrid cloud infrastructure so that it is highly available, resilient, and performant.

Before beginning to work with API Connect, it is a good time to introduce you to another valuable capability of API Connect. That is the ability to execute many of its features via the command line.

Introduction to the CLI commands

In this section, you will learn how to use the basic CLI commands. The examples shown here have been initiated from a Mac but will be the same for Windows environments.

The CLI commands come with the API Connect toolkit. You can download the toolkit from the web on Fix Central (`https://www.ibm.com/support/fixcentral/`). Alternatively, if your company has API Connect installed and you have access to API Manager, then you can download it from there. In fact, you are guaranteed to get the correct version if you do it in that manner:

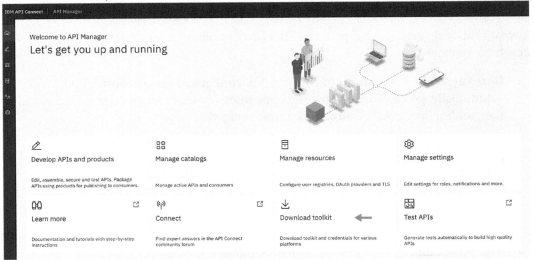

Figure 2.18 – Downloading the toolkit from API Manager

For what reasons would you want to use the CLI? Most of the time, the CLI is incorporated within your DevOps pipeline. However, before you start learning how to call the CLIs, you should try to understand the capabilities that are available within the CLI. One thing you should be aware of is the target of your CLI commands. When executing the commands, the location of the output can be either of the following:

- **Local filesystem**: This is for creating APIs and products locally. It uses general-purpose commands.

- **API management**: Here, commands are executed toward managing resources on the API Manager server, including draft APIs.

These distinctions are important because some commands might seem similar but have different implementations. On the one hand, you are generating APIs from the command line, and on the other hand, you are deploying and managing those same APIs against API Manager.

You have been briefly introduced to the development capabilities of API Connect. As a developer, you understand that you start by creating APIs. Then, you package them in a product, stage or publish them to a Catalog, and, finally, maintain a life cycle of revisions and updates. *Figure 2.19* shows the flow you could take to create an API and deploy it to API Connect.

The CLI API development flow has a series of steps that are required to promote your locally defined API (or an API that is checked into a source code manager such as Git) and published within API Connect's management component. These steps are very similar to what you need to introduce into your DevOps pipeline:

1. Determine which **Identity Provider (IDP)** you will be accessing.

2. Log in to the API Connect using the `login` CLI command.

3. Identify the Provider organization you will be updating.

4. Set a Catalog within the Provider organization as the target for the command.

   At this point, you have an established destination for publishing. Next, you can create your API.

5. Create an OpenAPI file using your favorite OpenAPI tool or the `create` CLI command.

6. Using the draft option in the CLI, create a draft API that moves the API into API Manager.

7. Create a draft product that establishes a relationship between the APIs and the plan(s) that will be contained within the product and moves that into API Manager.

8. Once your draft is ready to promote to a testing environment, you can use the CLI to establish the connection to the environment and publish the product.

The flow you just learned is shown in *Figure 2.19*:

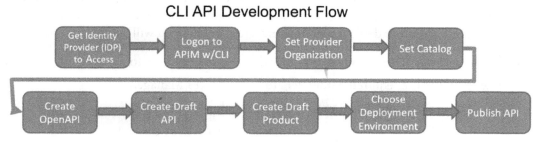

Figure 2.19 – Development process using the CLI

Your administrators have different tasks to perform. Often, their role is to provision new Provider organizations, update TLS certificates, and configure user registries. They will use other CLI commands to perform those activities. In the new world of automation, you can see these are perfect examples of tasks that can be streamlined for greater agility.

Greater agility leads to greater productivity, which leads to greater success in your digital transformation. You'll learn more about how the CLI participates in the DevOps process in *Chapter 14, Building Pipelines on API Connect*. With this background introduced, now is a good time to get you started with some simple CLI commands to whet your appetite.

Having fun playing with FHIR

FHIR is an emerging standard that many healthcare companies are moving to as they digitally modernize. FHIR is based on serving up REST APIs with JSON and is a good case study to begin our introduction into CLI.

To set the context of how the CLIs work, it's important to understand how an executed command integrates with API Connect. API Manager maintains the database of all the artifacts and resources that comprise the API Connect cloud. When you issue a command that will interface directly with API Connect, the call will pass through the **Platform-API** interface. For that to happen, you need to log in to API Manager. Additionally, your credentials need to have the appropriate role and permission to execute the CLI commands.

To make this introduction more interesting, we'll take a typical workflow scenario and issue the appropriate commands to achieve the goal. We'll assume you have downloaded the toolkit highlighted in the *Technical requirements* section and ensured that it's within the path.

You, as a developer, have been tasked to take an existing OpenAPI FHIR swagger file and create an API definition within the APIC healthcare organization. You must upload the FHIR API into API Connect and stage and publish it to the Sandbox Catalog. Your goal is to take what you have learned and teach the DevOps team, so they understand how to get started on building a future deployment pipeline. Here is what you will learn:

1. How to get your IDP; you need to know what registry to log in to.
2. How to supply the required parameters to log in to API Manager
3. Set the command configuration to specify a Catalog for deployment.
4. Create an FHIR API draft by using an existing OpenAPI document.
5. Run a stage deployment of your FHIR API to that Catalog.
6. Publish the FHIR API to the Catalog that also pushes it to the portal.

To identify the arguments that you will need to provide for each CLI command, you can use this guide to identify the required parameters. For these examples, we present arguments that can be used in a fictitious environment. As you run CLI commands that specify a server name, you will need to replace those with your server name. The use of the Sandbox Catalog will be throughout these examples because the Sandbox Catalog is always created for you, by default, when you create a new Provider organization. To provide some context regarding the target environment, *Figure 2.20* depicts the API Connect Manager's relationship with the Provider organization and Catalogs:

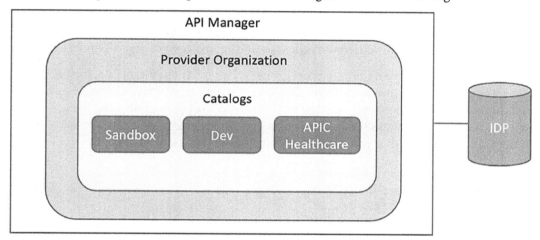

Figure 2.20 – An example target API Connect environment

The < > symbols signify a field, and you must supply a value for these fields. Inside the symbols will be example values to give you some context. Some examples include the following:

```
<api manager host> = api-manager-ui.apicisoa.com
<userid> = isoadeveloper
<catalog> = sandbox
<organization> = middleware
```

Now that you understand how to provide the variable arguments, you can begin with the CLI development flow. Open up a Terminal window/Command Prompt, and we will go ahead and learn how to interact with the IDP.

Determining realms and IDPs

There is a concept called **realms**. If you recall, the management server comprises two user interfaces: Cloud Manager and API Manager. When you use the CLI commands, you need to specify which target environment (either Cloud Manager or API Manager) you want the commands to apply to. Here are the realms:

- For Cloud Manager, the scope is called *admin*.

- For API Manager, the scope is called *provider*.

You need to find out which user registry points to the realm you desire. API Connect can have many user registries for different realms, so finding the correct one is important.

You will use the `apic identity-providers:list` command to obtain a list of registries used for Cloud Manager. To do that, you will have to pass the `scope` argument, specifying `admin`. Additionally, you need to provide the hostname, following the `-server` argument. As mentioned earlier, `<api-manager-ui.apicisoa.com>` represents the hostname that you will substitute with the hostname of your API Connect environment:

```
apic identity-providers:list --scope admin --server
<api-manager-ui.apicisoa.com>
```

```
default-idp-1
```

Notice that `default-idp-1` is returned. This means that when you log in to Cloud Manager, you must specify this user registry.

So, what about getting Providers? You can use the same command, but this type provides a scope of `provider`:

```
apic identity-providers:list --scope provider --server
<api-manager-ui.apicisoa.com>
```

```
default-idp-2
```

Once again, it returns the user registry. This time, it is telling you that `default-idp-2` is what you will be using when you log in to API Connect to work with API Manager. Now that you have that important data, it's time to log in.

Logging in to API Manager

The first thing you must do to interact with API Connect is to log in. You will be logging in as a Provider. The last command you ran provided you with the user registry to use (that is, `default-idp-2`). You will use that registry in the `login` command. You can run the `apic login` command by passing in all of the arguments or by issuing it interactively:

```
apic login --username <isoadeveloper> --password <password>
--server < api-manager-ui.apicisoa.com > --realm provider/
default-idp-2
```

You can choose either. Doing it interactively is shown here:

```
apic login
Enter your API Connect credentials
Server? <api-manager-ui.apicisoa.com>
Realm? provider/default-idp-2
Username? <isoadeveloper>
Password?
Warning: Using default toolkit credentials.
Logged into api-manager-ui.apicisoa.com successfully
```

All of the arguments are pretty straightforward. Only with realms can you specify it a little differently. Here, you provide the realm with the scope value and registry demarcated with a slash: `provider/default-idp-2`.

If done properly, you will get a successful message. If you are planning on working within a particular Catalog for multiple commands, it's useful to set your configuration to point to a Catalog as the default.

Setting the target Catalog and Provider organization for deployments

Often, you might be deploying several products to the same Catalog. By setting a configuration setting for Catalogs, you can omit the need to specify a Catalog when deploying. You can also do the same for Provider organizations:

1. To set configurations, you can use the `apic config:set` command. To set a Catalog, you specify a URL. This restful format is useful to remember for the future. The format is as follows:

    ```
    [Hostname of the API Manager]/api/catalogs/[name of
    organization]/[catalog name]
    ```

You must know the Provider organization and Catalog name ahead of time. Executing the command will set the Catalog in the configuration:

```
apic config:set catalog=<https://api-manager-ui.apicisoa.
com>/api/catalogs/<middleware>/<sandbox>
catalog: https://api-manager-ui.apicisoa.com/api/
catalogs/middleware/sandbox
```

2. Setting the Provider organization follows the same format, but the first argument changes from a Catalog to an organization. Now you can set the middleware organization as the default in the configuration:

```
apic config:set org=<https://api-manager-ui.apicisoa.
com>/api/orgs/<middleware>
org: https://api-manager-ui.apicisoa.com/api/orgs/
middleware
```

Since both of the commands returned successfully, everything is fine. However, you can validate that they are set if you wish to.

3. Check the configuration settings that you applied. To do so, you simply issue the `apic config:list` command:

```
apic config:list
catalog: https://api-manager-ui.apicisoa.com/api/
catalogs/middleware/sandbox
org: https://api-manager-ui.apicisoa.com/api/orgs/
middleware
```

So far, you have learned some general commands that are useful to you, the developer, and other roles, such as the DevOps engineer. They will be using the same commands to orchestrate the DevOps pipeline in the future.

With the basics out of the way, it's time to draft some FHIR APIs and Products. You will do this with the draft CLI commands. Those commands take the form of `apic draft-products:create` and `draft-apis:create`.

You can use the `apic draft-products:publish` command to publish a product that's already present in the drafts. First, create and upload a draft API.

Creating and uploading a draft FHIR API

You can view API drafts in API Connect when you are undergoing development. A draft is an API that is currently being worked on. You can find draft APIs on API Manager under **Develop API and products**. This is where you perform the development of API definitions and other functions such as deploying products, changing life cycles, and applying security to your APIs:

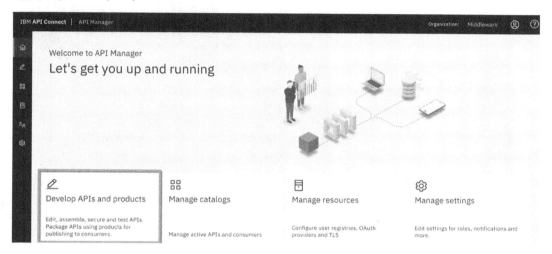

Figure 2.21 – Developing APIs and products

You can log in to API Manager to investigate how APIs are shown under **Develop APIs and products**. However, for now, we will continue with using commands to drive the flow:

1. On the GitHub site, as noted in the *Technical requirements* section, you will find two FHIR swagger files that you can play with. They fully support version 4.0.1 of the FHIR specification. Copy those files to your workstation so that you can try out the commands using these files.

2. Bring up Command Prompt and navigate to the folder where you copied the FHIR swagger files. You can view the files if you want to see more of what is contained, but for now, assume the files are fine. Let's convert one of the swagger files into an API Connect draft.

3. Creating a draft of an existing file is very easy. You will use the `apic draft-apis:create` command and pass in the appropriate arguments. There are only three required arguments:

    - The filename

    - The organization where the draft will reside

    - The server

4. Since we have already set the configuration for the organization, you only need to pass the filename and server. Run the following command:

```
apic draft-apis:create <Condition-swagger.json> -s
<api-manager-ui.apicisoa.com>
condition-api:4.0.1  https://api-manager-ui.apicisoa.com/
api/orgs/7c642f84-5028-43a7-89df-6cf2a5b7adc8/drafts/
draft-apis/17c10964-35a2-4010-ac73-fe92d312d8b4
```

The results are displayed along with a unique identifier. This identifier can be used to reference this API if we need to automate some behavior that can access this file in the future.

5. You can now log in to API Manager and review what just occurred:

Figure 2.22 – The draft created by the CLI

When you navigate to API Manager and select **Develop APIs and products**, you will find the newly created condition API, which is ready for customization. Next, you need to create a product so that you can publish it to a Catalog. You can certainly use the `apic draft-products:create` command to create a product and place it in the **Develop APIs and products** section within API Manager, but that will rely on a user working with the user interface, adding the APIs, and promoting them manually. That is one way many people do it, but you should be able to take APIs and products and deploy them directly from your build server. This is what you will learn how to do next.

Creating a Product with APIs

If you have an inventory of APIs and understand how you want to package them in a Product, using a command to do that is perfect. The `apic create:product` command works on your workstation or build server, not on the API Manager when using drafts.

When you run this command, it sets a version number of the Product and chooses the APIs you wish to package within it. Although we use the same `apic` command, the arguments are slightly different. You will provide a title of the Product, a specified version, and a list of APIs.

The `Observation-swagger.json` file was one of the two files you downloaded from GitHub. Run the command as demonstrated in the following code snippet. Notice that there is no server argument. The output will be on your local machine:

```
apic create:product --title "ObservationProduct" --version
1.0.0 --gateway-type "datapower-api-gateway" --apis
"Observation-swagger.json"
```
```
Created observationproduct.yaml product definition
[observationproduct:1.0.0]
```

> **Tip**
> Be mindful of the quotation marks in this command.

A successful creation will create the Product YAML file that you can now deploy. Let's do that next.

Publishing to a Catalog

When you publish your Product, you are publishing to a Catalog. You can publish it by using the `apic products:publish` command to pass the Product YAML filename, the Catalog, the organization, and the server. If you haven't set the configuration file within your organization and Catalog, you will need to supply them. Run the `apic config:list` command to verify:

1. Let's publish your product by running the `apic products:publish` command:

```
apic products:publish observationproduct.yaml -c
<sandbox> -o <middleware> -s <api-manager-ui.apicisoa.
com>
```
```
observationproduct:1.0.0 [state: published]
```
```
https://api-manager-ui.apicisoa.com/api/
catalogs/7c642f84-5028-43a7-89df-6cf2a5b7adc8/7aa487b3-
```

```
d215-48aa-a0a3-703053bb6a72/products/759842ca-2379-4b92-
8769-c4dbd58fb850
```

2. Once successful, you should be able to view the unique identities of the Catalog and Product. Now you can log in to API Manager and navigate to **Manage**. Then, click on the **Sandbox** tile and review the published Product:

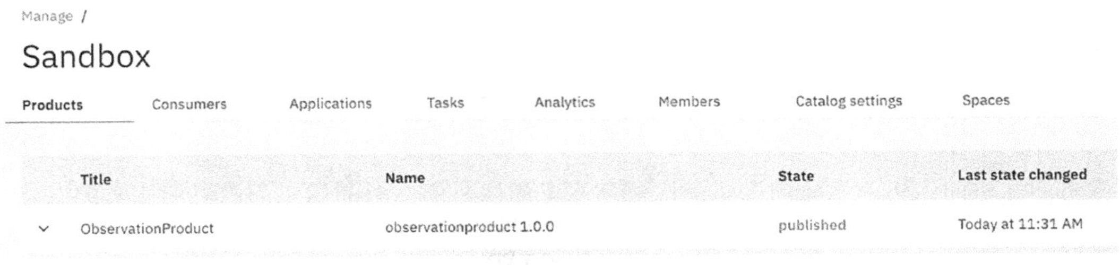

Manage /

Sandbox

Products	Consumers	Applications	Tasks	Analytics	Members	Catalog settings	Spaces

Title	Name	State	Last state changed
⌄ ObservationProduct	observationproduct 1.0.0	published	Today at 11:31 AM

Figure 2.23 – The published Product in Sandbox

As you can see in *Figure 2.23*, the Product is published with the version number specified. There is a lot more you can do with CLI, and you will learn more about that in *Chapter 14, Building Pipelines in API Connect*.

> **Tip**
> You might encounter errors when executing commands, and being able to troubleshoot them is important. There is the –debug argument that you can append to your commands to get information. But be prepared to get a lot of output.

Summary

In this chapter, you learned a lot about API Connect. You should now be familiar with the components of API Connect and the various deployment models. You have a lot of choices in terms of where to deploy API Connect, and it will certainly get you moving in the right direction on your digital journey.

You were also introduced to the API Connect CLI. The basic commands were provided so that you can utilize them and assist others such as the DevOps team. With the addition of some FHIR swagger files, you were able to walk through an API development flow from its setup to its publication.

You are now ready to learn more about preparing your API Connect topology and organizations so that you can begin building APIs for real. In the next chapter, you will learn how to configure your API Connect environment.

3
Setting Up and Getting Organized

Up to this point, you have learned the role that APIs play in our new digital transformation and modernization space. You have also seen how **API Connect (APIC)** can provide all of the tools you will need to realize these benefits in a seamless and secure manner. You are probably eager to dive right in and start creating and exposing your own APIs! You will get there but before doing that it is critical that you take some time to get organized and perform some housekeeping to ensure your digital transformation is following a framework for success.

If you think about it, an API is nothing new. It is simply a common piece of code or a module that can be called or consumed from some other program or application. The thing that sets APIs apart is how you organize, publish, and make them available to the outside world for consumption. The old *"if you build it they will come"* line of thinking really doesn't apply when we are talking about APIs. They need to be easily discovered by consumers and well organized. This is why taking the time to set up and get organized is critical. API Connect provides the framework for you to organize and publish your APIs so that consumers can discover them easily and know how to invoke them.

Of course, stating that you must take the time to plan out your organizational structure is easy to say, but if you are not aware of what each component is and how it could impact the final outcome, it might be a more difficult task than it should be and you will likely not get the desired outcome.

In this chapter, we will discuss and demonstrate how API Connect allows you to organize your API environments. This will include everything from who can create and manage products and APIs to how they are managed throughout their life cycle, and finally, how consumers can discover them. When you are finished, you should have a good understanding of how this all fits together, which will give you the knowledge to guide yourself through your own planning exercise. To accomplish this, we will cover the following main topics:

- The big picture
- Provider Organizations
- API Catalogs
- Utilizing spaces
- Developer Portal

Technical requirements

This chapter will start with the assumption that you already have an API Connect installation complete, along with the required components configured within the Cloud Manager. You should have all of your gateways services, SMTP server, and portal services already configured. Your hostnames being used for all of your services and components should be added to your DNS server and network communication between all components should be open.

The big picture

As you are introduced to all of the terminology and structure of how things are organized within API Connect, it could start to get a little overwhelming. So before diving into each piece of the setup that you will need to perform, let's take a look at the big picture and how things are organized. Hopefully, this will give you a high-level perspective on where each piece of the puzzle fits and provide some clarity on how you can structure your environment.

In *Figure 3.1*, you can see a very high-level overview of the main components and their hierarchy. You can see here that most of these components are required and you can have more than one defined. The one exception here is **spaces**. Spaces are not required but we will get into more detail on this in the *Utilizing spaces* section later in this chapter. Also, note that for simplicity, plans are not shown here as they will be covered in subsequent chapters:

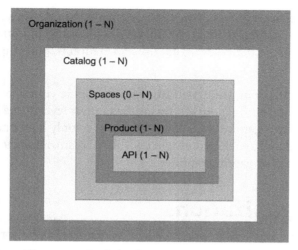

Figure 3.1 – High-level organization within APIC

As we go through each of the components shown in *Figure 3.1*, you can refer to it to see where and how each of these fits into the overall scheme. Although your implementation and organization will likely be more complex, this will give you the most simplistic view for your understanding.

For a more realistic illustration, *Figure 3.2* shows what a typical organization structure might look like. Your setup will likely expand upon this, but this will give you an idea of the basic structure:

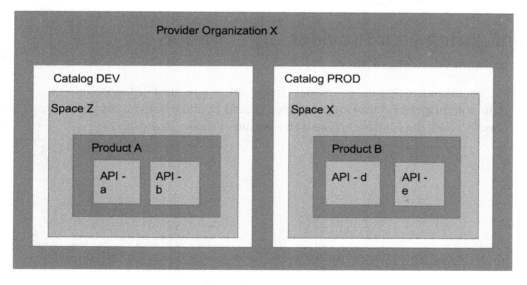

Figure 3.2 – Common configuration

The remainder of this chapter will take the time to explain the role of each of these components and how to configure them. Again, careful planning and understanding of these concepts will ensure that your digital transformation is following a framework for success!

Your first step when configuring your environment is to define your Provider Organization. As you can see in *Figure 3.1*, this is the highest level in the hierarchy. This is typically defined for your organization and it is common to only have one. It is, however, possible that you would like to define multiple Provider Organizations to represent different subsidiaries of your parent company.

Provider Organizations

A Provider Organization is just as it sounds. It is the organization that provides the APIs. This should not be confused with a consumer organization, which you might have guessed is the organization that consumes the APIs. Defining your Provider Organization is done within the Cloud Manager UI and must be defined before you or anyone else can access the API Management Console to develop and publish your APIs. It is here that you define a few important pieces of information, as follows:

- Organization title/name
- The user registry that will be used to authenticate the owner's user access
- The owner of the organization

Configuring your Provider Organization

Let's begin with the configuration:

1. To begin configuring your Provider Organization, you must be logged into the Cloud Manager UI. Once logged in, you should be at the home screen where you need to click the **Manage Organizations** icon as shown in *Figure 3.3*:

Figure 3.3 – Cloud Manager UI home screen

2. You should now see a screen that would show you all of your configured organizations. Since you don't have any yet, this list should be empty so you will need to add one. From this screen, in the top-right corner, you will see an **Add** button where you will click **Add | Create organization**.

3. This will bring you to a screen where you will enter all of the required information for your new Provider Organization.

4. The first field you will enter is the **Title** for your organization. This will be the name you will see when you view your existing Provider Organizations as well as the name that will be seen within the portal where your APIs will be discovered. For our example, we will create a fictitious healthcare Provider Organization with the title APIC Healthcare.

5. Once a value for **Title** is entered, you will see that the **Name** field will auto-populate with a slugified representation of the **Title** field. This means it is converted to lower case and spaces are replaced with the - character. For our example, our **APIC Healthcare** title was converted to **apic-healthcare**. This value will be used in the URL to access the Developer Portal and invoke APIs within this organization as well as to identify your organization within the Developer Toolkit CLI.

6. The next section, which will complete your Provider Organization configuration, will be to define the **Owner** of the organization. The owner of the organization will be the administrator of the organization who will have the authority to invite new members to the organization, author APIs and products, and manage the API and product life cycle. To set the organization owner, you have two possible options: you can select an existing user in one of the registries configured within the Cloud Manager, or you can add a new user to one of these pre-defined registries.

    All of your pre-defined public registries will be listed in the **User registry** field within a dropdown.

    > Info:
    >
    > User registries have visibility. When the cloud administrator sets a user registry to private, that user registry will not be available when creating a new Provider Organization.

    One of these must be selected before entering the owner's user ID. An important thing to note here is that once your Provider Organization is created, you cannot change this value to a different user registry.

7. Once the registry is selected, you can specify whether you will use an existing user from the registry selected, or you wish to add a new user. This will be specified via the radio button for the field **Type of user**. If you are using an existing user, you must specify the username in the **Username** field. If this is a new user to be added, you will see one additional field, **Email**, where you will enter the new user's email address in addition to **Username, First name, Last name, and Password**.

8. Once the organization title and the organization owner are set, you can click the **Create** button to create your organization. *Figure 3.4* shows our new Provider Organization, **APIC Healthcare**, being created. You can see that **Title** had spaces so **Name** was autogenerated as the title was transposed to all lower case and spaces were replaced with -. In this example, we chose to create a new user within our existing registry as the Provider Organization owner so we selected **New User** from the **Type of User** field and filled out the appropriate details:

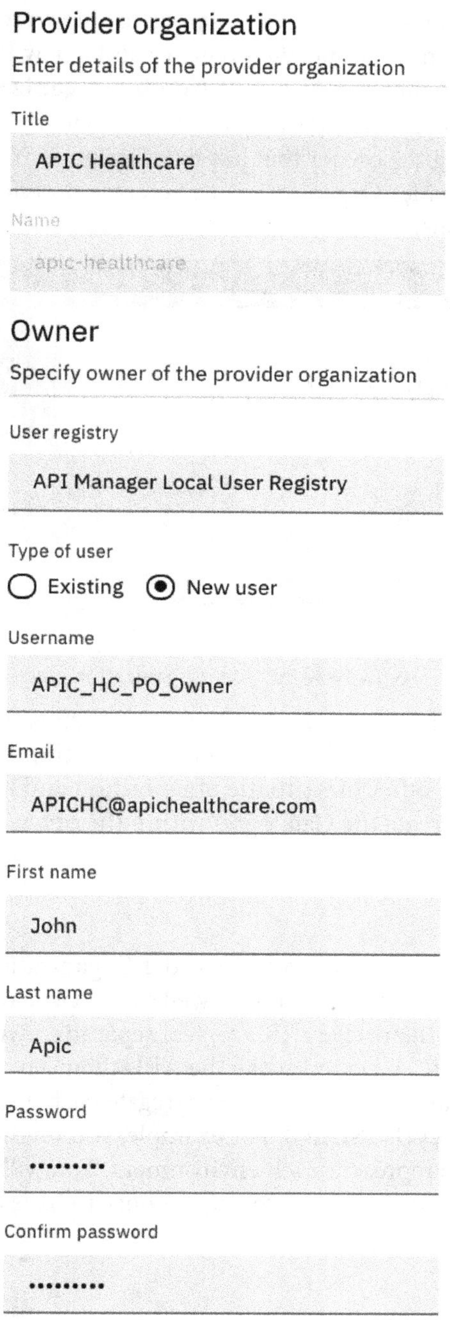

Provider organization

Enter details of the provider organization

Title

APIC Healthcare

Name

apic-healthcare

Owner

Specify owner of the provider organization

User registry

API Manager Local User Registry

Type of user

○ Existing   ◉ New user

Username

APIC_HC_PO_Owner

Email

APICHC@apichealthcare.com

First name

John

Last name

Apic

Password

••••••••

Confirm password

••••••••

Figure 3.4 – Create new Provider Organization

9. Once you complete the new Provider Organization step and click the **Create** button, a new user is entered into the specified registry and a new Provider Organization is created. You will be brought back to the **Provider Organizations** screen, where you will see a message indicating that your new Provider Organization has been created also indicating the owner that has been assigned. You can see in *Figure 3.5* that our example Provider Organization was created and now appears in the list of Provider Organizations:

Figure 3.5 – New Provider Organization created

Congratulations! You have now created your Provider Organization, which will be the foundation for all other components that you will define within it. The owner of this organization can now invite other users to the organization and move on to the next step of the process, which is to create the Catalog(s) within the organization.

API Catalogs

An API Catalog is a component within your Provider Organization that provides a logical partition within your Provider Organization as well as the Developer Portal and the gateway executing the APIs themselves. This logical separation will be apparent in the Developer Portal and the URL used to invoke the APIs themselves as they will be unique for each Catalog. Catalogs are typically used to segregate each environment within your **software development life cycle (SDLC)**. For example, you might configure DEV, STAGE, QA, and PROD Catalogs to represent each environment. You will see these as logical partitions within API Manager, however, you can choose to have different physical gateway services for each Catalog.

Now that you have your Provider Organization configured, you can move on to configuring your Catalog(s). The owner of the Provider Organization, or anyone that the owner has invited to the Provider Organization, can now log in to API Manager to perform this task. Before you begin to configure your API Catalogs, you should have already determined what Catalogs will be needed and what they will be named. This is all part of the careful, upfront planning you did before jumping right into your configuration effort.

Configuring your API Catalog

To begin configuring your Catalog(s), you will first log into API Manager using a credential that has been granted access to your particular Provider Organization. This could be the owner of the organization you previously set up, or some other person that was invited to join the organization. Let's move on to the next steps now:

1. Once logged into API Manager, you will be at the home screen with your newly-created Provider Organization shown at the top-right of the screen and several options on the main screen, as shown in *Figure 3.6*. From here, you will click on the **Manage catalogs** link:

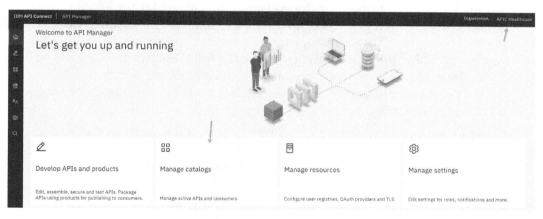

Figure 3.6 – API Manager home screen

2. This will bring you to the **Manage** screen where you will see all Catalogs currently configured for your Provider Organization. You will notice one Catalog named **Sandbox** is present, which is created by default when a Provider Organization is created. Since you just created your Provider Organization, you will only see the default Catalog named **Sandbox** as shown in *Figure 3.7*:

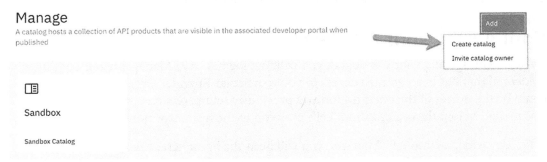

Figure 3.7 – Manage Catalogs

3. You can choose to use the default **Sandbox** Catalog to quickly test APIs and resources such as LDAP registries for API authentication. Using the Sandbox to initially test APIs has the benefit of not needing to create a Product nor acquire a ClientId or secret. Those are automatically generated for you in the Sandbox. The Sandbox should be limited to learning and testing APIs. You will most likely want to start creating your own, more meaningful Catalogs that you included while planning your overall architecture. As mentioned earlier, a practical use of Catalogs is to have one for each of your SDLC environments.

4. To start, we will create a Catalog named DEV to be used for our APIC Healthcare organization's development efforts and unit testing. To create a Catalog, from the **Manage catalogs** screen in API Manager, click the **Add** button at the top right of the screen, then click **Create catalog** as shown in *Figure 3.7*.

5. This will bring you to the **Create catalog** screen, showing the fields required to create a new Catalog where you will enter the **Title** for your new Catalog. In our example, we entered DEV as the title. Once the title is entered, you will see that the **Name** field is automatically populated with the slugified representation of the **Title** field. This will be used as part of the URL used to access the Developer Portal as well as to invoke your APIs published to this Catalog. This will be evident as we configure the Developer Portal.

6.   You will notice that the **Select user** field is prepopulated with the user logged into API Manager. This user will be the owner of the new Catalog. You can select from the drop-down list any user that has been added to the Provider Organization as the owner of this Catalog. If you wish to add a user to the registry that you would want to be designated as the owner of this Catalog, you will need to return to the previous **Manage** screen and select **Add | Invite catalog owner** as you can see in *Figure 3.7*. Once all fields are completed for your new Catalog, you will click the **Create** button to create the new Catalog. *Figure 3.8* shows this screen with the new Catalog being created named **DEV**:

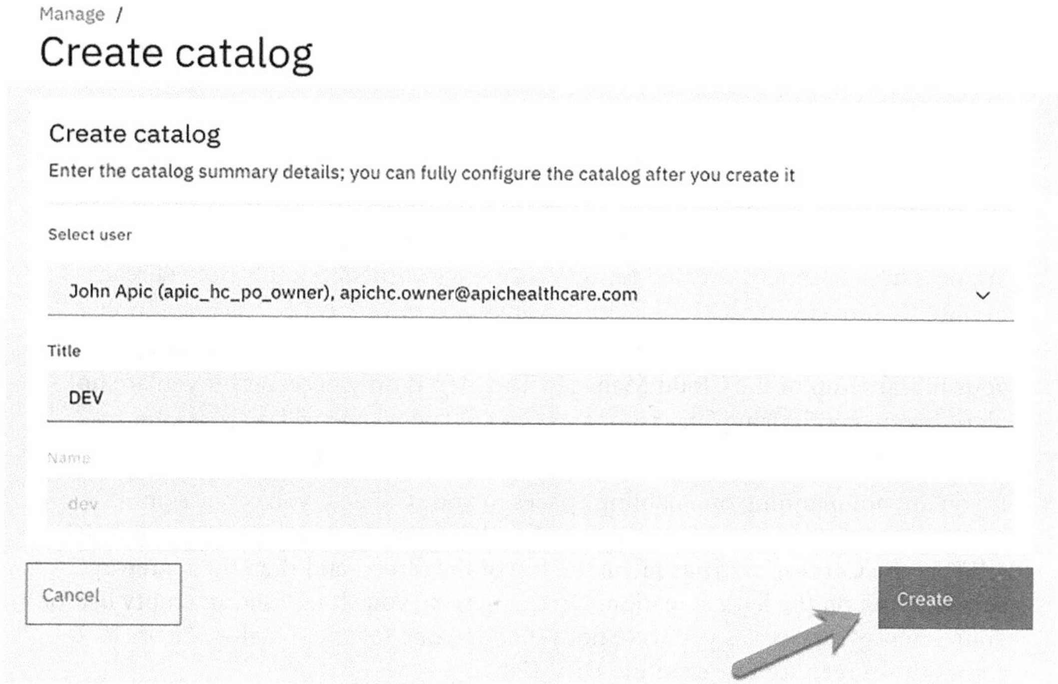

Manage /
Create catalog

Create catalog
Enter the catalog summary details; you can fully configure the catalog after you create it

Select user

John Apic (apic_hc_po_owner), apichc.owner@apichealthcare.com ⌄

Title

DEV

Name

dev

Cancel   Create

Figure 3.8 – Create a new Catalog

7. After you click the **Create** button, the Catalog is created and you will be returned to the **Manage** screen where you will see the Catalog you just created listed on the main screen with a message at the top right, as seen in *Figure 3.9*:

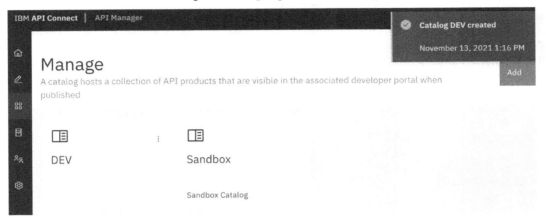

Figure 3.9 – New Catalog created

We now have a Catalog created for our DEV environment but it is not yet ready to publish and execute APIs. For this, a gateway service must be assigned to this Catalog. This will be one of the gateway services that you would have configured in your initial setup in the Cloud Manager. This step is only necessary if you are not planning on configuring Spaces for this Catalog, which we will discuss later in this chapter.

8. If you are not planning on enabling Spaces in your Catalog, you can assign a gateway service to it by selecting your Catalog from the **Manage** screen and then clicking the **Catalog settings** tab at the top of the screen and then the **Gateway services** link on the left navigation. On this screen, you should see an empty list for your gateway services as you have not yet added one for this Catalog. *Figure 3.10* shows this screen for our example DEV Catalog:

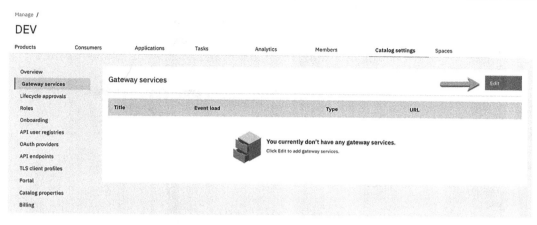

Figure 3.10 – Catalog settings | Gateway services

9. From this screen, you can add a Gateway service by clicking the **Edit** button in the upper-right corner of the screen. This will then bring you to a screen with a list of all gateway services configured that you can select from. Select the checkbox next to the gateway service you want to add to this Catalog and click the **Save** button. *Figure 3.11* shows a gateway service selected for the DEV Catalog in our example:

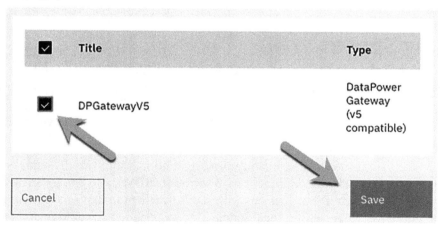

Figure 3.11 – Select gateway service

10. After clicking the **Save** button, you will be returned to the previous screen where you will see the gateway service added to the list of **Gateway services** along with a message at the top right indicating that it has been added to the Catalog as shown in *Figure 3.12*:

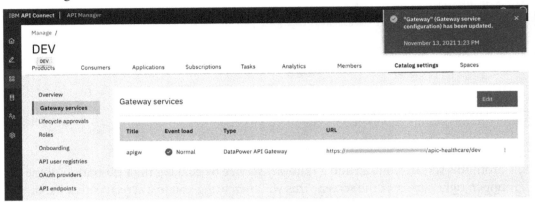

Figure 3.12 – Gateway service added

At this point, you should have a good understanding of what a Catalog is and how it could fit into your API Connect Provider Organization structure. In our example, we walked you through the creation of one Catalog named **DEV**. You can repeat this process for every Catalog you need to create for your organization. Once that has been completed, you can move on to the next steps, which we will describe in similar detail.

Utilizing Spaces

Spaces are a part of the API Connect syndication feature, which will allow you to subdivide your Catalog into partitions. Of all of the configurations we have covered within this chapter thus far, this is the first optional configuration we will demonstrate. Spaces could be useful if you have different divisions within your organization that you need to segregate within a Catalog.

For example, our sample healthcare organization might be split into two divisions within the company. One for **prescription benefit management** (**PBM**), and one for **insurance**. We might want to segregate these divisions within our Catalogs so that each division will be independent of the others within the Catalog. This independence would include the separation of authorized users, product life cycle, and runtime environment altogether.

Spaces would be the ideal feature to accomplish this because it provides this logical partition from all other spaces within the Catalog as well as a segregated runtime environment. In addition, Spaces will provide its own analytics for each space configured. Keep in mind that this only provides this segregation for development, management, analytics, and runtime, which are all internal. Spaces will not be evident within the Developer Portal to an outside consumer. To the outside consumer who will discover and consume your APIs, they will all appear to be within the same Catalog.

> **New with v10.0.1.5**
>
> There is a new feature to apply Catalog actions to Spaces. You can read more on this capability in the What's New section: `https://www.ibm.com/docs/en/api-connect/10.0.1.x?topic=overview-whats-new-in-latest-release-version-10015`

If you decide that your organization will require the use of Spaces, you must configure them at the Catalog level within each individual Catalog configuration.

Configuring your spaces

Here is how we configure the spaces:

1. To accomplish this, you will click the Catalog you wish to configure from the **Manage catalogs** screen. If you are at the **Home** screen from within the API Manager, you will navigate to **Manage catalogs | DEV**.

2. You will now be on the configuration screen for your Catalog. From here, you will click on the **Catalog settings** tab at the top of the screen and toggle the **Spaces** slider to **On** as shown in *Figure 3.13*:

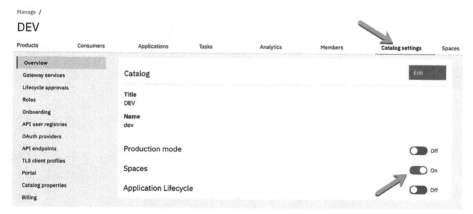

Figure 3.13 – Enable Spaces for a Catalog

Once you enable Spaces within a Catalog, you can no longer configure a gateway service for the Catalog. At least one space must be configured and assigned a gateway service. Products and APIs can only be published to a space and not the Catalog itself.

3. You can now begin to configure the individual spaces for your Catalog. While in **Catalog configuration**, click on the **Spaces** tab at the top of the screen. By default, you will see one Space listed with the same name as your Catalog. To create a new space, click the **Add** button in the top-right corner of the screen and then **Create space** as shown in *Figure 3.14*:

Figure 3.14 – Create a new Space

4. This will now bring you onto the configuration screen to configure your new space where you will enter the details required. These details should be starting to look familiar to you as they are similar to the ones entered for your Provider Organization and your Catalogs. Like these, you will select the owner for space and provide a title that will then be converted to lower case and used to auto-populate the **Name** field.

If you wish to invite a new user to be the owner of the space, you will go to the previous screen and click **Add | Invite space owner**. Once these fields are populated, you can click **Create** to create your new Space. *Figure 3.15* shows the creation of a new space for our healthcare organization with a **Title** of **PBM**:

Manage / Spaces /

Create space

Create space

Enter the space summary details; you can fully configure the space after you create it

Select user

John Apic   ⌄

Title

PBM

Name

pbm

Cancel   Create

Figure 3.15 – Create a new Space

5. Once you click the **Create** button, the space is created and you are brought back to the **Catalog configuration** page. Here, you should see your new space listed and a message at the top right. You can see in *Figure 3.16* the new Space we created within the **DEV** Catalog named **PBM**:

Figure 3.16 – New Space created

6. Once you have created all of your spaces for your Catalog, you must repeat the process for each Catalog in your Provider Organization that you want to use spaces within. In our fictional healthcare company, we configured two spaces for each Catalog, one for PBM and one for Insurance as we want each division to have its own management, life cycle, analytics, and runtime. When we publish our products and APIs, this isolation is never seen in the Developer Portal. *Figure 3.17* shows our **DEV** Catalog with these two spaces created:

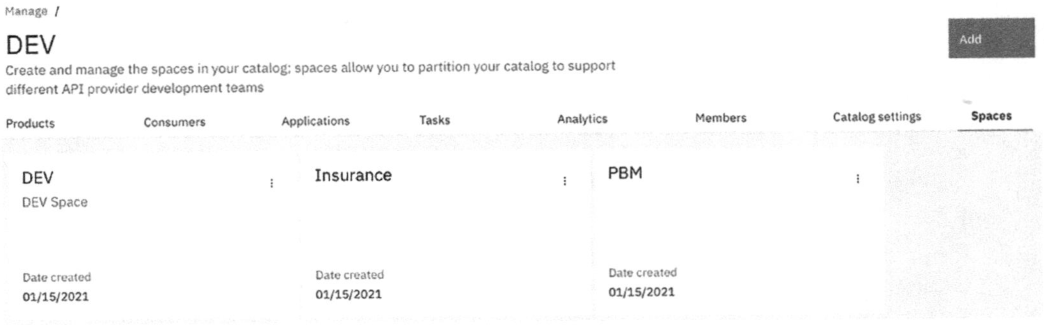

Manage /

DEV

Create and manage the spaces in your catalog; spaces allow you to partition your catalog to support
different API provider development teams

[Add]

Products	Consumers	Applications	Tasks	Analytics	Members	Catalog settings	**Spaces**

DEV		Insurance		PBM	
DEV Space					
Date created		Date created		Date created	
01/15/2021		01/15/2021		01/15/2021	

Figure 3.17 – DEV Catalog Spaces

7. If you recall when we configured our Catalogs, you had to add a gateway service but only if you were not planning on utilizing spaces within the Catalog. That is because each space will have its own runtime, or gateway service, configured. To configure the gateway service for each of your spaces, you will click on the space from within the **Spaces** tab of your Catalog. From here, you will navigate to the **Space settings** tab and click the **Gateway services** link on the left navigation menu. From here, you can add a gateway service to the space the same way you would have for a Catalog by clicking **Edit**, selecting the gateway service to add, and then **Save**.

The structure of our new organization is now starting to take form. Up to this point, we have defined our Provider Organization, Catalogs, and spaces. We now have the ability to start publishing products and APIs, which we will cover in later chapters.

Developer Portal

Once you have configured your carefully planned out Provider Organization, Catalogs, and spaces, you can technically begin developing and publishing your APIs and products. Although you might be tempted to jump right into this, there is one final piece we must configure so that the consumers can discover your APIs—the Developer Portal. You might develop the most useful APIs going, but as we said earlier in this chapter – *if no one can find them, they aren't much good.*

The Developer Portal is the place developers will go to discover and register for your APIs that are available. This is where you will socialize your APIs to the outside world. This is their window into your API organization. In addition to socializing your APIs, the Developer Portal also provides analytics, forums, blogs, and rating facilities.

Since this is where the consumers of your APIs will go to find them, the Developer Portal will be a representation of your company to the outside world. For this reason, the Developer Portal is fully customizable so that you can brand and customize it to best represent your organization. The level of customization is entirely up to you, though. You can put as little or as much time into customizing it as you see fit.

Before we think about customization and branding, we must first configure the Developer Portal for each Catalog.

Configuring the Developer Portal

This is done at the Catalog level for each Catalog you have configured within your Provider Organization.

1. To begin this configuration, you must be logged into API Manager and be on the **Home** screen.

2. You will then navigate to **Manage catalogs | DEV** and click the **Catalog settings** tab. From this screen, you will click the **Portal** link on the left navigation menu as shown in *Figure 3.18*:

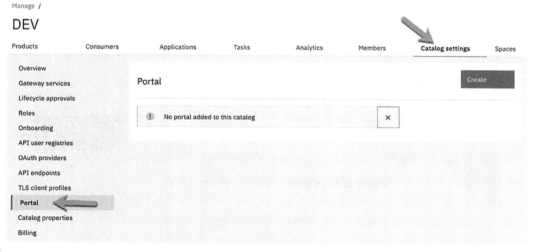

Figure 3.18 – Catalog settings

3. From this screen, you can see that there are no Developer Portals added to this Catalog so we will need to create one by clicking the **Create** button in the top-right corner.

    This will bring you to a screen where you will simply need to select one of the Portal services that were created in the Cloud Manager during your initial setup. You will see that the URL field will be automatically populated for you, which will include the Provider Organization and the Catalog name in the URI.

4. Clicking the **Create** button then completes the Developer Portal creation. *Figure 3.19* shows this screen where we are creating a new Developer Portal for our DEV Catalog. You can see that we have chosen our previously defined portal service named **Developer Portal** and the URL was automatically populated with the base URL defined in the Portal Service and the URI made up from the Provider Organization name and the Catalog name:

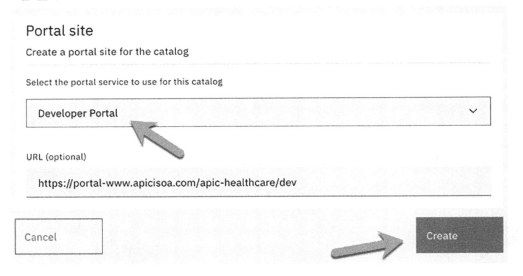

Figure 3.19 – Create a new Developer Portal

5. Once you click the **Create** button, you will be brought back to the previous **Catalog settings** page. That page indicates that your new Developer Portal creation has been initiated along with the new Developer Portal details showing the Portal Service being used, the Portal URL, and the User Registry used. *Figure 3.20* shows this confirmation page for the creation of our new Developer Portal for our DEV Catalog in our APIC Healthcare Provider Organization:

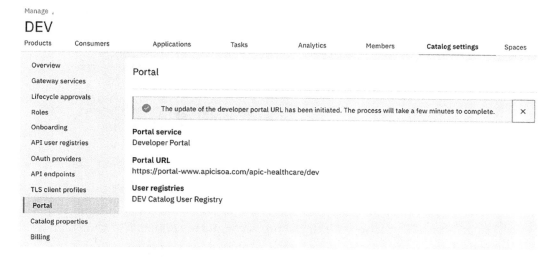

Figure 3.20 – Developer Portal creation confirmation

6. The Developer Portal creation can take some time to complete, so be patient. Once it completes, you will receive an email that will be sent to the email address listed for the user that you have logged into API Manager with. The text within this email will be addressed to **Administrator (admin)** because this email will contain a link that will bring you to the Developer Portal page to reset the password for the **admin user**. An example of this email is shown here:

Administrator (admin),

Your Developer Portal site has been created. You can log in as the 'admin' user by using the following one-time log in link. After you have logged in with the following link, you must change the password for the 'admin' user account immediately.

https://portal-www.apichealthcare.com/apic-healthcare/test/user/dev/1/1636831498/7og7bulz6iCcoGy8nORVrn0Qai2AAC925Ca25xGRq6Y/new/login

For more information on setting up and customizing your new site see:

http://www.ibm.com/support/knowledgecenter/SSMNED_v10/com.ibm.apic.devportal.doc/capim_devportal_admin.html

Note: The 'admin' user is a special system user and cannot also be used as an API Consumer.

Figure 3.21 – Developer Portal creation email

This is a one-time-use-only link and will bring you to the page shown in *Figure 3.22*, which will make it clear that this link can only be used once. From here, you will click the **Sign in** button to change your admin password:

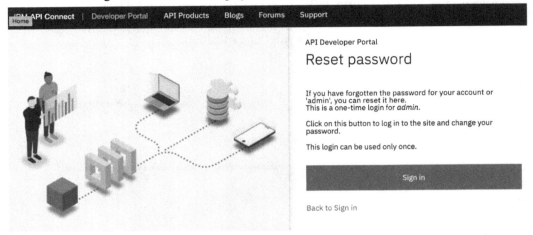

Figure 3.22 – Change admin password

7. The next screen will again make it clear that this is a one-time login link and you will enter your new password for the admin account and confirm it as shown in *Figure 3.23*:

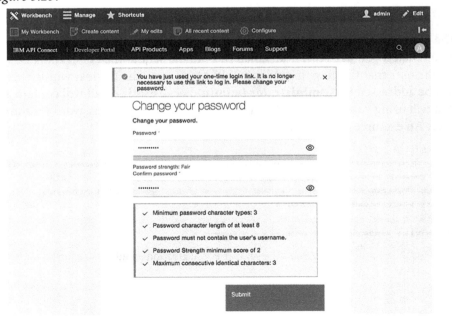

Figure 3.23 – Enter new admin password

8. After entering your new admin password and confirming it, click the **Submit** button to complete the process. If the passwords are valid, your admin password will be changed and you will be redirected to the Developer Portal home page as shown in *Figure 3.24*:

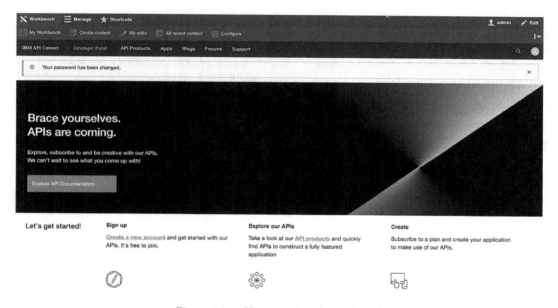

Figure 3.24 – Your new Developer Portal

You have now successfully added a Developer Portal for your Catalog. You would need to repeat this process for each Catalog within your Provider Organization the same way. As you complete this process for each Catalog, you will notice that the URL to access the Developer Portal for each Catalog will change to reflect the Catalog that it belongs to. As you publish your products and APIs to each of these Catalogs, you will see them appear in the respective Developer Portal for consumers to discover.

At this point, you can move on to customize your Developer Portal to personalize and brand it to represent your organization, or you can simply leave the default look. The choice is entirely up to you. You can find more information on customizing the look of the Developer Portal in *Chapter 15, API Analytics and the Developer Portal*:

```
https://www.ibm.com/support/knowledgecenter/SSMNED_2018/com.
ibm.apic.devportal.doc/capim_portal_controlling_appearance_de-
veloper_portal_drupal8.html
```

You have now completed the configuration of your entire Provider Organization and can now begin to create, publish, and socialize your APIs!

Summary

In this chapter, we have taken you on a long journey and introduced you to many terms and concepts. You were introduced to configuring your first Provider Organization, Catalogs, spaces, and finally, the Developer Portal. All of this may seem a bit overwhelming at first, but as you go through your planning sessions to determine the structure and layout of your own organization, this will become clearer. It cannot be stressed enough how critical careful upfront planning and organization are to the success of your digital framework transformation. This will become more evident as you go through the API life cycle from development, to publishing, to retirement.

Now that you have learned about the foundation for setting up your API environments, our next chapter will expand on this by explaining and demonstrating the process of creating your own APIs.

Section 2: Agility in Development

API Management is more than creating APIs. In this section, you will learn how to deliver an API that is performant, secure, measurable, and consumer-centric.

This section comprises the following chapters:

- *Chapter 4, API Creation*
- *Chapter 5, Modernizing SOAP Services*
- *Chapter 6, Supporting FHIR REST Services*
- *Chapter 7, Securing APIs*
- *Chapter 8, Message Transformations*
- *Chapter 9, Building a GraphQL API*
- *Chapter 10, Publishing Options*
- *Chapter 11, API Management and Governance*
- *Chapter 12, User-Defined Policies*

4
API Creation

A software platform's success is often defined by the ease with which engineers can build solutions using that platform. What can lead to an elongated **Software Development Life Cycle (SDLC)** and is a genuine gripe among experienced engineers (such as yourself, of course!) is if the software of their choice does not allow for compartmentalized development and testing.

Apart from support for a compartmentalized development platform, other factors that are extremely important for engineers are ease of tool navigation, support for logic constructs (if, switch, and so on), pre-packaged capability modules (built-in policies), the ability to create custom modules (user-defined policies), and ease of testing. **API Connect (APIC)** recognizes these developer demands and meets them by providing a comprehensive and feature-rich development toolset.

This toolset's main components are a desktop-based API development tool, **API Designer (Designer)**, a container-based runtime execution platform, **Local Test Environment (LTE)**, and a comprehensive cloud-based tool, **API Manager**. API Manager, through its web interface, provides administration, configuration, development, and unit testing (based upon **DataPower Gateway**) capabilities. Designer and LTE can enable you to get started on your APIC journey in a much faster manner by removing dependencies on centralized API Manager and DataPower Gateway.

The goal of this chapter is to take you through a journey of these tools, help you set up your localized development environment, and introduce you to the development aspects of API development for you to build and test great APIs. In this chapter, we will cover the following topics:

- Introduction and installation of Designer and LTE environments on a local workstation
- Working with Designer's interface and the key commands of LTE
- OpenAPI design
- Methods to extract data from a message payload
- Logic constructs and policies available in APIC and using them for your API development

After you have finished going through each of the preceding topics, you should be comfortable with the following:

- Installing Designer and LTE on a Windows workstation
- Navigating around Designer and executing some important LTE commands
- Creating a RESTful API
- Unit testing your APIs
- Enhancing your APIs with different policies, logic constructs, and variables

Technical requirements

The steps discussed in this chapter are based on Windows 10 Pro, hence you will require Windows-specific software. The setup of the Designer and LTE components requires the following prerequisites:

- Download and install Docker Desktop for Windows: `https://www.docker.com/products/docker-desktop`.
- Download APIC toolkits and LTE (requires IBM ID and a proper API Connect license): `https://www.ibm.com/support/fixcentral/swg/select-Fixes?parent=ibm%7EWebSphere&product=ibm/WebSphere/IBM+API+Connect&release=10.0.1.2&platform=Windows&function=all`. Download the following files:

```
toolkit-loopback-designer-windows_10.0.1.5.zip
apic-lte-images_10.0.1.5.tar.gz
apic-lte-win_10.0.1.5.exe
```

> **Note**
> You can search for the preceding files in the search results without the filename extensions.

- Download three sample JSON files from the GitHub site: `https://github.com/PacktPublishing/Digital-Transformation-and-Modernization-with-IBM-API-Connect/tree/main/Chapter04`.

- Network connectivity from your workstation to the FHIR server: `https://stu3.test.pyrohealth.net`.

You will use the downloaded prerequisites to set up the development environment on your workstation. But before that, you should understand the various development tools available as part of the APIC toolset. APIC provides development tools for various kinds of requirements and preferences. These tools can be broadly classified into two categories:

- **API development tools**: Tools used to build the APIs

- **API runtime/execution environment**: Tools used in the execution of the APIs

You will learn in detail about each of these categories and their installation procedures.

Development tools

APIC development tools accommodate various developer preferences. These preferences could be development through a command-line environment, a desktop-based UI-driven tool, or a web-based tool. These accommodations are made with the support of the following toolset options:

- **Command-line toolkit option (CLI) also known as toolkit**: This toolkit provides commands for the cloud (called the Cloud Manager user interface) administration and API development and management. Typically, you will not be using the toolkit for API development purposes. It is commonly used for scripted cloud management (user management and gateway server management) and API management (API, Product, and application management).

- **Designer user interface option (CLI + LoopBack + Designer) also known as Designer**: You will use this for most of your API development work and this is going to be the focus of this chapter.

- **API Manager**: You can also develop APIs using the web-based API Manager tool. API Manager provides a more comprehensive development platform (as compared to Designer) where you can create new catalogs and manage API life cycles, for example. These advanced facilities are not available in Designer.

You can see that APIC provides many development tools to cater to all kinds of developer preferences. The next section will take you through the installation of Designer on your workstation.

Installing Designer

Before you start with the installation of the Designer component, ensure that you have downloaded the file `toolkit-loopback-designer-windows_10.0.1.5.zip` as specified in the *Technical requirements* section of this chapter. The following steps assume that you are carrying out the installation on a Windows platform. If your environment is either macOS- or Linux-based, please consult the IBM documentation for your environment-specific steps. You will probably have to execute (`chmod +x`) permissions for the `apic` file on a non-Windows environment:

1. Extract the contents of `toolkit-loopback-designer-windows_10.0.1.5.zip` to a folder of your choice. You should see two files:

    `apic.exe`: Supports the CLI option discussed earlier

    `api_designer-win.exe`: Supports the desktop-based UI-driven option

2. It is advisable to set the location of `apic.exe` in the `PATH` variable.

3. Go to the folder where you have extracted the contents of the `toolkit-loopback-designer-windows_10.0.1.5.zip` file.

4. To accept the toolkit (CLI) license, execute the following command:

    ```
    apic licenses --accept-license
    ```

5. Install Designer. Execute the following command and follow the standard installation steps:

    ```
    api_designer-win.exe
    ```

6. Launch Designer by either clicking on the API Designer icon on your desktop or by executing the `API Designer.exe` file in the directory where you installed Designer.

7. Accept the license.

8. Create a **workspace** directory on your workstation where Designer will store various artifacts. This is the directory where Designer will store the API and Products that you create using Designer. You will need to provide the path to this directory after the initial login screen.

9. Designer requires credentials to connect to an API Manager instance or to an LTE. You will provide these credentials after the installation of your LTE.

10. You can exit Designer for now. You will come back to it after performing the installation and configuration of the LTE (covered in the next section).

You just completed the setup of the CLI toolkit and Designer. As was mentioned earlier, these two tools will help you build your APIs (by using Designer) and manage your cloud environment (by using the toolkit).

You now need a local API execution environment where APIs can be published and tested. The environment to publish and test the APIs is provided by the LTE component. It is discussed next.

Installing the LTE

One of the vital requirements of a development-friendly platform is its ability to support local/desktop-based testing. The APIC platform meets this critical requirement by providing a local runtime execution environment called LTE for you to publish and test your APIs. This platform is based upon the Docker container framework. It is worth mentioning that the API Manager web-based tool uses DataPower Gateway for API execution.

There are multiple components of this containerized LTE platform. These components are an API management service, two DataPower gateways (API and v5 compatible), an authentication gateway, a local user registry, and a Postgres database. APIC does an excellent job of hiding the many components' installation complexity and provides a more straightforward installation path via a container deployment model. We will cover this installation in this section.

Prerequisites

Before performing the installation of LTE on your local workstation, ensure that the following prerequisites are taken care of for a successful installation of LTE:

1. Install the toolkit and Designer components as covered in the previous sections.

2. Create a directory. Place the following files in that directory:

   `apic-lte-images_10.0.1.5.tar.gz`: Container images for various LTE components.

   `apic-lte-win_10.0.1.5.exe`: This is the LTE binary. This is required to execute the LTE commands. It is advisable to have this file's directory in your PATH variable.

3. Ensure that the Designer and the LTE components are from the same APIC fix pack release for them to work together. For your setup, you use `toolkit-loopback-designer-windows_10.0.1.5.zip`, `apic-lte-images_10.0.1.5.tar.gz`, and `apic-lte-win_10.0.1.5.exe` (also mentioned in the *Technical requirements* section of this chapter). As is apparent from these filenames, they are from the same APIC fix pack release.

Installation

Having ensured that the prerequisites are in place, go ahead and execute the following commands. These can be executed on the Windows command shell, Windows PowerShell, or a Linux terminal (if you are performing this setup on Linux):

1. Test that Docker is correctly installed on your machine. Execute the following command:

    ```
    docker version
    ```

2. Start Docker by executing this command:

    ```
    docker run -d -p 80:80 docker/getting-started
    ```

3. Load LTE container images to your Docker repository. You will need to specify the path to the `apic-lte-images_10.0.1.5.tar.gz` file in this command:

    ```
    docker load -i <images-tar-directory>\apic-lte-images_10.0.1.5.tar.gz
    ```

> **Note**
>
> Ensure that previous command successfully loads images `apiconnect/ibm-apiconnect-management-apim`, `apiconnect/ibm-apiconnect-management-lur`, `apiconnect/ibm-apiconnect-gateway-datapower-nonprod`, `apiconnect/ibm-apiconnect-management-juhu`, `postgres`, and `busybox`.
>
> These images are required for the LTE environment to function. If there are any errors in the loading of the container images, then you will need to resolve the errors before performing the rest of the steps.

4. Verify that the images are loaded in the Docker repository by executing this command:

```
docker image ls
```

```
REPOSITORY                                        TAG
apiconnect/ibm-apiconnect-management-apim          v10.0.1-395-171f74e88
apiconnect/ibm-apiconnect-management-lur           v10.0.1-263-90f1379
apiconnect/ibm-apiconnect-management-juhu          2021-07-31-14-09-30-v10.0.1-0-gf2c95e7
apiconnect/ibm-apiconnect-gateway-datapower-nonprod 10.0.1.4.334153
postgres                                           12.2
busybox                                            1.29-glibc
```

Figure 4.1 – The Docker image ls command results

5. Make sure that the `apic-lte-win_10.0.1.5.exe` directory is in your environment's PATH.

6. Ensure that all LTE container components are installed (as shown in *Figure 4.2*) by executing this command:

```
apic-lte-win_10.0.1.5.exe status
```

This will get the following output:

```
Container                        Status
---------                        ------
apic-lte-apim                    Not Running
apic-lte-datapower-gateway       Not Running
apic-lte-datapower-api-gateway   Not Running
apic-lte-db                      Not Running
apic-lte-juhu                    Not Running
apic-lte-lur                     Not Running
```

Figure 4.2 – LTE components

7. Start Docker containers for APIC by executing this command:

```
apic-lte-win_10.0.1.5.exe start
```

> **Note**
>
> There are a number of switches that are applicable to the `start` command. Some of the main switches are discussed next:
>
> `--keep-config`: By default, `apic-lte-win_10.0.1.5 start`, without the `--keep-config` switch, deletes all the previously published API/Product information from the configuration DB and re-initializes the LTE configuration. Essentially, LTE starts with an empty backend DB that does not contain the APIs and Products that may have been published during an earlier run of LTE. To retain the previously published configuration, use the `--keep-config` switch. When using `--keep-config`, any other switches (for example, `--datapower-api-gateway-enabled` `--datapower-gateway-enabled`) specified for the start are ignored, and instead the same flags that were used during the earlier start are used. The **start** switch starts the DataPower API Gateway by default.

8. After the successful start of the LTE environment, you should see a response similar to *Figure 4.3* in your console:

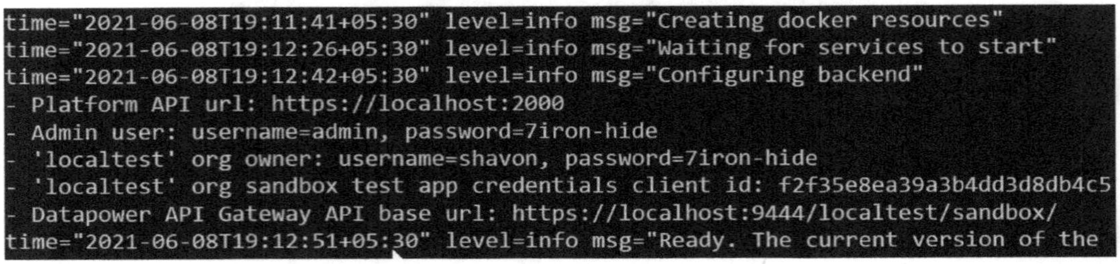

```
time="2021-06-08T19:11:41+05:30" level=info msg="Creating docker resources"
time="2021-06-08T19:12:26+05:30" level=info msg="Waiting for services to start"
time="2021-06-08T19:12:42+05:30" level=info msg="Configuring backend"
- Platform API url: https://localhost:2000
- Admin user: username=admin, password=7iron-hide
- 'localtest' org owner: username=shavon, password=7iron-hide
- 'localtest' org sandbox test app credentials client id: f2f35e8ea39a3b4dd3d8db4c5
- Datapower API Gateway API base url: https://localhost:9444/localtest/sandbox/
time="2021-06-08T19:12:51+05:30" level=info msg="Ready. The current version of the
```

Figure 4.3 – LTE environment start results

You will be using a number of the values that are displayed in *Figure 4.3*. Copy these values to a notepad. They will be used later. There are two important values not displayed here. These are DataPower web admin URLs for DataPower API Gateway and DataPower Gateway (v5 compatible). Sometimes, during your API testing and debugging, you might have to refer to the gateway logs. In such cases, having access to these URLs will be beneficial. These URLs are as follows:

*   `https://localhost:9090/` (v5 compatible)
*   `https://localhost:9091/` (API Gateway)

You can log in using admin/admin credentials on both these URLs.

LTE verification will be discussed next.

Verification

It is important that you perform some basic verification of the LTE installation that you just completed:

1. Check the status of all the services to make sure that all the containers are running by executing the following command:

    ```
    apic-lte-win_10.0.1.5.exe status
    ```

    This will give the following output:

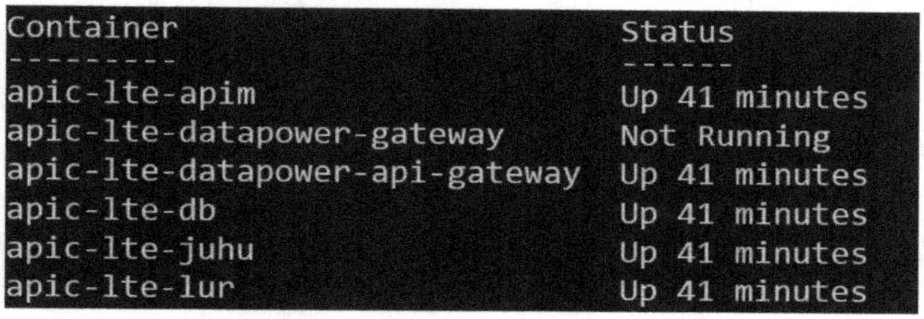

    ```
    Container                          Status
    ---------                          ------
    apic-lte-apim                      Up 41 minutes
    apic-lte-datapower-gateway         Not Running
    apic-lte-datapower-api-gateway     Up 41 minutes
    apic-lte-db                        Up 41 minutes
    apic-lte-juhu                      Up 41 minutes
    apic-lte-lur                       Up 41 minutes
    ```

    Figure 4.4 – LTE environment status results

2. You can also consult the log file for each of these LTE containers by using the following command. Consulting the container log files might be required if you face any problems/errors while publishing and testing your APIs later:

    ```
    docker logs -f -n 10 container-name
    ```

    Replace container-name with the name of the container whose logs you want to monitor. Container names can be retrieved from the command (apic-lte-win_10.0.1.5.exe status) described in the previous step. They are provided in the Container column of *Figure 4.4*.

This concludes the verification of your LTE.

Now that you have installed the Designer and the LTE environments on your workstation, it is time for you to connect these two tools together. As you will recall, Designer helps you in the development of your APIs and the LTE provides an execution runtime for your APIs. You will see these tools in action.

Connecting Designer to the LTE

Designer allows you to connect to multiple environments. You can connect Designer to the LTE or to the API Manager cloud environment. In all cases, you should maintain the version compatibility between Designer and the execution environment (the LTE or API Manager) being connected to. You will now connect Designer to the LTE:

1. Open Designer. Click on **Open a Directory** (*Figure 4.5*). You will provide the path to a local directory where your APIs and Products definitions can be stored. This is typically your development workspace.

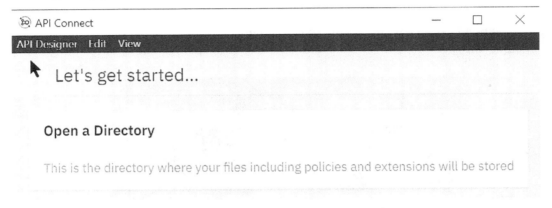

Figure 4.5 – Local workspace directory path

2. On the **New API Connect connection** screen (*Figure 4.6*), provide information about connecting to your LTE. This information was gathered earlier, during the setup of the LTE environment. You can always retrieve that information by executing the `apic-lte-win_10.0.1.5.exe status` command. Click **Submit**.

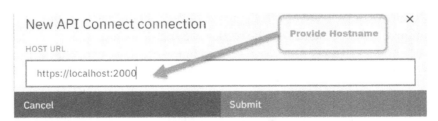

Figure 4.6 – LTE environment host URL

3. On the login screen, provide the username (`shavon`) and password (`7iron-hide`) information. Click **Log in**. Your Designer should now be connected to the LTE!

4. Before going further, it is important for you to quickly familiarize yourself with some key navigation areas of the Designer window. Key areas of Designer are marked with numbers in *Figure 4.7*:

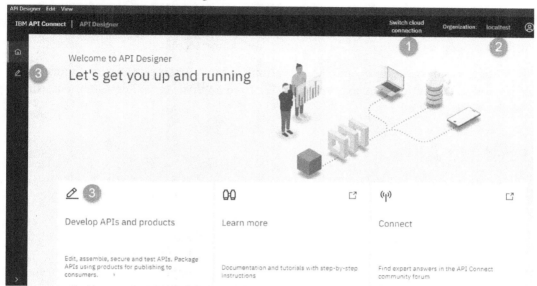

Figure 4.7 – API Designer interface

The details of the key areas of the Designer are provided in the following table:

Area	Purpose
1	Allows for the ability to switch to another APIC cloud. Designer allows the flexibility to connect to multiple APIC clouds.
2	Allows for the ability to switch Designer's context to different organizations, if the logged-in user is part of multiple organizations under a single cloud.
3	Allows access to the main development work area, allowing the development of APIs and Products. There are two methods to launch it.

Table 4.1 – Designer Interface's key navigation areas

It is time to take a quick recap of the things you have accomplished thus far. By now, you have set up on your workstation the Designer component (for API development) and the LTE (for API execution). You have also connected Designer to your local LTE. Then you took a quick overview of Designer's main screen and understood the main navigation areas.

The Designer and LTE components provide a great way to perform localized API development. These two tools offer a perfect starting point for many developers. They should be sufficient for you to start API development. Yet, many developers choose to use a more comprehensive web-based API Manager tool for their API development needs. API Manager comes with multiple functionality enhancements, such as working with OAuth providers, user-authentication registries, and a unit test harness.

With Designer and the LTE now in place, it is time for you to start putting these tools to use. The following sections will expose you to the fundamentals of API development using the tools you have set up.

Creating APIs

APIC supports OpenAPI design, multiple security standards, has an extensive repository of **Out-of-the-Box (OOTB) Policies**, and supports various logic constructs. This really enables developers to develop advanced APIs.

It is practically impossible for a single book to cover every feature and function provided by APIC and all the use cases that it can solve; the concepts discussed in the following sections should make it easier for you to undertake API development with much more confidence. And for the preparation of that journey, you will learn about the following in this part of the chapter:

- OpenAPI design
- Creating a simple proxy API
- Testing APIs
- Variable usage
- Enhancing APIs by applying policies
- Error handling

Let's explore each topic in detail.

What is an OpenAPI design?

The **OpenAPI Specification (OAS)** is the grammar to the language of the API. Grammar is a sign of a language's maturity and advancement. Mature APIs use OAS grammar rules and syntaxes to describe themselves. Any API written following these grammar rules is called an **open API**.

In other words, an **OpenAPI** document is a document (or set of documents) that defines or describes an API using **OAS**. **OAS** specifies a standard format for describing a REST API interface in a standard, vendor-neutral, and language-neutral way. The OAS standard is necessary because it helps in the API economy's growth through increased API consumers' participation (after all, standards-driven communication leads to more participants).

OAS is the most widely adopted specification today when it comes to describing RESTful APIs. It is worthwhile mentioning that OAS is not the only specification out there. Some notable mentions are **RESTful Service Description Language** (**RAML** – `https://raml.org/`) and **API Blueprint** (`https://apiblueprint.org/`).

An **API definition** (or an OpenAPI document) is written in a file that can either be a JSON or YAML formatted document. The main sections of an OpenAPI document from an OAS 3.0 perspective are Info, Servers, Security (Scheme and Enforcement/Security), Paths, Operations, Parameters, and Responses.

> **Security Definition and Security Enforcement:**
>
> **Security Scheme:** This is where you can specify the security implementation details such as authentication URLs, OAuth settings, and API keys' header names. It is worth highlighting that creating a scheme does not enforce that scheme on an API consumer. You can define multiple schemes in your API Proxy but have only one of those schemes enforced upon that proxy's consumers. The enforcement of the scheme is done by Security Enforcement.
>
> **Security Enforcement** or simply called Security: This is where you enforce the Security Scheme on the API consumer.

At the time of writing this chapter (September 2021), OAS 3.1.0 is the latest candidate release. These are the following things that you should be aware of:

- APIC 10.x supports OAS 3.0 with certain limitations on both the user interface (Designer and API Manager) and DataPower API Gateway. Refer to IBM documentation for limitations: `https://www.ibm.com/docs/en/api-connect/10.0.1.x?topic=definition-openapi-30-support-in-api-connect`

- OAS 3.0 is not supported by DataPower Gateway (v5 compatible). OAS 3.0 API support is provided by the DataPower API Gateway only.

- APIC has extended the OAS and has added some of its own message processing policies to support complex message processing.

Since APIC supports OAS, any API definitions you build using OAS can be imported directly into APIC. You can view the OAS design of your API by clicking the **Source** icon on the **Design** tab. Refer to *Figure 4.8*:

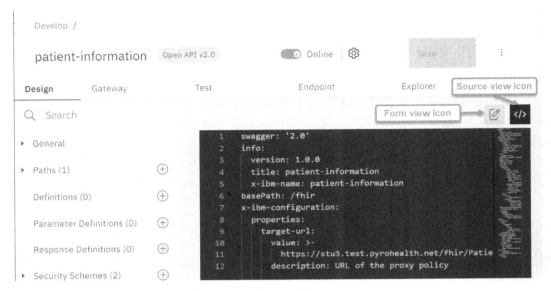

Figure 4.8 – OpenAPI Source icon

You can update the source code within the source view or use the more user-friendly form view.

> **Note**
>
> There are two important distinctions to be made here:
>
> **API Proxy**: An API proxy, in general, concerns itself with the implementation of an API definition in an API Management tool such as APIC. Though it is certainly possible to write an API proxy completely within APIC, without it having to connect to a backend service provider, you would typically implement an API proxy that proxies a backend service.
>
> **API**: An API will mostly be related to the implementation of the service definition by a backend service provider, outside of the API management tool.
>
> It is certainly possible for an API to deviate from the API definition that is proxied by an API Proxy. And that is where various Policies and logic constructs of APIC come into play. Policies and Logic constructs map the API definition of an API Proxy to the definition provided by the actual API.

You will now begin the process of creating an API Proxy using APIC.

Creating an API Proxy

The previous section was about understanding the API's grammar. This section will be about building the sentences and starting to put those grammar rules into practice. You will do that by creating an API Proxy using Designer and then test that on the LTE. It is worth mentioning that you can follow the steps provided in this section on the API Manager interface as well. Steps on the API Manager are quite similar to the steps provided for the LTE environment.

Prerequisites

- Make sure that you have Designer and the LTE installed and configured on your workstation.

- Ensure that you can access the `https://stu3.test.pyrohealth.net/fhir/Patient/d75f81b6-66bc-4fc8-b2b4-0d193a1a92e0` resource from your workstation. The API Proxy that you will soon build uses this backend service URL.

API development

It is time for you to start with the API development:

1. On the Designer's home screen, click on the **Develop APIs and products** tile. Click **Add | API (from REST, GraphQL or SOAP)**.

2. Stay on the default **OpenAPI 2.0** tab.

3. Notice the various API types (*Figure 4.9*) that can be created. You will learn how to implement the SOAP Proxy and REST proxy using a WSDL file in *Chapter 5, Modernizing SOAP Services*. In *Chapter 9, Building a GraphQL API*, you will learn how to create a GraphQL proxy.

    You will be creating your first API utilizing an existing target service.

In the **Select API type** view, select the **From target service** option and click the **Next** button.

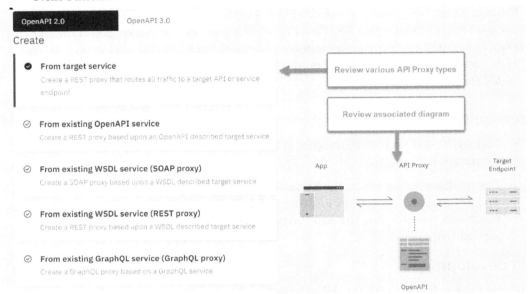

Figure 4.9 – API types

4. In the **Info** section, specify the following information (*Table 4.2*):

Field	Value
Title	`patient-information`
Version	`1.0.0`
Base path	Replace any values generated and enter `/fhir`
Description	`Proxy API from a target service`
Target Service URL	`https://stu3.test.pyrohealth.net/fhir/Patient/d75f81b6-66bc-4fc8-b2b4-0d193a1a92e0`

Table 4.2 – Info section values for the new API

5.   Click the **Next** button. In the **Secure** section, keep the options of **Secure using Client ID** and **CORS** selected.

6.   Click the **Next** button, where APIC will display the summary of the API generation process.

7.   Click the **Edit API** button to further configure your API definition.

8.   Review the various sections (**General, Paths, Definitions,** and so on) under the **Design** tab (*Figure 4.10*).

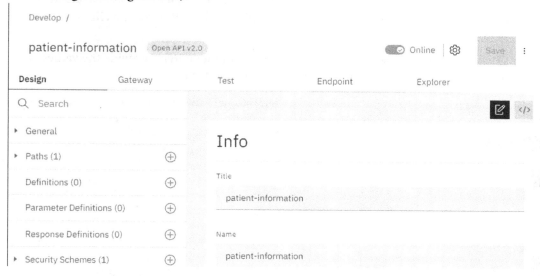

Figure 4.10 – Design tab user interface

Figure 4.10 shows how APIC has organized OAS under various key sections. By clicking on each section of the **Design** tab and reviewing various parameters of each section, you will see how easy APIC makes it to create the OAS source, instead of typing it all into a file. Of course, you can click on the source view icon to review the OAS in a file format. As you click through the various sections under the **Design** tab, take time to correlate much of the information in these sections (**General | Info, General | Schemes List, General | Security, Paths**) with the OAS that you learned about in the *OpenAPI design* section of this chapter.

APIC provides comprehensive security coverage for APIs. As was discussed earlier, in the *OpenAPI design* section of this chapter, APIC supports multiple security schemes. You will be getting a comprehensive tour of APIC's API security features in *Chapter 7, Securing APIs*. For now, you will keep it simple and secure your API using the API key security scheme. Go ahead and review the security definitions of your API. Go to the **Design** tab | **Security Schemes** section. You will note that it already has an **apiKey** with the name `clientID`. This was created because you kept the option of **Secure using Client ID** selected in the earlier step. Go ahead and further strengthen your API's security by adding a definition for the **client_secret** type **apiKey**.

9. Click the **Plus** icon next to the **Security Schemes** menu item. Provide the values highlighted in *Figure 4.11*. Once complete, click on **Save**.

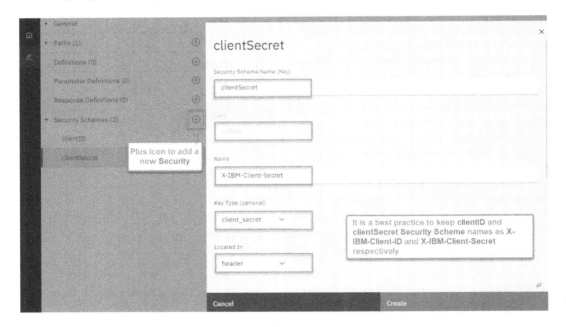

Figure 4.11 – client_secret apiKey

> **Note**
>
> The `X-IBM-Client-Id` and `X-IBM-Client-Secret` header names are not APIC specific. You can substitute these with any other names, for example, `X-ABC-Client-Id` or `X-ABC-Client-Secret`. It is still a good practice to use the standardized names of `X-IBM-Client-Id` and `X-IBM-Client-Secret`.

Enabling API security

Now that you have the security schemes for `client_id` and `client_secret`, it is time to use them in your API's Security. You will remember from our earlier discussion that Security Scheme creation and security enforcement are separate steps. Security scheme creation is followed by the security enforcement step. In the previous step, you created the security scheme. Now you will apply that scheme to your API's security.

1. Go to the **Design** tab | **General** | **Security** | **clientID**. On the **Security Requirements** screen, select the `clientSecret` scheme. Click the **Submit** button.

2. Go to **Paths** in the navigation pane. Designer has a default **Path** (/) definition created. You will be modifying this default definition.

3. Click on / under **Paths**. Click on the **Update** button. Change **Path** to /`Patient` and click **Save**. Refer to *Figure 4.12*. This path will be appended to the **Base path** defined in the earlier step (refer to *Table 4.2*), for example, the URL for invoking this API will be `https://localhost:9444/localtest/sandbox/fhir/Patient`.

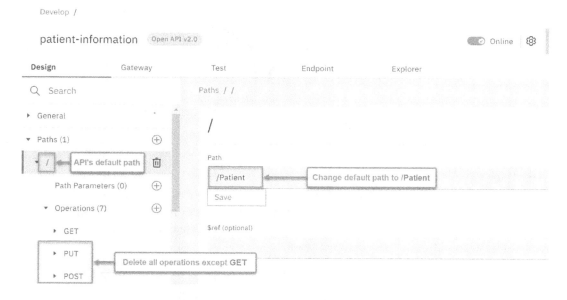

Figure 4.12 – Setting Path

4. Delete all **Operations** except GET.

5. Click the **Save** button.

6. Locate and choose the **Gateway** tab. You will notice an **invoke** policy as shown in *Figure 4.13*:

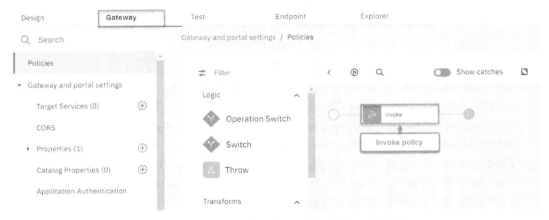

Figure 4.13 – An Invoke policy in the Gateway Policies

The invoke policy instructs APIC to proxy the API to the endpoint you provided earlier (`https://stu3.test.pyrohealth.net/fhir/Patient/ d75f81b6-66bc-4fc8-b2b4-0d193a1a92e0`).

Congratulations! You have just created an **API Proxy** that has been deployed for testing.

Now that you have created your first API Proxy on APIC, it is time to run some tests and see this API Proxy in action.

Testing APIs

There are multiple methods to test an API Proxy. You can either use external tools such as Postman, cURL, or SoapUI, or you can use the test capability provided by the APIC platform. You will learn how to use the test facility in APIC so that you can understand the test suite capabilities of the APIC platform.

> **Testing Note**
>
> A more comprehensive testing facility will be introduced in *Chapter 13, Using Test and Monitor for Unit Tests*, where you can build unit tests that can be executed in your DevOps pipelines.

Before you can test your API, it needs to be put online. To put your API online, simply click on the **Test** tab as shown in *Figure 4.14*:

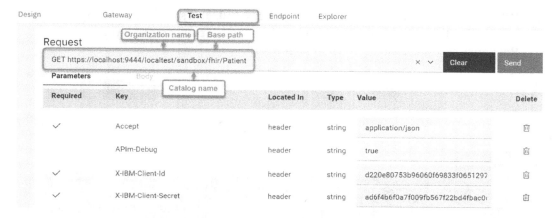

Figure 4.14 – Using the built-in test feature

You will notice in *Figure 4.14* that the **GET** operation you created is already displayed. Some other interesting points in the URI are as follows:

- localtest in the URI is the organization name.
- sandbox in the URI is the catalog environment where the API was deployed.
- The base path you defined earlier is also in the URI followed by the REST target.

Other features you will see are the headers that are already available. The generated client id and client secret that were provided when you started the LTE are automatically inserted. The **APIm-Debug** header is set to provide you with detailed debugging information.

Testing your API Proxy

It is time to test the API Proxy. Here's how:

1. While still on the **Test** tab in *Figure 4.14,* click **Send**. You should now receive an HTTP 200 successful response with the associated payload displayed in the **Response** section. That is shown in *Figure 4.15*:

Figure 4.15 – HTTP 200 response with payload

2. You can click on the **Headers** tab to see the header used or the **Trace** tab to see the tracing of your policies. In this case, you only had the invoke policy, but in future cases, you will be able to see how APIC traverses through multiple policies.

You have completed the testing of your API Proxy. That was easy.

> **Note**
>
> The test capability you just completed only works on the DataPower API gateway service. This is the default gateway when you create your APIs. If you are supporting the previous version 5 edition of APIC and you want to specify DataPower Gateway (v5-compatible), go to the **Gateway** tab | **Gateway and portal settings** in the navigation pane of your API. Scroll to the **Catalog Properties** section and you will find the **Gateway** selection option. Choose `datapower-gateway`.

By default, when you test an API with the **Test** tab, a range of test parameters are pre-configured, for example, a default Product is automatically generated, to which the API is added, and that Product is published to the `sandbox` catalog. You can configure testing preferences such as the target Product and target Catalog. Some of these configuration options are not available in LTE though.

You should now feel sufficiently prepared to embark on the journey of enriching your API Proxy with Policies and Variables. **Policies** and **Variables** are the building blocks for adding functionality to your Proxies. These are often used to add custom logic, transformation, routing, logging, and security capabilities to the Proxies. Variables and Policies are important to understand so that you can use them to enhance an API Proxy's functionality. You will begin by learning about variables.

Using variables

Like any mature application development framework, APIC also supports various variable types to help manage/manipulate the data being passed in an API and control API's Policies' behavior. There are two types of variables supported by APIC: **Context variables** and **API properties**.

Context variables

These are the variables that are relevant during the context of an API call. Therefore, their scope is also limited to the context of a specific API call on the runtime gateway. There are several context variables available. You can refer to the complete list at `https://www.ibm.com/docs/en/api-connect/10.0.1.x?topic=reference-api-connect-context-variables`.

Context variables are categorized into the following categories:

Each category contains information about a particular aspect of an API call. Here are some of the most used categories:

- `api`: Provides API metadata, for example, API name, version, operation path, and provider organization information. One of the important elements of interest in this category is `api.properties.propertyname`. `propertyname` can be any custom property (catalog specific) defined for the API. API properties will be discussed in detail in the next section.

- `client`: Provides information about the client's organization information and the client's application that issued the request to the API.

- `message`: One of the most heavily used categories. This is the payload of the request or response message. Note that this is different from the **request** category (discussed next). Loosely, the `message` category represents a message as it moves in processing policy. You can read/write data to the `message` category. If you need to make any modifications to request and response payloads, then the message category is your playground.

- `request`: This represents the original HTTP request, its headers, URI, verb, parameters, and so on. This is read-only. And this is the fundamental difference between the message and the request category. HTTP request headers can be modified in the preflow policies.

- `system`: Provides system data and time information of the gateway.

Of course, you may not need to use all the different categories, but you should be aware of some of the more useful context variables. These are shown in *Table 4.3*.

Name	Description
`api.properties. propertyname`	This will be covered in the next section.
`message.body`	Discussed above. This represents the payload of the request or response message. It is read/write. You would typically set `message.body` if you are constructing a request that would be sent to a backend service via an invoke policy.
`message. headers.name`	You can extend `message.headers.headername`, for example, `message.headers.test` in a Gateway script policy and then access `message.body.test` during the latter part of the processing flow to retrieve the test value. You can also set custom headers in the message to be passed to backend services.
`request.body`	The payload of an incoming request. All request category variables are read-only.
`request. headers. headername`	Value of the named header (`headername`) of the HTTP request.
`request. parameters`	Provides incoming parameters from the `path` and `query` parameters.
`request. querystring`	Provides the request query string without the leading question mark.

Table 4.3 – Important context variables

You can read more about context categories and some associated variables of the message and request categories at https://www.ibm.com/docs/en/datapower-gateways/10.0.x?topic=object-context-api-gateway-services.

Now that you know some of the useful context variables, you should become familiar with how API properties are used.

API properties

API properties are used by the gateway to control the behavior of policies. One of the most common uses for custom API properties is to specify environment-specific URLs. In a typical APIC configuration, each catalog generally represents a deployment environment, for example, Development, Test, and Production.

Further, each catalog has a specific gateway URL assigned to all the APIs executing in its context. You can use API Properties to specify environment-specific backend URLs (for each catalog) in the API definition and then use that property (instead of hardcoding the URL) in the invoke policy.

This will result in a dynamic API configuration. You will soon build an example to get a better understanding of this. You will use the API Proxy that you created earlier to create a Catalog specific property, assign it a backend URL, and then use that property through inline referencing (more about this shortly) in the Invoke Policy:

1. Open your patient-information API in Designer.
2. Go to the **Gateway** tab | **Gateway and portal settings** in the navigation pane | **Properties**.

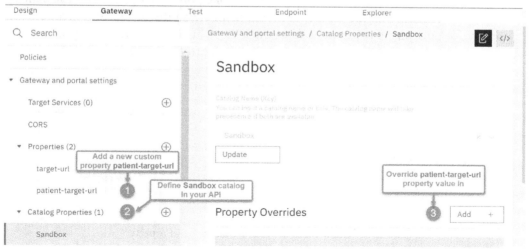

Figure 4.16 – Adding API properties

You will notice there already is a **target-url** property defined in *Figure 4.16*. You will create a new one for demonstration purposes.

3. Click the plus icon button to create a new custom API property.

4. In the new property view, provide the **Property Name** field with a value of `patient-target-url`.

5. Now go to **Catalog Properties** in the navigation menu. Click the plus icon button.

6. Select `Sandbox` in the **Catalog Name (Key)** field. Click **Create**. Now you will be able to override properties for the `Sandbox` catalog. Scroll down to the **Properties Overrides** section and click the **Add** button.

7. Provide the following values in the new property override view:

   - **Property Name to Override (Key)**: `patient-target-url`

   - **patient-target-url (optional)**: `https://stu3.test.pyrohealth.net/fhir/Patient/d75f81b6-66bc-4fc8-b2b4-0d193a1a92e0`

8. Click the **Create** button and save your API. You just created a custom API Property with the name `patient-target-url` and have overridden its value in the `Sandbox` catalog. Since you are working with LTE, `Sandbox` is the only catalog available to you. Typically, in API Manager, you will have access to multiple catalogs (if you have been granted access by your administrator). In that case, you can add more catalogs to your custom property and assign them different URLs.

9. Now that you have created a custom API Property in your `Sandbox` Catalog, it is time for you to use this property in your API. Go to the **Policies** view in the **Gateway** tab of your API. Click on the only **Invoke** policy that you see on the panel. A panel will display on the right that shows parameters intended for the **Invoke** policy. Remove any existing value from the **URL** field of the **Invoke** Policy and replace it with `$(patient-target-url)`. Refer to *Figure 4.17*:

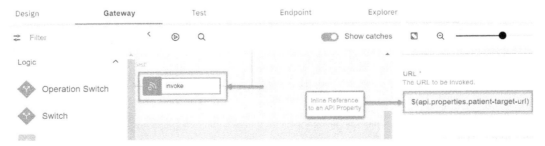

Figure 4.17 – Inline referencing of a custom API Property

Inline referencing is a variable or property referencing technique using the $(variable) format. Here, variable is the name of a context variable or an API property that is utilized. Often this method is used to construct dynamic URLs in an invoke policy and for building switch conditions. You will see another example of an Inline referencing method when we introduce you to the switch policy later in this chapter.

10. **Save** the API, click on the **Test** tab, and run the test again. You should see the same result as before.

You have learned about the concept of variables and API properties that are utilized in APIC. Many of these variables are system variables. They allow you to access meta-information about an API call, for example, API version, base path, request payload, and headers. You also reviewed the concept of API Properties that you can leverage to manage environment-specific configuration.

You have just seen an example of an Invoke policy using API Properties. There is more to learn about policies, so you will do that next.

Adding policies

What are policies? Policies are pieces of configurations that invoke a specific type of action (depending on the type of policy) on a message. There are Built-in policies that come pre-packaged in the APIC solution. They are divided into five categories:

- **Logic**: These policies control the flow of the API by providing conditional coding aspects to your APIC such as if, switch, Operation-switch, and catch.

- **Transforms**: Provide the ability to transform payloads using various methods or manipulate headers. In *Chapter 8*, *Message Transformations*, you will work exclusively with these policies.

- **Policies**: Provide the ability to route, inspect, validate, log, set limits, and apply GatewayScript coding. This chapter will introduce you to some of these policies.

- **Security**: Apply client and user security as well as applying JWT capabilities. In *Chapter 7*, *Securing APIs*, you will work with these policies in detail.

- **User-defined policies**: Build your own policies. The user-defined policies feature is available only with the on-premises offering of APIC. In *Chapter 12*, *User-Defined Policies*, you will learn how to create your own custom policies.

These policies support your API use cases with data transformation, message routing, securing, validation, and logging. Essentially, these Policies provide building blocks to truly enhance your API's capability. For instance, you may use these Policies to **Map** (transformation), **Invoke** (message routing), **Validate** (schema validation), apply **Client Security/Validate JWT** (for security), **Log** (to control logging information to the analytics server), and **Throw** (for exception handling).

It is worth mentioning that **Catch** is not a policy. Catch is implemented by IBM extensions to OAS. You will be introduced to the interplay of the Throw policy and Catch IBM extension through an example later in this chapter.

You should review the documentation for details of all the different Built-in Policies: `https://www.ibm.com/docs/en/api-connect/10.0.1.x?topic=constructs-built-in-policies`.

Policies are gateway type-specific. All Built-in policies, except for the Validate Username token policy, are available as part of DataPower API Gateway. There are several policies that are not available for DataPower Gateway (v5 compatible). You can review the Gateway specificity of the policies at `https://www.ibm.com/docs/en/api-connect/10.0.1.x?topic=constructs-built-in-policies`.

A sample screenshot to show you how policies are organized in an API is shown in *Figure 4.18*:

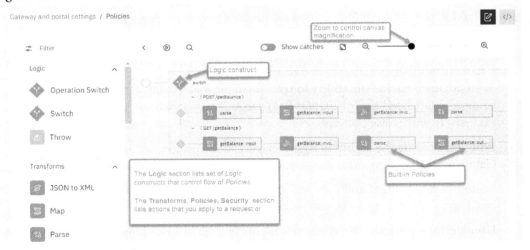

Figure 4.18 – Policies and Logic constructs

As you can see, you can apply multiple policies to a single API. To help your understanding of policies, you will be introduced to a few of the policies and how they supplement your API. In this chapter, we are focusing on the Invoke policy. It is the most important policy so you will learn about it now.

The Invoke policy

The Invoke policy is one of the most heavily used policy types and probably the most feature-rich too. It can often be used (depending upon the nature of the backend service provider) by keeping its default settings and by only specifying the URL property. But other than its basic usage, the Invoke policy provides many advanced properties to control the execution of the backend service call. You will learn about some of these advanced properties by implementing some new behavior.

Handling error conditions on a failed invoke

Within the Invoke policy, there are capabilities that instruct how the Invoke policy will behave on error conditions. A typical error condition would be a backend service that is experiencing slow response times. It is possible that your proxy needs to respond back to the consumer with a response (success or failure) within a defined **Service-Level Agreement (SLA)**. If the SLA is shorter than the Invoke policy's default Timeout value of 60 seconds, an error occurs. This scenario can be easily handled by leveraging the **Timeout** and **Stop on error** properties found within the Invoke policy setup. Here is a brief description of these properties:

- The **Stop on error** property provides an option to stop the message flow's execution when one or more pre-defined errors (for example, `ConnectionError`, `SOAPError`, `OperationErrror`) is encountered during an Invoke policy's execution. If a catch flow is configured to handle these errors, then that catch flow is executed. If there is no catch flow configured, then the message flow processing continues, thereby ignoring the error. You add a catch flow by first enabling the **Show catches** slider. That will display an empty **Catches** box. When you click on the **Catches** box, it opens the **properties** box to allow you to add a new `catch` or add a default `catch`. This is shown in *Figure 4.19*:

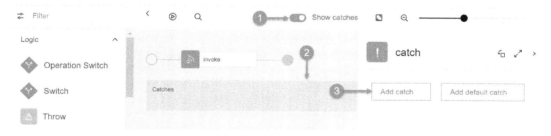

Figure 4.19 – Enabling Catches

When you click on **Add catch**, you have various catch conditions to choose from. When you choose one of the 11 catch types, it updates the **Catches** section with the selected catch type and allows you to configure how to handle the catch. This is shown in *Figure 4.20*:

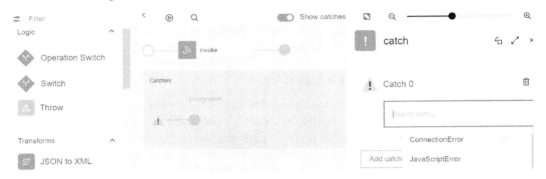

Figure 4.20 – Adding a catch

You can add more catches to handle other conditions.

- The **Timeout** property is the time (in seconds) that the Gateway waits to receive a reply from the backend service.

You will now build an example to see the **Stop on error** and **Timeout** properties in action:

1. Open the `patient-information` API in Designer. Open the **Gateway** tab.

2. Enable the **Show catches** section of your API as shown in *Figure 4.19*. Click anywhere on the **Catches** block. A **catch** view will open. Add a `ConnectionError` catch by clicking on the **Add catch** button and selecting `ConnectionError`.

3. Drag a **GatewayScript** policy onto the **ConnectionError** catch. This is shown in *Figure 4.21*:

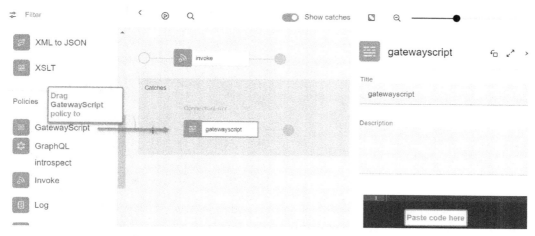

Figure 4.21 – Applying error coding to handle error

4. Enter the following code into the editor of your newly added **GatewayScript** policy. You can also fetch this code (errormsg.js) from this chapter's GitHub repository:

```
var errorName = context.get('error.name');
var errorMessage = context.get('error.message');
var errorResponse =
{
    "name": errorName,
    "message": errorMessage
};
context.message.header.set('Content-Type',
   "application/json");
context.message.body.write(errorResponse);
```

Click **Save** to save your changes.

5. Click on the **Invoke** policy, scroll down the **properties** page, and check the **Stop on error** checkbox.

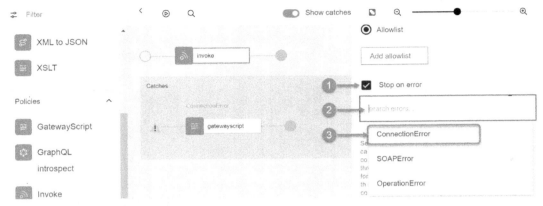

Figure 4.22 – Configuring Stop on error

When you click on **search errors ...**, a dropdown with error types is displayed as shown in *Figure 4.22*. Choose ConnectionError to match up with the catch you just configured.

This completes your configuration of handling ConnectionError when such an issue is encountered in the API Proxy. With your API now configured to handle the connection error issues, you will next do a temporary configuration to create this error condition. This error condition will be created by utilizing the **Timeout** property of the Invoke policy.

6. The **Timeout** property (*Figure 4.23*) has the default timeout value of 60 seconds. You should set it to 10 seconds.

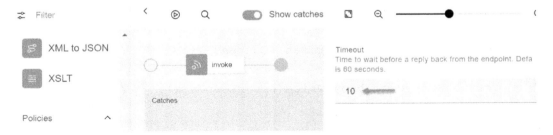

Figure 4.23 – Invoke policy's Timeout property

7. The last thing that you need to do is force a timeout. This will be done by updating the `patient-target-url` API property. Use a URL that will time out. Change the API property value from `https://stu3.test.pyrohealth.net/fhir/Patient/d75f81b6-66bc-4fc8-b2b4-0d193a1a92e0` to `https://stu3.test.pyrohealth.com/fhir/Patient/d75f81b6-66bc-4fc8-b2b4-0d193a1a92e0` (note the change from .net to .com in the server name). **Save** the changes to your API.

8. Next, click on the **Test** tab. **Send** the request. After about 10 seconds, you should get the backend server timeout and observe the **ConnectionError** catch in action.

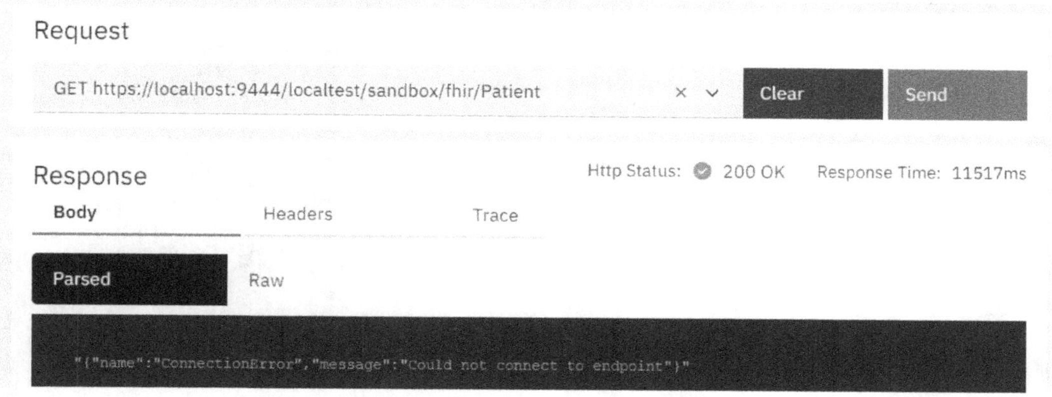

Request

GET https://localhost:9444/localtest/sandbox/fhir/Patient   × ∨   Clear   Send

Response   Http Status: ● 200 OK   Response Time: 11517ms

Body   Headers   Trace

Parsed   Raw

`"{"name":"ConnectionError","message":"Could not connect to endpoint"}"`

Figure 4.24 – Error caught – "Could not connect to endpoint"

Figure 4.24 shows the **ConnectionError** response `"Could not connect to endpoint"` in JSON. You have successfully tested the **Stop on error** and **Timeout** properties within an Invoke policy.

Remember to change the `patient-target-url` property back to `https://stu3.test.pyrohealth.net/fhir/Patient/` for future testing.

Next, you will learn about another important property of the Invoke policy that is frequently used, and that is changing the HTTP method of the outgoing request (to the backend service).

Changing the HTTP method in the Invoke policy

Typically, an API proxy's operation's method should match the backend service's HTTP method. Sometimes you might come across a scenario where the backend service that you are proxying supports an HTTP method that is different from what is exposed by your API proxy. A typical example is exposing a REST API (**GET**) method that proxies a SOAP backend service. It is common for SOAP services to be exposed via a **POST** method irrespective of whether they fetch the results (typically **GET**) or perform other updates (**POST/PUT/DELETE**). In such a case, you can use the **HTTP Method** property (*Figure 4.25*) of the Invoke policy to match the HTTP method that is supported by the backend service.

Figure 4.25 – Invoke policy's HTTP Method property

You may be wondering about the **Keep** option. **Keep** is the default value and it means that the incoming HTTP method should be continued as the HTTP method to the backend service. You will learn more about these options in, *Chapter 5, Modernizing SOAP Services*.

Often there are strict requirements around the header and parameters that can be passed to the backend service. The Invoke policy provides a couple of properties that can be used to meet such requirements. You'll see an example of this next.

Header Control and Parameter Control

The **Header Control** and **Parameter Control** properties control the request's headers and parameters that get passed to the backend service.

Often, you will have a requirement to block certain header values from being passed to the backend service called by your Invoke policy, for example, blocking the API request's X-IBM-Client-ID/X-IBM-Client-Secret values from being passed to the backend service. You can easily handle such a requirement using the Header Control property:

1. Open the patient-information API in Designer. Open the **Gateway** tab | invoke Policy | **Header Control** property. Click **Add blocklist** as shown in *Figure 4.26*:

Figure 4.26 – Header Control

2. Add the value ^X-Environment$ to **Blocklist**. This instructs the gateway to copy all headers except for X-Environment to the target service.

3. **Save** and **Publish** your API.

4. Click on the **Test** tab to run your API. Add the following headers to the request (refer to *Figure 4.27*):

    `X-Environment` header with a value of `Skip Me`

    `X-Book-Reader` header with a value of `Allow Me`

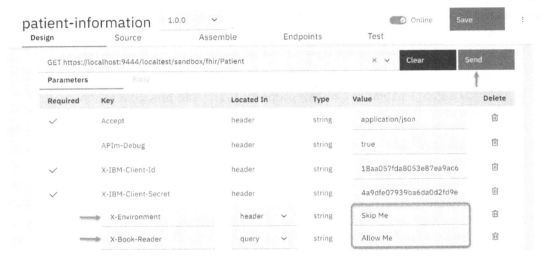

Figure 4.27 – Adding new headers in the Test facility

5. Click **Send** to send the request. When the request completes, click on the **Trace** tab | **Invoke** policy | **Advanced** slider. Refer to *Figure 4.28*:

Figure 4.28 – Reviewing Trace for headers

This will allow you to see the entire flow from the initial request to the call to the backend and the resulting response.

6. Scroll through the output and expand the headers to look for your X-Environment and X-Book-Reader headers. Notice that all headers except for X-Environment are passed, including your X-Book-Reader header, to the target backend service as shown in *Figure 4.29*:

Figure 4.29 – Header omission verification

This example demonstrated the use of the header control property of the Invoke policy and how it can be used to block certain headers from being passed to a backend service. You can use the same method to filter or allow query parameters using the Parameter Control property of the Invoke policy.

Another important capability of the Invoke policy is its ability to utilize variables and properties inside its URL property. This allows for the dynamic building of the backend URL. You will review this capability through an example in the next section.

Accessing variables and properties in an Invoke Policy

An Invoke policy's target URL is often built using variables and properties. Earlier, in the *Create an API Proxy* section, you created a proxy that utilized a static backend URL containing a hardcoded `patient-id` value (`d75f81b6-66bc-4fc8-b2b4-0d193a1a92e0`). You will now improve upon that proxy by removing the hardcoding of the `patient-id` value. Instead, you will now define `patient-id` as a parameter in your API's definition and pass the `patient-id` value with the request message. You will then access this `patient-id` parameter using the **Inline Referencing** technique and dynamically build the backend service URL of the Invoke policy using the captured value of `patient-id`.

Before diving into the example, it is important for you to understand key Parameter types. Parameters represent the values that come as part of an API's request. The following are the main types:

- **Path type**: Parameters defined at the path level are the variable parts of a URL path. These parameters cannot be omitted from the URL path. They are used to point to a specific resource. For example, a parameter defined as `/customers/{customer-id}` can be accessed as `/customers/123` by an API consumer. In this case, API consumers will get details of the customer resource represented by `customer-id=123`. You learned about path-based Parameter usage in the example *Accessing variables and properties in an Invoke Policy*.

- **Query type**: Parameters defined as a query type are typically used to sort, filter, or paginate the resources. For example, a query parameter defined as `/customers?firstName=Drew` can apply a filter of `firstName=Drew` to the complete dataset before returning the subset data results to the API consumer.

- **Header type**: Header type Parameters can also be used to pass values to an API flow. You can use this type to pass custom values, such as security headers, to an API flow.

Now that you understand the key parameter types, it is time to create a new example of a REST API proxy:

1. On the Designer tool's home screen, click the **Develop APIs and products** tile. Click **Add | API**.

2. Ensure that **OpenAPI 2.0** is selected.

3. In the **Select API type** view, select **From target service** and click **Next**.

4. In the **Info** section, specify the following information:

Field	Value
Title	`inline-access`
Version	`1.0.0`
Base path	`/inline-access`
Description	`Proxy API from a target service`
Target Service URL	`https://stu3.test.pyrohealth.net/fhir/Patient/`

Table 4.4 –API proxy details

5. Click **Next**. In the **Secure** section, unselect **Secure using Client ID** but keep **CORS** selected.

6. Click **Next** to create your API definition.

7. In the **Summary** section, click **Edit API** to further configure your API definition.

8. Go to your API's **Design** tab | **Paths** section in the navigation pane. Change the default path (/) to `/patient/{patient-id}`.

9. Add a new path parameter by clicking on the plus icon next to **Path Parameters (0)**. Create a new parameter with the name `patient-id` and assign it the values as per *Figure 4.30*.

10. Delete all the operations except GET. Click **Save** to save the definition.

A completed screen is shown in *Figure 4.30*. It is important to understand all the things that are highlighted in this figure. There is a lot happening here. It is important to note that the parameters **Located In** path are always **Required**. Another thing to note is that the parameter defined in the path's **Name** {patient-id} is inside curly braces and that it matches the value in the **Name** field.

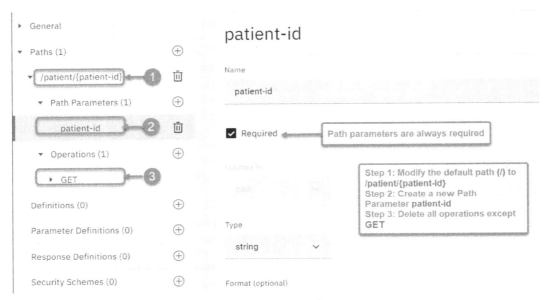

Figure 4.30 – Path, Operation, and Path parameter creation

Now you need to set the target-url property to the backend endpoint.

11. Go to API **Gateway** tab | **Gateway and portal settings** | **Properties**. Click on the target-url property. Remove the value from the **Property Value (optional)** textbox.

12. Override this target-url property value by creating a catalog property in the sandbox catalog. Name this catalog property target-url and assign it a value of https://stu3.test.pyrohealth.net/fhir/Patient/. Click **Save** to save the definition.

13. With this setup done, let's use the Inline referencing technique to construct the target URL for your Invoke policy. Go to the **Gateway** tab of your API and click on the only invoke policy on the canvas.

14. Remove the **URL** property $(target-url) of the Invoke policy and enter the following value:

```
$(api.properties.target-
url)/$(request.parameters.patient-id)
```

> **Note**
>
> You should understand the format of $(api.properties.target-url)/$(request.parameters.patient-id). This technique is called Inline referencing. You are using Inline referencing to access an API property (target-url) and a request **Path** parameter (patient-id) that you just define. You are also combining the two values to construct the final backend target URL.

15. Click **Save** to save your API. Enable the **Online** slider to publish your API.

16. Click on the **Test** tab.

17. You can now test the API proxy. Note the absence of the "X-IBM-Client-Id" and "X-IBM-Client-Secret" headers:

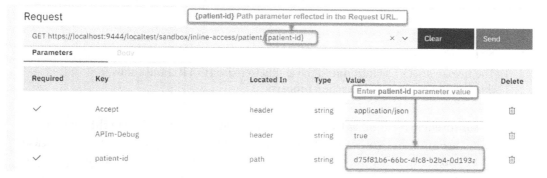

Figure 4.31 – Adding the value of patient-id

As you can see in *Figure 4.31*, the value of `patient-id` is required. Enter the value `"d75f81b6-66bc-4fc8-b2b4-0d193a1a92e0"` in the highlighted area. This value is the `patient-id` path parameter that you defined earlier.

18. Click the **Send** button and review the response.

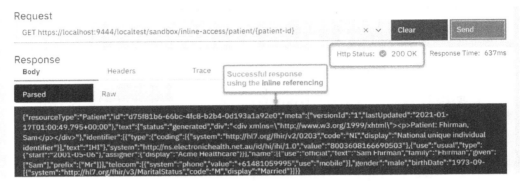

Figure 4.32 – A successful test using Inline referencing

The preceding steps demonstrated the method of using inline referencing to dynamically build a target URL. As mentioned earlier, Inline Referencing coupled with API Properties is among one of the most used methods to dynamically build the target URL. Make sure that you develop a good understanding of this technique.

> **Note: v10.0.1.5 enhancements**
> The Invoke policy has introduced the ability to reuse HTTP connections through the persistent connection property.

That was a comprehensive look at the Invoke policy and some of the crafty things you can do with it. We are sure that you can think of many more use cases. Some of those use cases might lead to you using other policies provided with APIC. One other important policy in the APIC arsenal is the switch policy. You will be introduced to this next.

Building a Switch policy

You will make changes to the `inline-access` API that you recently created. You will use a Switch policy and make a determination based upon a header variable (Environment) to process different branches of the message processing flow. Begin by opening the `inline-access` API in Designer. Navigate to the **Gateway** tab to review the single invoke policy:

1. Click on the Invoke policy to bring up the property pane. Notice that the title is `invoke`. Replace the title `invoke` with the more readable title `Production invoke`. Making this change helps with the identification of the responsibility of the policy. You must follow the practice of giving meaningful names as per the identification of the responsibility of your policies and variables. You see that this new title is represented on the canvas as shown in *Figure 4.33*. Click **Save** to save the change.

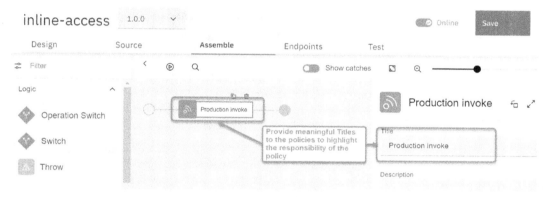

Figure 4.33 – Adding readable code

2. Next, we want to add our switch logic policy. Drag the switch logic policy to the left of the `Production invoke` policy. You will drop it when a small square appears. Refer to *Figure 4.34*:

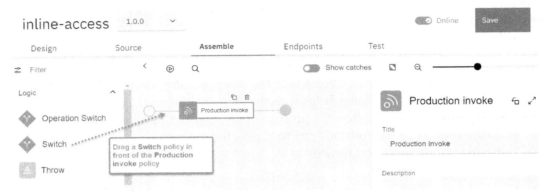

Figure 4.34 – Dragging switch before the Production script

You will now see a **switch** policy on the pane and the property screen on the right with **Case 0** set to a condition of true.

3. Change the **Case 0** condition using an Inline reference of `$header('Environment')='Production'`:

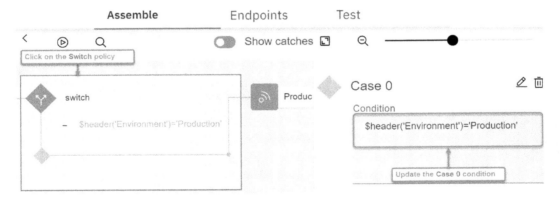

Figure 4.35 – Adding case to switch

4. Click on **Add case** to add our other leg to the switch. Using the same process as in step 3, change the value of **Case 1** from **true** to `message.headers.Environment = 'Test'`.

5. In *Figure 4.36*, you can see how the switch has changed to show the two possibilities:

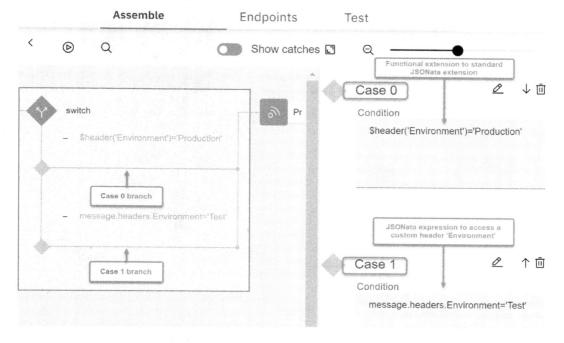

Figure 4.36 – Adding a second case

You will notice that conditional logic for each case is built a bit differently using Inline references to the `Environment` request header. Notice the use of the `$header('Environment')` and `message.headers.Environment` expressions. `$header` is just a functional extension of the variable `message.headers.name` in standard JSONata notation. You can use either of these techniques.

You will want to add a second Invoke policy to represent the Test environment.

6. Drag the `Production invoke` policy to **Case 0** as shown in *Figure 4.37*. Drag
    and drop another Invoke policy for the **Case 1** branch and change its title to `Test
    invoke`. Make sure that the URL for both `Production invoke` and `Test
    invoke` policies is set to `$(api.properties.target-url)/$(request.
    parameters.patient-id)`. Click **Save** to save your API.

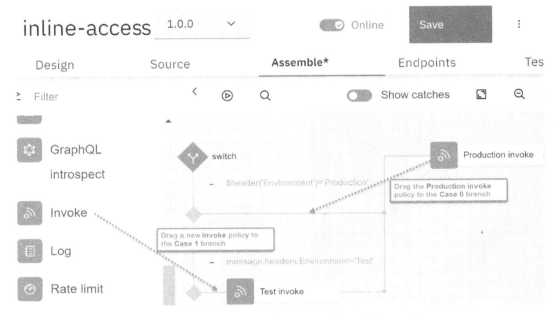

Figure 4.37 – Dragging Production invoke and Test invoke

You have configured your first switch policy. Your screen should look like *Figure
4.38*. This example assumes that the client application is passing a custom header,
`Environment`, with the request.

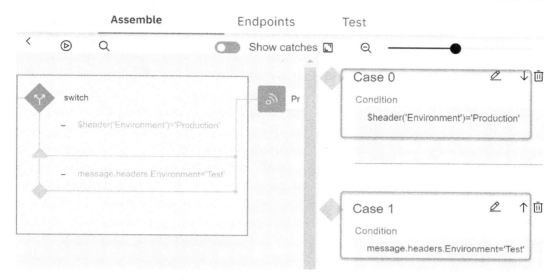

Figure 4.38 – Use of Inline referencing in Switch conditions

Review the highlighted sections in *Figure 4.38*. You will notice that conditional logic for each case is built using inline references to the `Environment` request header.

It is time to test your API. You already know that the two Invoke policies point to the same service. So how will you know which branch your Switch policy took? You are going to learn how to use the **Trace** facility to answer this question.

7. Click on the **Test** tab and add the `Environment` header. Set the header to `Test`. Also set **Patient-ID** to `d75f81b6-66bc-4fc8-b2b4-0d193a1a92e0`. Click **Send**.

You should get the same results as before. But since the Invokes looked identical, how do you know whether it really executed **Case 1**? The next step shows you how.

8. After you run your test, click on the **Trace** tab. You will see the graphic that shows you which case statement was run. The other statement is grayed out. *Figure 4.39* shows that result:

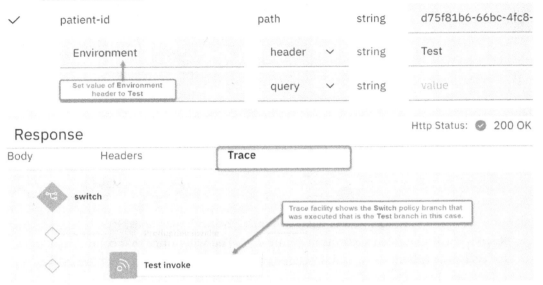

Figure 4.39 – Using Trace to determine the execution of Switch

You have now learned another valuable technique for accessing Parameters and API properties through the Inline Referencing method, as well as how to take advantage of the built-in testing and tracing capabilities of API Connect.

The last feature on the Invoke policy will be how to capture response objects and merge the information. Service chaining is common with APIs.

Service chaining using double Invokes

Under-fetching is a common symptom when services are created where the consumer needs to make multiple calls to obtain the results they desire. Service chaining in APIC can be created to handle such situations. You can build an API Proxy that acts as an aggregator of more than one backend services. You can implement the aggregator API Proxy by chaining calls to the backend services and combining the response data from each of these services into a single response to the consumer. Another solution for under-fetching is using GraphQL. You will learn about GraphQL in *Chapter 9, Building a GraphQL API*.

To build the aggregator response, you will use a combination of Invoke Policies and a GatewayScript Policy. For this use case, we will use the same backend service we have been using: `https://stu3.test.pyrohealth.net/fhir/Patient/`.

Previously, you have learned how to modify the titles of invoke policies, set path parameters for patient-id, and pass values to the Test capability of APIC. You will be doing that again with some slight modifications and the assistance of GatewayScript code. You can download **patient-response.js** from the GitHub site.

The approach you will be taking is constructing a new API definition and implementing two Invoke policies that return a different patient payload, followed by a GatewayScript that merges the responses of the two Invokes. The new skill you will learn is how to capture the response objects individually from each Invoke as query parameters. Once captured, you will filter specific fields from these responses and send a custom aggregated response back to the API consumer.

The patient-id values you will use are as follows:

- 9df54eeb-a9ac-47ec-a8f6-eb51dd7eb957
- 97e47441-b8b9-4705-bda7-248ae6ae2321

Here are the steps you should follow:

1. Create a new REST API Proxy (from a target service) using the URL https://stu3.test.pyrohealth.net/fhir/Patient/. Name this proxy aggregation-service. Keep the default **Base path** value. Remove all the security definitions.

2. In **Design | Paths**, create a path (/Patient) that takes two query parameters, patient-id1 and patient-id2. It should support a GET operation. Refer to *Figure 4.40*:

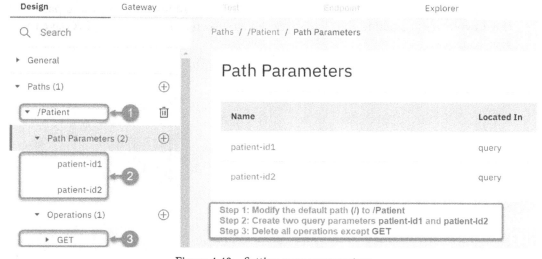

Figure 4.40 – Setting query parameters

3. In the **Gateway** tab | **Policies**, update the existing Invoke and set **Title** to `get patient 1` and the URL to `$(target-url)/$(request.parameters.patient-id1)`. Scroll to the bottom of the invoke policy's properties and enter `patient1-response` in the **Response object variable** property. Refer to *Figure 4.41*:

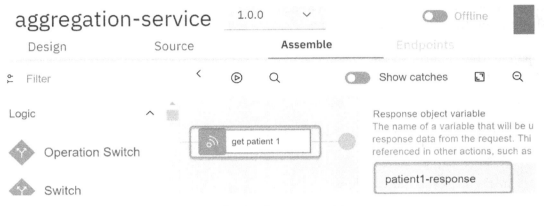

Figure 4.41 – Setting the response object name

4. Drag and drop a new Invoke to the right of `get patient 1`. Perform step 4 again but instead use the title of `get patient 2`, set the URL to `$(target-url)/$(request.parameters.patient-id2)`, and set up the response object to `patient2-response`. Next, you will add the aggregation code to filter data.

5. Drag and drop a **GatewayScript** policy to the right of `get patient 2`. Set the title to `aggregate results`. Download the GatewayScript's code (`patientResponse.js`) from this chapter's GitHub repository. Copy and paste the downloaded code into the GatewayScript as shown in *Figure 4.42*. Click **Save** to save your API.

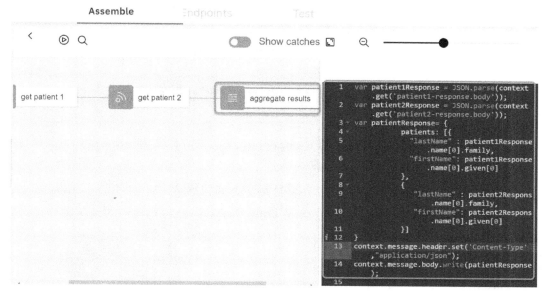

Figure 4.42 – Aggregation GatewayScript

A quick review of the JSON code would be helpful. The code uses the `context.get('response object name')` method to retrieve the response objects you designated. It utilizes the technique of converting the response returned by the backend service to a JSON object using the `JSON.parse()` method. The converted JSON object can then be navigated using the dot notation method. For example, to access the `family` name of `patient 1`, you can use `patient1Response.name[0].family`. The aggregated response is stored in the `patientResponse` variable. This aggregated response is finally sent as the output to the API caller using the `context.message.body.write` method:

```
var patient1Response =
   JSON.parse(context.get('patient1-response.body'));
var patient2Response =
   JSON.parse(context.get('patient2-response.body'));
var patientResponse= {
         patients: [{
            "lastName" :
               patient1Response.name[0].family,
            "firstName":
               patient1Response.name[0].given[0]
         },
```

```
        {
            "lastName" :
                patient2Response.name[0].family,
            "firstName":
                patient2Response.name[0].given[0]
        }]
    }
context.message.header.set('Content-
    Type',"application/json");
context.message.body.write(patientResponse);
```

You are now ready to test. Again, you will use the **Test** tab to execute the API.

6. Click on the **Test** tab and provide the query parameters:

    Patient1 = 9df54eeb-a9ac-47ec-a8f6-eb51dd7eb957

    Patient2 = 97e47441-b8b9-4705-bda7-248ae6ae2321

    Click **Send** to test your API. Your results should appear as shown in *Figure 4.43*:

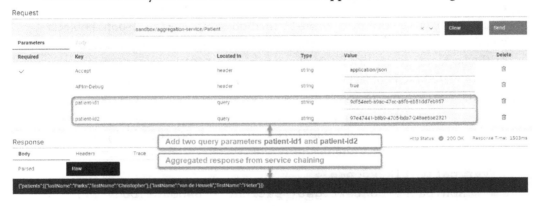

Figure 4.43 – A successful test of aggregation of response

Your results should be as follows:

```
{"patients":[{"lastName":"Parks","firstName":"Christop
her"},{"lastName":"van de Heuvel","firstName":"Pieter"
}]}
```

Well done!

The previous example covered some important aspects of API development. You were introduced to the concept of service chaining and techniques to access an Invoke Policy's response objects in a GatewayScript policy to create an aggregated response. Capturing the backend service response in a **Response** object variable is just one of many advanced features provided by the Invoke Policy.

One of the most important aspects of any software development is to build software that handles exception conditions that arise during that software's execution. APIC's development framework also provides you with a catch IBM extension and a throw policy to handle exceptional scenarios. The following example will demonstrate the technique of managing such error conditions within an API Proxy.

Error handling

Error handling involves handling custom error conditions and pre-defined error conditions. Examples of custom error conditions could be data validation faults (`missing last-name`, `missing first-name`, and so on) and business validation faults (typically arising from backend services such as `account number not available`). You gained some experience in error handling earlier in the *Invoke policy section*.

Examples of pre-defined error conditions are `ConnectionError`, `JavaScriptError`, and `TransformError`. A complete list of pre-defined errors is available at `https://www.ibm.com/docs/en/api-connect/10.0.1.x?topic=reference-error-cases-supported-by-assembly-catches`. A **Throw Policy** is available to throw an error reached during the execution of a message flow. Handling of the thrown error is done via an IBM **Catch extension**. This extension allows you to catch the thrown exception, build a processing flow to take remedial measures, and generate an appropriate error message for the API consumer. In this section, you will learn techniques to do the following:

- Throw a custom error using a Throw Policy
- Throw a pre-defined error in the GatewayScript Policy
- Catch an error
- Access the Error context variable in a GatewayScript Policy

You will use the `inline-access` API proxy that you built earlier to apply some error handling.

The Switch Policy has two execution paths. These two paths will be used to demonstrate the throwing of a custom error using the Throw Policy and the throwing of a pre-defined error in the GatewayScript Policy:

1. Open your `inline-access` API Proxy in Designer and click on the **Gateway** tab | **Policies** in the navigation menu.

2. Drag and drop a Throw Policy to the `Production` path of the execution flow, as shown in *Figure 4.44*:

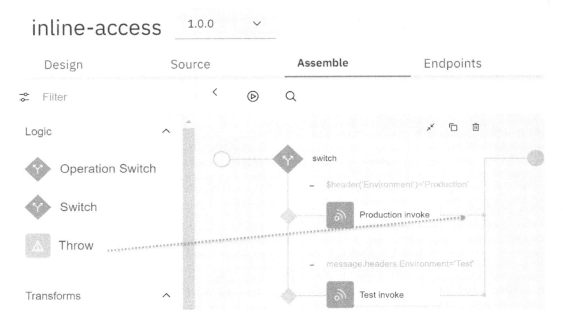

Figure 4.44 – Adding a Throw to the Production branch

3. Provide the following values for **Error name** and **Error message**:

Error Name: `UnsupportedEnvironment`

Error Message: `Set 'Environment' header to 'Test' because 'Production' environment is currently unsupported.`

> **Note: v10.0.1.5 Enhancements**
>
> When configuring a throw policy, you can now specify the HTTP status code and reason phrase in the **Error status code** and **Error status reason** properties, respectively.

4. Next, drag and drop a GatewayScript Policy at the beginning of the `Test` path. Name it `Validate patient-id`.

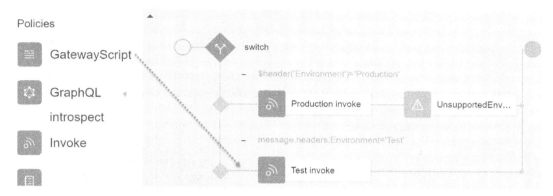

Figure 4.45 – Add the PID validation check

5. You can use the code downloaded from GitHub called `pidcheck.js` to copy and paste in the code block of your GatewayScript policy.

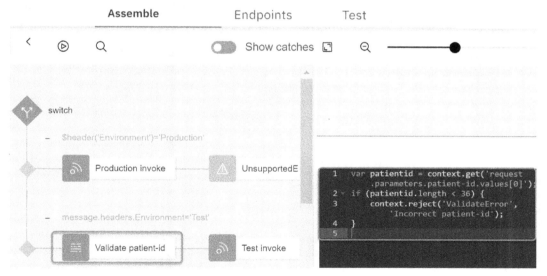

Figure 4.46 – Applying error handling to GatewayScript

Review the following code. You will see that it checks for the length of the `patient-id` parameter value and throws an error based on the length of the `patient-id` value:

```
var patientid =
    context.get('request.parameters.patient-
```

```
      id.values[0]');
if (patientid.length < 36) {
      context.reject('ValidateError', 'Incorrect
        patient-id');
}
```

The code uses the `context.get()` method to access context variables. Parameters that you receive as part of the request are also considered context variables.

> **Inline Referencing and GatewayScript Referencing of Parameters/Variables**
>
> Techniques for accessing Parameters using Inline Referencing and Context variables are slightly different. Path, query, and header type Parameters can all be accessed using `request.parameters.[parameter-name]` using the Inline referencing technique. But to access Parameters in the GatewayScript Policy, you will need to use `request.headers.[header-name]` to access the header type Parameters or `request.parameters.[parameter-name]` to access the query/path type Parameters.

6. Toggle the **Show catches** option. This will enable the **Catches** block. Click inside the **Catches** box to enable the **Properties** panel. Click **Add catch**.

    You can build error-specific execution paths or a single execution path to handle multiple errors. Notice that the `UnsupportedEnvironment` exception that you recently created is in the list of errors to catch. This is shown in *Figure 4.47*:

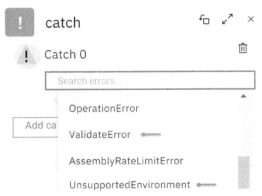

Figure 4.47 – Throws you create are listed in available Catches

7. Select `UnsupportedEnvironment` and `ValidateError` errors in `Catch 0`. You will be handling both these errors under a common Catch flow.

8. Drag and drop a GatewayScript Policy on the `Catch 0` execution path and name it `Create Error Response`. Copy the following code from GitHub, named `errormsg.js`, into this policy:

```
var errorName = context.get('error.name');
var errorMessage = context.get('error.message');
var errorResponse =
{
    "name": errorName,
    "message": errorMessage
};
context.message.header.set('Content-Type',
   "application/json");
context.message.body.write(errorResponse);
```

The code demonstrates the method of accessing the error context variable in a GatewayScript policy. During the execution time, when an error is encountered, the `error.name` and `error.message` variables contain the name and the message of the thrown error. The other critical element to notice is the use of the `message.header.set` and `message.body.write` methods to set the `Content-Type` header and actual response, respectively. You can use the `message.body.write` method to create a message body for an invoke policy's request as well. Use of the `message.body.write` method is not limited to generating custom API responses.

Your screen should look like *Figure 4.48*:

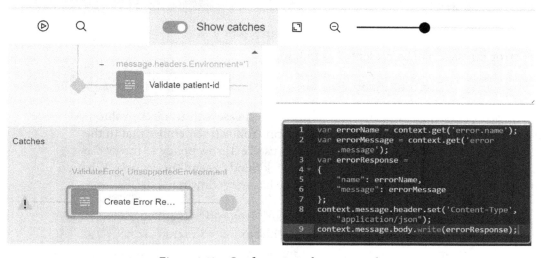

Figure 4.48 – Configuration of custom catch

9. The final step is testing. As you have done in the past, click **Save** to save your API. Now click on the **Test** tab to test your new error features.

    `ValidateError` testing: Add an `Environment` header and assign it a value, `Test`. Put in a random `patient-id` that is less than 36 characters and see how the API behaves. You should see the error message you configured in your error routine `pidcheck.js`.

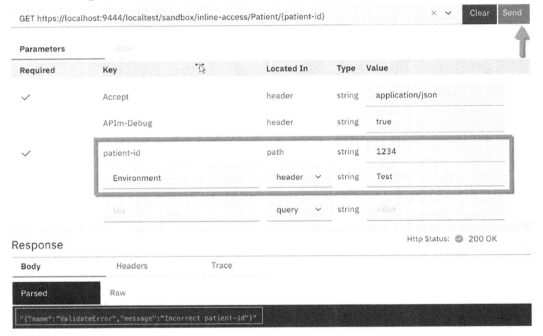

Figure 4.49 – Incorrect patient-id

The error message will be presented as follows:

```
{"name":"ValidateError","message":"Incorrect patient-id"}
```

`UnsupportedEnvironment` testing: In your test case, substitute the value of the `Environment` header for `Production`. You will remember that in the `Production` branch of your API proxy, you used a Throw policy to throw the `UnsupportedEnvironment` exception. Click **Send** to send the request. What do you observe? You should see an error message like the following:

```
{"name":"UnsupportedEnvironment","message":"Set
 'Environment' header to 'Test' because 'Production'
  environment is currently unsupported."}
```

The preceding section demonstrated ways to handle API Proxy error conditions. You can build upon these examples to handle different error conditions, both custom and APIC runtime specific.

You have now built a solid foundation of various APIC development features. Soon you will discover many other rich features of the APIC framework, starting with its support for modernization patterns, building RESTful services using FHIR (a healthcare standard), API security, GraphQL, advanced transformations, and much more. This chapter was just the beginning of a rich journey that lies ahead of you. This chapter got you to "buckle up." Now is the time for you to "enjoy the ride."

Summary

You started this chapter by installing and configuring a local development (Designer) and a testing environment (LTE). After getting a brief introduction to the **OpenAPI Specification (OAS)**, you put your local environment to good use by developing a simple API Proxy. This simple proxy creation exercise should have helped you get your feet wet and prepare you for a long swim in the pool of APIC.

This chapter then took you into a deep dive into the extensive development framework features provided by the APIC platform. You were introduced to many Policy types (Built-in and Custom) and Logic constructs, with a particular focus on the Invoke policy. With the introduction to variables, you learned about various context variables (request and message, especially) available to you for the fetching and manipulation of data flowing through the API. API properties and their usage in an Invoke Policy taught you methods for building environment-specific dynamic target URLs.

As you worked through many step-by-step examples, you were introduced to the vast capabilities of the Invoke policy. You covered the capabilities of applying blacklists and the setup of error handling for timeouts. You even learned how to execute more than one call to multiple backend service providers as part of the same flow. You were also briefly introduced to GatewayScript (as part of numerous examples) and the switch policy. The switch policy introduction contained examples for building switch conditions using scripts and variables. These capabilities will be highly beneficial to you as you begin building more APIs with varying use cases. There was an extensive repository of hands-on examples that covered essential topics of error handling, customizing an API's response, and advanced features of the Invoke Policy.

This chapter also explored multiple testing techniques available to you as a developer to test your APIs. Speaking of testing, you learned how to use the built-in test capability of APIC that includes the powerful feature of tracing. API Connect provides a one-stop shop experience by keeping development, testing, and tracing proxies in a single toolset – an ask of so many engineers!

You have seen just how quickly you can take an existing endpoint and create, promote, and test an API proxy. And with all the available features, **Policies/Variables/Referencing**, you as a developer can truly build robust, complex, and flexible APIs. But with *great power comes great responsibility*. The idea of API development is rooted in speed, agility, and simplicity. On your API development journey, if you observe yourself using all these features in developing a single API, it might be an opportunity to reflect on the design of that API itself. Food for thought!

With all this information and a solid base, you are now ready to take it up a notch and start with the modernization journey. The next chapter will teach you the techniques to expose hidden IT services to the outside world.

5
Modernizing SOAP Services

SOAP (short for **Simple Object Access Protocol**) is an XML-based protocol for accessing web services over HTTP. SOAP web services started gaining traction somewhere in the early first decade of the 21st century. Numerous standards were developed, especially in the following decade, which allowed the overall web services to mature, especially the SOAP-style web services. Some of the notable ancillary standards designed to support these SOAP web services are **Universal Description, Discovery and Integration (UDDI), Message Transmission Optimization Mechanism (MTOM), WS-Attachments, WS-Security, WS-Trust,** and **WS-Policy.** Many organizations still support SOAP web services because of SOAP's long history (after all, history is relative!) and the development and support of these standards by major software vendors.

Though the new development of web services has given way to REST-style services, SOAP services continue to provide many essential business functions. There are often external consumers who access these business functions.

Meanwhile, as the migration of these SOAP-based assets to REST-based assets continues to happen in the background, there is still a large requirement to expose SOAP services as **APIs**. **API Connect (APIC)** can help do precisely that, with considerable ease. In this chapter, you will learn various techniques using APIC to accomplish the goal of exposing SOAP web services as APIs. There are two main patterns for exposing SOAP services as APIs. You will learn both patterns in this chapter. These patterns are as follows:

- Exposing a SOAP service as a SOAP proxy: This use case is mostly for the service consumers/clients that continue to use the SOAP protocol.

- Exposing a SOAP service as a REST proxy: This use case handles the scenario where consumers/clients have already modernized their services and are using XML/JSON with a REST architectural style pattern.

After you have finished going through this chapter, you should become comfortable with the following:

- Understanding APIC's capabilities with respect to SOAP web services

- Creating a SOAP Proxy API using a **Web Services Description Language (WSDL)** using **out-of-the-box (OOTB)** features of APIC

- Creating a REST Proxy API using a WSDL using OOTB features of APIC

- Developing an understanding of the many APIC policies that are automatically generated because of exposing a SOAP service as a SOAP/REST Proxy API

Technical requirements

The steps discussed in this chapter are built upon the installation of the APIC Designer and APIC LTE environment covered in *Chapter 4, API Creation*. They also use two publicly available SOAP services: the **Calculator service** (used for building a SOAP API proxy) and the **Account service** (used for building a REST API proxy). You should ensure the following:

- Access to these services is not blocked by your machine's firewall or corporate network. The service endpoints are as follows:

  - Calculator service: `http://www.dneonline.com/calculator.asmx`

  - Account service: `https://apictutorials.mybluemix.net/AccountService`

- Service definitions (WSDL files) for these services are downloaded to your local workstation. You can download service definitions from the following URLs:

  - Calculator service: `http://www.dneonline.com/calculator. asmx?wsdl`

  - Account service: `https://www.ibm.com/support/knowledgecenter/ SSMNED_v10/com.ibm.apic.apionprem.doc/AccountServicing. txt?view=kc`

You can test these services using a SOAP testing tool such as **SoapUI**. Go ahead and import these WSDL files in SoapUI and send example requests. Your results should look like *Figure 5.1* and *Figure 5.2*.

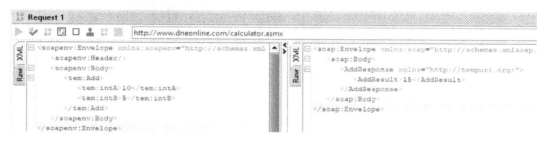

Figure 5.1 – Example test case for the Calculator service

And here is the output for the Account service:

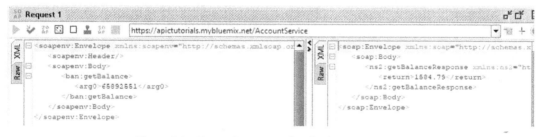

Figure 5.2 – Example test case for the Account service

Before we go into building SOAP and REST API proxies on top of our simple Calculator SOAP service, let's discuss the support for the SOAP protocol in APIC V10.

SOAP capabilities in APIC

APIC supports exposing existing SOAP services as either SOAP or REST proxy APIs. The basic requirement for the services to be exposed is that those services should support Web Services Basic Profile 1.1 – Second Edition.

To create a SOAP API definition, you will need access to an existing SOAP web service. This existing web service can have a WSDL that can be defined either by a single standalone WSDL (we will be using this in our examples) or by a WSDL file that references other WSDL files and/or schemas.

> **Note**
> For standalone services definitions that have no external dependencies, such WSDLs can be directly loaded from a directory to create the SOAP API definition.

For service definitions that have dependencies on other WSDLs and schemas, you will need to create a ZIP archive of the main service definition and all its dependencies and then load the ZIP file to create the SOAP API definition.

APIC provides API generation patterns that take care of generating either SOAP or REST API proxies from a provided SOAP service definition (standalone or otherwise). You will cover how to use these patterns and techniques to generate SOAP/REST API proxies in the following sections, starting with generating a SOAP API proxy from an existing SOAP service. But before deep diving into a couple of these patterns, let's quickly look at all the proxy patterns supported by APIC v10. To do that, log in to your API Designer and after performing the initial steps of selecting your workspace and logging into your LTE cloud environment with `shavon/7iron-hide` credentials, go ahead and **Add** a new API. Your screen should look like *Figure 5.3*.

Let's take a quick walk-through of the **Select API type** landscape and all the options that are made available by APIC here. We will stay with the **Open API 2.0** tab for now.

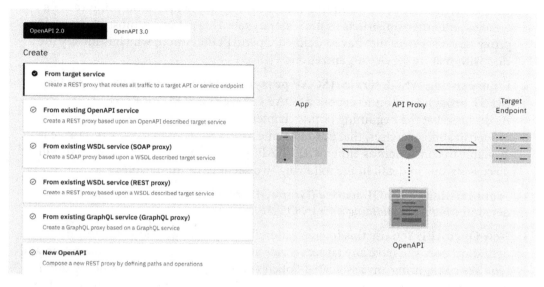

Figure 5.3 – API creation patterns

As is apparent, there are a number of options made available by APIC to accommodate various API patterns. Let's take a quick overview of some of the important ones before we go into discussing a couple of them in detail:

- **From target service**: Use this to build an API proxy definition from the ground up. The most common use case is when you have a target service URL available, and it is to be exposed through a custom schema that is different from the schema exposed by the target service. The target service could be a REST, SOAP, MQ, or database endpoint. This option is most suitable when custom processing logic needs to be applied before invoking the target service and/or the target service is not described by a standardized schema.

- **From existing OpenAPI service**: Use this to expose an OpenAPI-based REST service. You can import an existing YAML, JSON, or XML OpenAPI definition. Of course, you will still be building the API proxy implementation using various policies and other constructs in the Gateway tab. This pattern is most often used to proxy target services that have a defined OpenAPI definition, which is mostly the case with mature service organizations.

- **From existing WSDL service (SOAP proxy)** and **From existing WSDL service (REST proxy)**: Use this to expose a SOAP service as a SOAP or REST proxy. As discussed at the beginning of this chapter, this is the primary focus of service modernization exercises, that is, to use an existing SOAP-based asset and make it available to consumers as either a REST API proxy or a SOAP API proxy. We will be discussing this in detail in the following two sections of this chapter.

- **From existing GraphQL service (GraphQL proxy)**: You will learn about this in detail in *Chapter 9, Building a GraphQL API*.

- **New OpenAPI**: You use this to create a draft API definition. Note that this draft definition does not have any target service attached to it. It is simply a definition and requires configuring any associated Policies and target services. Depending upon what tool is used to create the API definition, the definition is either stored on the management server (if API Manager is used) or the workspace directory specified when you logged into API Designer.

> **Note**
>
> The **From target service**, **From existing OpenAPI service**, and **New OpenAPI** options are available for both OpenAPI 2.0 and OpenAPI 3.0 Proxy types.
>
> For the **From existing OpenAPI service** option, you can only create an OpenAPI 2.0 proxy if your target OpenAPI is described by OpenAPI 2.0 specification. Similarly, you will only be able to create an OpenAPI 3.0 proxy if your target OpenAPI is described by an OpenAPI 3.0 specification. Creating an OpenAPI 2.0 proxy using an OpenAPI 3.0 specification or vice versa is not permitted. You can experiment with various example definitions at `https://github.com/OAI/OpenAPI-Specification/tree/master/examples`.

Now that we have explored the various API proxy patterns available in APIC, it is time to look at the patterns that will help us in our service modernization efforts. You will soon appreciate the ease with which APIC helps us proxy our SOAP-based services either as a SOAP or REST proxy.

Creating a SOAP proxy that invokes a SOAP service

Consider a scenario where you want to expose an existing SOAP service where requests from API consumers are forwarded, as-is, to that SOAP service. This pattern of API/service interaction is depicted in *Figure 5.4*:

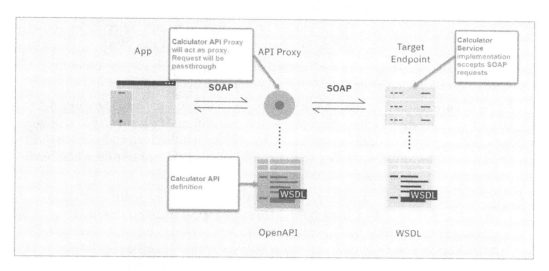

Figure 5.4 – API proxy/SOAP service interaction

We will now explore this pattern in detail by creating a simple SOAP API proxy that will expose a SOAP-based target service (**Calculator service**). To complete the rest of this section, ensure that you have read the *Technical requirements* section of this chapter. We will be using a service definition that is completely defined in a single file.

Creating a SOAP proxy

By now, you should already have become comfortable with starting your LTE environment, opening the API Designer, connecting to your workspace, and finally, connecting your API Designer to your running LTE server. We will hence start with the process of creating a new API:

1. In the **Develop APIs and products** tile of API Designer, click **Add** then select **API**.

2. In the **Select API type** view, select **From existing WSDL service (SOAP Proxy)**. Review the diagram displayed to the side of the **Select API type** view. Click **Next**.

3. In the **File upload** view, upload the Calculator service definition that was downloaded and saved by you as part of the technical requirements. The system should validate the service definition and display a success message. Click **Next**.

4. In the **Select Service** view, the Calculator service should be auto-selected for you as that is the only service defined in the service definition file. Click **Next**.

5. In the **Info** section, review all the details that are populated by APIC. You can change **Base path** in this section. For now, accept the default values and click **Next**.

6. In the **Secure** section, unselect **CORS** (more about this in *Chapter 7, Securing APIs*) and select **Activate API**. The effect of the **Activate API** option is that APIC will automatically create a draft Product, add this API to that Product, and publish the Product to the sandbox Catalog so that the API is available for testing. Click **Next**. This should create your API definition.

7. On the **Summary** page, you should see important information about your API proxy. Make a note of the **Client ID** and **Client secret** values. You will be using them shortly to test your API proxy and if for some reason you forget to make a note of these values, we will explore a way to recover them from the API definition.

Before we go about testing our SOAP API proxy, let's review all the considerable work that was done behind the scenes by APIC to expose the provided SOAP service as an API.

Review SOAP Proxy configuration

To fully appreciate the work done by APIC in mapping the WSDL definition to an OpenAPI-based definition, it is imperative that you review the various mappings. In brief, the mappings are done so that each WSDL `portType` is mapped to API Path, each path has a **POST Operation**, and each Operation contains Parameters that are mapped to **Input message type**, and Response that are mapped to **Output message type**. You will now review the `calculator` API's configuration. This will give you an in-depth understanding of the APIC's API generation process:

1.   APIC creates a set of **API Operations** based on the `portType` section in the WSDL document. For each operation in the WSDL `portType`, the API definition creates a matching API operation. The API **Paths** correspond to the names of the operations. You can review this mapping of WSDL `portType` to API **Paths** in *Figure 5.5*:

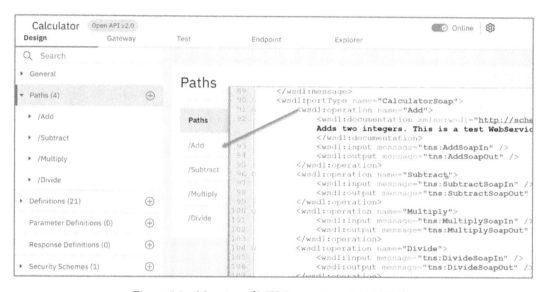

Figure 5.5 – Mapping of WSDL operations to API Paths

The HTTP method is always set to **POST** (SOAP services do not use HTTP methods of **GET**, **PUT**, or **DELETE**). Go ahead and click on one of the **Paths** and review the **Operations** section of **Edit Path** view. You will notice an operation with the name **POST**. Click on **Operation POST**.

APIC has done tremendous work in the background. For each API operation, it has mapped WSDL input and output types to request and response messages. Navigate to the `Calculator API` | **Design** | **Paths** (in the left menu) | **/Add** path | **Operations** | **POST** | **Path Parameters** type `AddInput` and **Response** schema `AddOutput` (*Figure 5.6*). **Path Parameters** are related to API requests and **Responses** are related to API responses. You will review the schema definitions for `AddInput` and `AddOutput` objects in the next step.

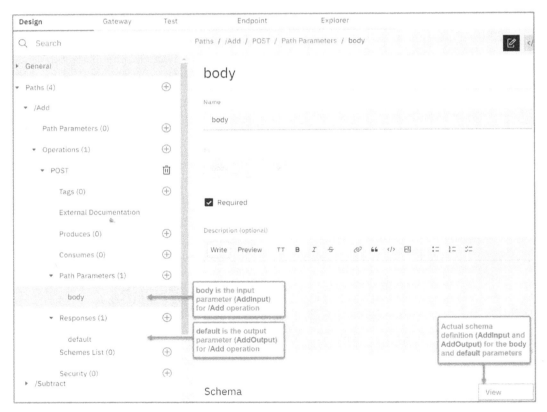

Figure 5.6 – Mapping WSDL Message Types to API Parameters/Responses

2. Navigate to the `calculator` API | **Design** | **Definitions** (in the left menu). Review the `AddInput` and `AddOutput` schemas, as shown in *Figure 5.7*.

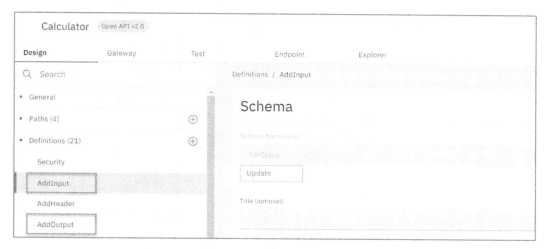

Figure 5.7 – Path Parameters/Responses Definitions

These steps should have helped you develop a good understanding of the behind-the-scenes work done by APIC in exposing your SOAP-based target service as a SOAP API proxy. Before we proceed to test the API, we should review two more important items. These are the **Gateway** tab and **Endpoint** tab of our API proxy.

Go ahead and click on the **Gateway** tab. You should see nothing but a single **Invoke** policy. This is because we are exposing our SOAP target service as a SOAP API proxy. There is no translation of data required. Requests from the API consumer are passed as-is to the backend service and the responses from the backend service are sent to the API consumer, again without any translations. All this work can be carried out by a single **Invoke** policy. You will soon see the difference (in the next section) in the **Gateway** tab of a REST API proxy that proxies a SOAP backend service.

Lastly, you should look at the **Endpoint** tab. Earlier in this section, we mentioned that **Client ID** and **Client secret** values can be retrieved from the API definition. You can retrieve these values from the **Endpoint** tab.

Having now built your SOAP API proxy and reviewed the configuration generated behind the scenes, it is time to test this API proxy.

Testing the SOAP Proxy

There are multiple ways to test your API. You can either use the SoapUI tool or the test tool built into API Designer/API Manager. You will be using the SoapUI tool for this test. Through this test, you will explore how the API's base path and path values are mapped to the actual URL. You will also be able to make a comparison between the API URL and the service URL. This test is divided into two parts. Part one will test the SOAP service by directly calling the service URL. This will ensure that there is required connectivity to the backend SOAP service and give us a glimpse of the response (to make a later comparison) returned by the SOAP service. Part two will test your SOAP API proxy. The highlight of your API proxy test will be to compare the responses produced by the two test cases and to see the Client ID and Client Secret security in action.

Testing the SOAP service

Open your SoapUI tool and create a new SOAP project. You can give this project any name. The initial WSDL value will be `http://www.dneonline.com/calculator.asmx?wsdl`. SoapUI should create a project based on this service.

Run a quick test by invoking the SOAP service directly. Refer to *Figure 5.1*. You should be able to get a valid response from the **Add** operation.

Testing the SOAP API Proxy

Using the same SoapUI project, change your testing endpoint URL.

Your URL consists of the following parts:

- **API base url**: `https://localhost:9444/localtest/sandbox/` (provided to you when you started your LTE environment).

- **API base path**: `/calculator` (available in the **Design|General|Base Path** section).

- **API path**: `/Add` (available in the **Design|Paths** section).

- **Complete URL**: You can construct the complete API URL as **API base URL + API base path + API path**, that is, `https://localhost:9444/localtest/sandbox/calculator/Add`.

Add this new endpoint to your SoapUI project as shown in *Figure 5.8*. But before you can send a request to this APIC endpoint, you have to do a couple of things.

1. First, add Client ID security to your SoapUI request. This can be done by adding a new header, **X-IBM-Client-Id** (it's worth mentioning that the Client ID is mapped to parameter name **X-IBM-Client-Id** in the **Security Schemes** section of your API.

You can give any header name, but it is recommended to use **X-IBM-Client-Id** for the sake of standardization). The value for this header is available in the **Endpoint** tab of your API.

2. To enable rapid testing, APIC auto-publishes your Product to a **Target Product** (`Calculator auto product`) and a **Test application** (`Sandbox Test App`). Refer to *Figure 5.8*.

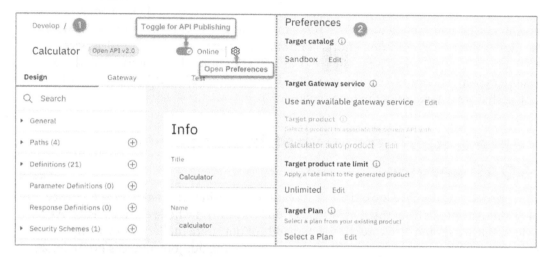

Figure 5.8 – Auto-publishing Preference

3. Now you are ready to test your API. Send a request to this new endpoint. You should see a response, as shown in *Figure 5.9*.

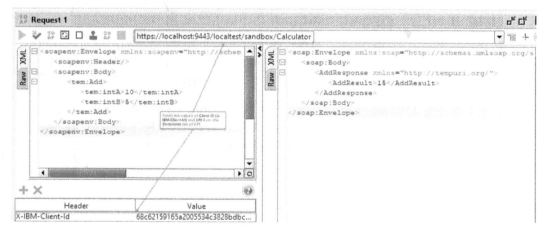

Figure 5.9 – SOAP API proxy test

4.   You should take time to test other operations as well. Make sure that you change the API URL as you test other service operations. With this test you should now be able to identify the key elements of an API URL, a simple method to enforce security on your API, and an API proxy that sits in front of a SOAP service.

This section covered in detail the facilities available in APIC to expose an existing SOAP-based service as a SOAP API proxy. This ability to expose an existing service as an API, without much effort, can help you on your modernization journey. Many of your API consumers who continue to use SOAP services will be able to consume your APIs with ease. But as their modernization efforts continue and they move from SOAP services to REST services, so will you have to evolve your API endpoints. To accommodate these consumers, you will need to expose your SOAP services as REST API proxies. This is exactly what we will cover in the next section.

Create a REST proxy that invokes a SOAP service

In the previous section, we looked at the method of exposing an existing SOAP service as a SOAP API proxy. That method works for our consumers who are still using SOAP services. But we should also cater to the demands of our more modern or evolving consumers who have shifted to REST services. Continuing our journey, let's now explore the methods in APIC to make available a sample SOAP service as a REST proxy API and satisfy our more modern consumers. APIC provides simple methods to expose an existing SOAP service as a REST API. Let's dive into learning these methods.

Creating a REST proxy

Similar to the starting steps that you took while creating a SOAP proxy earlier, go ahead and start your LTE environment, open API Designer, connect to your workspace, and connect to your LTE environment. After that, follow these steps:

1.   In the **Develop APIs and products** tile of API Designer, go to **Add | API**. In the **Select API type** view, select **From existing WSDL service (REST Proxy)**. Review the diagram displayed on the side of the **Select API type** view (*Figure 5.10*).

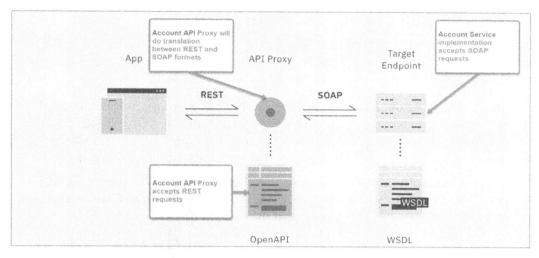

Figure 5.10 – REST proxy to SOAP service pattern

There are two main things to note. One is that requests received by API Gateway are RESTful in nature; that is, they will accept either JSON or XML payloads. Second, the requests going to the backend service are of the SOAP type. Click **Next**.

2. In the **File upload** view, upload the `Account service` definition that was downloaded and saved by you as part of the *Technical requirements section*. The system should validate the service definition and display a success message. Click **Next**.

3. In the **Select Service** view, **Account service** should be auto-selected for you as that is the only service defined in the service definition file. You will notice that there is only a single operation defined in this service. Compare this with the **Calculator service,** which had four operations (Add, Subtract, Multiply, Divide) defined in its WSDL file. Click **Next**.

4. In the **Info** section, review all the details that are populated by APIC. You can change **Base path** in this section. You can change the **Base path** value to / `accounts`. For now, accept the other default values and click **Next**.

5. In the **Secure** section, as done previously, unselect **CORS** and select **Activate API**. Click **Next**. You have now created your proxy API definition.

It is time to review the configuration that is generated by APIC to make this conversion between REST and SOAP possible. You are just going to observe how this differs compared to the SOAP proxy to SOAP service configuration.

Review the REST proxy configuration

The configuration of the newly created REST proxy is quite different from the SOAP proxy. This is obvious from the fact that our proxy is no longer a SOAP passthrough type. There are a number of extra actions that our proxy now performs:

- **Message translation**: It now consumes a JSON REST message instead of an XML SOAP message, and then does a JSON request translation to SOAP. It then performs the same actions in reverse, that is, a SOAP response coming from the backend service is translated to a JSON response sent to the proxy consumer.

- **Support for multiple HTTP methods (GET and POST)**: It now supports a GET HTTP method along with a POST method. We will shortly see how it does that and all the configuration required to support these methods.

Through the next steps, you will now review the generated REST proxy and observe the important elements of its configuration:

1. Open your Proxy API in API Designer. Click on the API's **Design** tab and then on the **Paths** section (in the left menu). Since there is only one operation defined in your WSDL, there is a single **Path** that is, **/getBalance**. You use this **Path** value while constructing your Proxy's URL and use it for your Proxy testing.

2. In the **Paths** section, click on /getBalance and explore the **Operations** section. You will observe two HTTP methods, **POST** and **GET**.

3. Click on the **GET** operation. It is important to carefully go through the **Path Parameters** and **Responses** sections.

   The **Path Parameters** section is where the request's incoming parameters are defined. Incoming requests can contain data in **query**, **path**, **header**, **form data**, and **body**. You will see a single arg0 **query** parameter defined. This matches our simple service that only accepts one parameter, arg0, as part of the SOAP request body.

   The **Responses** section is where the response schema is mapped to various HTTP status code values. The system has mapped **HTTP 200** code to **getBalanceResponse_element_tns** schema and **HTTP 500** to the **CommonFault** schema. It is important that you go through these schemas. You can review these schemas by going to the API's **Design** tab and clicking on **Definitions**.

Navigate to the /getBalance **Path | POST Operation**. You will observe that
the main difference (compared to the GET operation) is in the **Path Parameters**
section. The POST operation's **Path Parameters** contains a request parameter
that is in the **body** of the request and its type is **getBalance_element_tns**. This is in
line with the nature of the HTTP method. Data should now be sent in the request
body instead of the query parameter (as was the case with the GET method request).
There are no changes to the **Responses** section.

Having now reviewed the **Paths** configuration, let's take a tour of the **Gateway** tab
(*Figure 5.11*) of the generated Proxy. It is important to carefully review the *Table 5.1*.
It will give you an in-depth understanding of the generated configuration that facilitates
the mapping of a REST proxy to a backend SOAP service. You will immediately observe
that this generated configuration is much different and more complex than the SOAP
proxy configuration. The SOAP proxy configuration was a simple passthrough with no
translation of schema or the protocol itself. On the other hand, the REST proxy needs to
perform a schema translation from JSON/REST to XML/SOAP and a method translation
from GET/POST (REST) to POST (SOAP), hence the complexity.

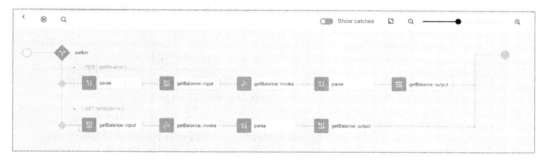

Figure 5.11 – Gateway tab of REST to SOAP Proxy

Policy	Details
S	A **Switch** policy to differentiate between POST and GET REST requests. Please review the *Tip* after this table. This Switch policy sends the execution of the flow either to the POST method path (P1-P5 policies) or to the GET method path (G1-G4 policies).

Policy	Details
P1	A **Parse** policy to perform two things: one, to create an XML DOM or JSON object (in this case) from the input string data (available in `request.body`); two, to copy the parsed value (that is, a JSON object) to `message.body`. You might want to take time to compare the **Parameters** section (**Paths | Operations | POST**) of your API. You will see the name of the input variable defined as `request`. After this action, the incoming data (JSON) is available in `message.body` and ready to be transformed into SOAP format.
P2	A **Map** policy to map incoming JSON request values to a SOAP format XML. Note the setting of **content-type** and **SOAPAction** values in the Map policy. Make sure to click the pencil icon and review both the **Input** and **Output** sections of the Map policy.
P3	An **Invoke** policy to call the actual backend service. In the auto-generated configuration, the target service URL is directly entered into the **URL** field. You can also use **Catalog Properties** to avoid hardcoding the URL values in the policy configuration. Using **Catalog Properties** will make your configuration portable. Another thing to note in the Invoke policy is the setting of **HTTP Method**. We will soon see how this value will be important in the GET execution branch.
P4	A **Parse** policy to parse/serialize the string contents of the incoming `message.body` variable into an XML DOM object.
P5	A **Map** policy to map the SOAP response from the backend service to JSON. Review the setting of **content-type** in both **Input** and **Output** sections of this Map policy. You will notice that `content-type` in case of **Input** is `application/xml` and **Output** is `application/json`.
G1	A **Map** policy to map the `arg0` JSON query parameter to a SOAP message. You will notice that there is no incoming parsing policy required in for GET methods. This is because of the presence of the incoming parameter (`arg0`) in the query parameter. (Refer to the **Parameters** section of the GET method in **Paths | Operations | GET**). For the REST proxy, the incoming `arg0` value is captured via `request.parameters.arg0`. Note the setting of the **content-type** and **SOAPAction** values in this Map policy. This is similar to what we saw for P2. After all, we are sending the SOAP request to the same backend service.

Policy	Details
G2	An **Invoke** policy that is exactly the same as P3. Similar to the P3, the setting of the HTTP Method to POST is most critical. The incoming request is using GET as the HTTP Method.
G3	A **Parse** policy. Same as P4.
G4	A **Map** policy. Same as P5.

Table 5.1 – Policy details of the Account service

> **Tip**
>
> You can write conditions scripts in the **Switch** policy using JSONata functions, numeric operators, comparison operators, and functional expressions. This is applicable for API Gateway only. There are many expressions supported, such as $httpVerb() and $operationPath(). $httpVerb() that map to the request.verb context variable and $operationPath() that maps to the api.operation.path context variable.
>
> You will see more examples of **Switch** condition scripts in *Chapter 6, Supporting FHIR REST Services.*

That was a great deal of insight into all the nut and bolts of proxy configuration generated for you by APIC. You can now try to expand on this information by changing this configuration to use Catalog Properties and redesigning the schema for proxy JSON responses. Catalog Properties are available in your API's **Gateway** tab. As another learning exercise, you can generate a more complex configuration by using the Calculator service to generate a REST proxy. With all this learning behind you, it is time to test the REST proxy.

Testing the REST Proxy

It is time to test the REST proxy. You will use the **Test** tab to test the REST endpoint. The endpoint will remain the same for GET and POST method invocations:

1. Go to your proxy **Test** tab.

2. Select the GET method URL in the URL selection drop-down box. You will notice that the testing facility has automatically populated various **Parameters**, such as **Accept**, **APIm-Debug**, and **X-IBM-Client-Id**. Refer to *Figure 5.12*.

3. Create a query parameter, `arg0`, in the same request and assign this parameter a value of `12345`. Refer to *Figure 5.12*.

4. After you have done the required preparation, go ahead and **Send** the request. You should receive an `HTTP 200` message and a valid response.

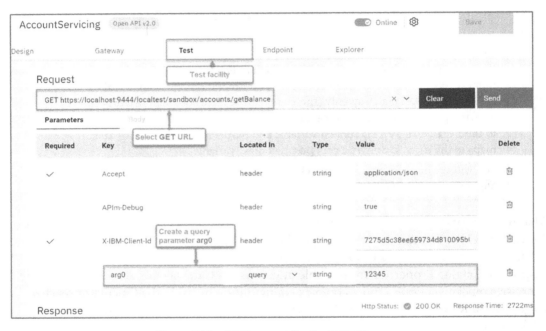

Figure 5.12 – GET request for the REST Proxy

5. You will test the `POST` method now. Change the **Request** URL to the `POST` method URL. Delete the `arg0` query parameter from the **Parameters** tab. Construct a simple JSON request payload in the **Body** tab. Refer to *Figure 5.13*.

6. Click **Send**. You should receive an HTTP 200 message and a valid response. This test helped you to understand the method of testing a REST proxy, along with the method of passing the request data to a POST method of a REST proxy.

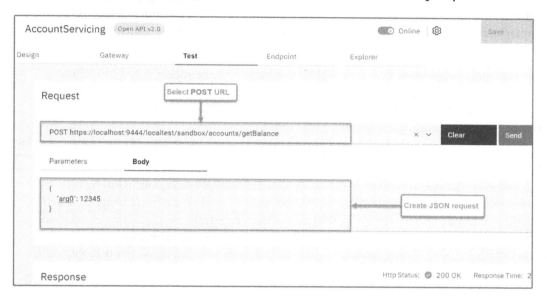

Figure 5.13 – POST request for REST Proxy

This section covered in detail the process of creating a simple REST proxy, configuring it, and testing it. You also reviewed the difference in complexity of REST proxy configuration and SOAP proxy configuration. Your configuration exploration took you into the depths of how the **Path Parameters** section is mapped to various incoming request attributes (query, headers, form data, body) and the ways of accessing those attributes in the **Map policy** of the **Gateway** tab.

Chapter 4, API Creation, and this chapter have given you a good overview of various development concepts of API development. They should have also laid a solid foundation for working with APIC LTE and API Designer. The next couple of chapters will build upon this foundation. *Chapter 6, Supporting FHIR REST Services*, will utilize the concepts covered so far to teach you how to build advanced FHIR (an extremely important healthcare interoperability standard) services. *Chapter 7, Securing APIs*, will cover one of the most important aspects of software development, and that is security. But before we take on those advanced concepts, let's wind up this chapter with a short summary.

Summary

The focus of this chapter was to cover in detail two different methods of modernizing your SOAP-based legacy assets. Method one was about exposing these assets via a SOAP proxy. Method two was about exposing them as a REST proxy. The techniques discussed in this chapter helped you quickly expand your company's API repository while making it easier for your existing customers to make use of your SOAP-based business functionality.

There are other legacy assets or non-SOAP assets that can be modernized too. These non-SOAP assets could be your messaging infrastructure (MQ) or your databases. You can use advanced APIC policies such as the GatewayScript policy to make connections to your backend servers.

Though it is now easier, with all the tools and flexibility at our disposable, to create APIs, the question that you should answer before embarking on your API'fication is if the functionality that you intend to make available as an API is even qualified to be one. API candidates (only from a modernization perspective) should be carefully weighed. Each organization builds their own API candidate selection criteria. Some common ones worth mentioning are the reusability score, granularity of business function, and consumer maturity index. Not all integration modernization problems can be solved using an API.

Make sure you remember *Maslow's hammer*:

> *"I suppose it is tempting, if the only tool you have is a hammer, to treat everything as if were a nail."*

APIC is most definitely a hammer, but not everything is a nail. Use the hammer judiciously!

In the next chapter, we will see how to develop REST APIs that support FHIR.

6
Supporting FHIR REST Services

In *Chapter 4*, *API Creation*, and *Chapter 5*, *Modernizing SOAP Services*, you learned how to set up the **Local Testing Environment** (**LTE**) and began developing APIs with the designer. You also learned how to take advantage of your existing SOAP assets and convert those into APIs.

In this chapter, you will learn about how to begin implementing APIs that utilize the **Health Level Seven International** (**HL7**) **Fast Healthcare Interoperability Resources** (**FHIR**) standard (pronounced *fire*) that is now being mandated by the **Centers for Medicare & Medicaid Services** (**CMS**).

FHIR is an HL7 standard that supports providing resources using REST APIs and JSON. This chapter introduces how to develop REST APIs that support the FHIR specification. You can read more on FHIR by visiting the website: `https://www.hl7.org/fhir/overview.html`.

In this chapter, we will cover the following main topics:

- Introducing FHIR API
- Creating a RESTful FHIR API
- Applying logic policies in your API

When you complete this chapter, you will know how to create RESTful services using API Connect with a focus on delivering FHIR APIs and embedded logic. You will also have a good understanding of how FHIR is a good example of digital transformation.

Technical requirements

With this chapter, you will be referencing a number of Swagger files to assist you with the learning experience. You will find these files in GitHub using the following URL: `https://github.com/PacktPublishing/Digital-Transformation-and-Modernization-with-IBM-API-Connect`.

You should copy the files for this chapter to your development environment. We will be utilizing the LTE and API Manager to perform the development tasks. If you prefer to use some of the skills you learned from *Chapter 2, Introducing API Connect*, you can use the CLI commands to push the sample files up to API Connect. The assumption going forward is you will begin utilizing the API Manager to draft new APIs.

Now that you understand what you will be attempting, let's begin with understanding the business use case for FHIR.

Introducing FHIR

Interoperability between healthcare organizations has been like the search for the holy grail. The **HL7** organization has been around since 1987 and has tackled the task of establishing standards for sharing healthcare information. There have been many levels of maturity of the standard over the years, with the most prominent being **HL7v2**.

Version 2 is an exchange of data that is delimited with pipes. Unfortunately, at the time, this method was too lenient, and some fields were left up to the provider to declare. These fields then created issues for consumers because they didn't have the consistency needed to receive the same type of file from different sources. Fields were of different types and values. Version 3 was introduced by HL7 to improve the data exchange quality by using XML and XML schemas. The adoption of version 3 was not as successful as they had hoped. XML schemas were difficult to digest and made implementations complicated. The goal was still to develop an improved version, but there needed to be something more digital that took advantage of Web 2.0, APIs, and JSON. That is how **Fast Healthcare Interoperability Resources (FHIR)** got started. The current version is FHIR v4.0.1. Although you can implement FHIR in various ways, the predominant way is by creating an FHIR server and accessing the FHIR resource data with APIs. Remember that the goal is to exchange data. Let's explore how that is accomplished by creating FHIR resources.

FHIR resources

A resource in FHIR represents small and locally distinct units of exchanged data. Each resource has a defined behavior and meaning. Contained within the resource are **elements** and extensions of elements, some narrative, and other extensions to the resource.

It also has **metadata**, which has elements such as known identity and the location of the data. Of course, each resource has some interest in healthcare (patients, observations, diagnostic reports, and so on). *Figure 6.1* shows the relationship between resources and elements:

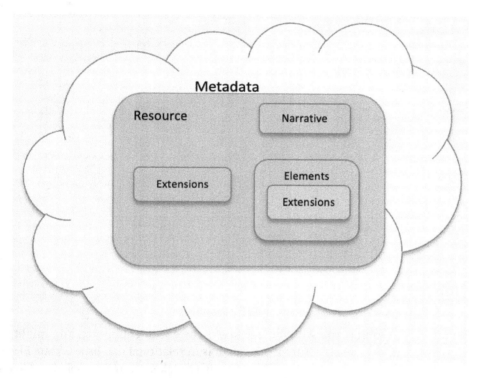

Figure 6.1 – A FHIR resource examined

There are hundreds of resources defined by FHIR. You can visit the HL7 FHIR website, `https://www.hl7.org/fhir/resourcelist.html`, to peruse the various resource definitions.

Figure 6.2 – FHIR resources

When you review a resource, you will find a data structure that it supports. You might wonder how you support this structure if your data is in relational databases. Your FHIR architecture may vary based on the approach you wish to take, given the tools you have at your disposal. You have various options for implementing your interoperability interface:

- **Broker adapter for FHIR**: If you have already used HL7 schemas from previous versions, you can set up a broker transformation into FHIR.

- **Proprietary mixed APIs**: Perhaps you are using multiple methods to exchange FHIR data. It can be a combination of FHIR and other APIs.

- **Vendor-neutral clinical repository**: In a clinical situation where you have multiple backend systems, you can establish a vendor-neutral repository to hold all your FHIR resources.

Something to keep in mind is the number of API endpoints you may be exposing. Let's look at the patient operations:

- Get lists of all patients.
- Get a single patient resource.
- Create a patient record.
- Update a patient record.
- Delete a patient record.

There are approximately 144 FHIR resources. If we support the 5 endpoints listed on all the resources, your number of endpoints totals 720, and that doesn't include custom endpoints based on consumer/business requirements.

Assuming you may be only addressing a small number of resources, let's discuss where the data resides after you have utilized one of the various options.

FHIR server

The FHIR server is where you will initiate your API calls to begin the exchange of data. You might think of it as the repository of the resource data accessed via the APIs you create. Many of the FHIR servers are built on Java but other languages exist. There are a number of test FHIR servers on the web and many others that you can download and use as models. The **HL7 application programming interface (HAPI)** FHIR server is one of them:

Figure 6.3 – The HAPI FHIR server can be used for testing APIs

In this chapter, we will be running tests on a HAPI FHIR server that is prepopulated with test resources. You can access the HAPI FHIR test server at this URL:

```
http://hapi.fhir.org
```

Before you start developing your first FHIR API, it will be beneficial to understand the motivation of using API Connect to implement your RESTful FHIR APIs.

Government-mandated FHIR interfaces

As mentioned in the chapter introduction, CMS has mandated payers (insurance companies) to implement a Patient Access API and a Provider Directory API for consumption by third-party entities using the FHIR standard. While the implementation date was moved from January 1, 2021, to July 1, 2021, nevertheless, payers will need to comply.

The benefits proposed by these two requirements are as follows:

- **Patient Access**: This is the government's response to putting patients first, giving them access to their health information when they need it most, and in a way that they can best use it. During a pandemic such as COVID-19, this would be very beneficial.

- **Provider Directory**: The government is requiring an FHIR interface to a directory of a health insurer's in-network providers and pharmacies for patients/consumers. Again, during a pandemic such as COVID-19, this would be very beneficial.

These two benefits to consumers are just what the doctor ordered. Having information readily available is critical to healthcare and patient care. Now that we will be learning how to create RESTful FHIR APIs, it is important to note the reasons why using API Connect is a good choice.

Figure 6.4 highlights reasons for building APIs, managing the FHIR API lifecycle, and maintaining the integrity of data:

Build APIs quickly

- No code model-driven methodology
- Generate Swagger and REST
- Pre-built connector
- Access data securely using OAuth, TLS and other security methods

Manage your FHIR API lifecycle

- Secure, manage, and share APIs
- Manage FHIR APIs across multiple clouds and environments
- Version manage your FHIR APIs
- Socialize your APIs on a Developer Portal customized for your organization

Maintain integrity of data

- Sync data across many applications and systems across clouds and environments
- Retrieve FHIR data securely

Figure 6.4 – FHIR and API Connect synergy

You now have a good overview of FHIR and how CMS mandates implementing capabilities for the Patient Access and the Provider Directory using FHIR. It's time to start playing with FHIR.

Creating a RESTful FHIR API

As you may recall, your APIs will be accessing an FHIR server that provides healthcare resources. Given that, when you create your FHIR RESTful API, you will need an endpoint. For the examples provided in this section, you will be referencing the online HAPI Test FHIR server that is at this location:

```
http://hapi.fhir.org/baseR4/swagger-ui/
```

You have learned to use the LTE and have been introduced to the API Manager Drafts Designer. You'll be using the LTE initially and then switch to the API Manager so that you can learn the other capabilities of the API Manager.

If you haven't already downloaded the chapter files from the URL mentioned in the *Technical requirements* section of this chapter, you should do that now and place them in a directory/folder on your local device.

Now that you have the files, you can log in and begin developing APIs.

Playing with FHIR

Everyone has a lot of fun utilizing FHIR in puns, but when it gets down to interoperability in healthcare, utilizing FHIR APIs is just what the doctor ordered. What you will be creating is the FHIR Patient resource as an API and retrieving data from an FHIR server on the web. The endpoint you will be calling is `http://hapi.fhir.org/baseR4/Patient?_pretty=true`.

If you were to call this from your browser, you would see all the patients in JSON data format returned, as shown in *Figure 6.5*:

```json
1   {
2     "resourceType": "Bundle",
3     "id": "461ec8ed-9168-4b8f-8259-10853f243d78",
4     "meta": {
5       "lastUpdated": "2021-06-30T23:58:00.913-05:00"
6     },
7     "type": "searchset",
8     "link": [ {
9       "relation": "self",
10       "url": "http://hapi.fhir.org/baseR4/Patient?_pretty=true"
11     }, {
12       "relation": "next",
13       "url": "http://hapi.fhir.org/baseR4?_getpages=461ec8ed-9168-4b8f-8259-10853f243d78&_getpagesoffset=20&_count=20&_pretty=true&_bundletype=searchset"
14     } ],
15     "entry": [ {
16       "fullUrl": "http://hapi.fhir.org/baseR4/Patient/618772",
17       "resource": {
18         "resourceType": "Patient",
19         "id": "618772",
20         "meta": {
21           "versionId": "1",
22           "lastUpdated": "2020-02-08T09:31:48.074+00:00",
23           "source": "#q305rtGyo6eF8xCL"
24         },
```

Figure 6.5 – Patient resource data

You will be working with the `Patient-swagger.json` file that you downloaded from the book's GitHub repository.

To create the FHIR, perform these steps:

1. Start the LTE environment:

    ```
    apic-lte start
    ```

    With a successful start of the LTE, you will receive some valuable information that you will need to log in to API Connect. They are shown as follows:

    ```
    INFO[0000] Creating docker resources
    INFO[0011] Waiting for services to start
    INFO[0117] Configuring backend
    - Platform API url: https://localhost:2000
    - Admin user: username=admin, password=7iron-hide
    - 'localtest' org owner: username=shavon, password=7iron-hide
    - 'localtest' org sandbox test app credentials client id: ea5a0964c30af6ba4f02ebeb4f4da940
    , client secret: 6f103704ae8515a3a180cdc66a627352
    - Datapower API Gateway API base url: https://localhost:9444/localtest/sandbox/
    ```

    Figure 6.6 – apic-lte startup output

    For now, that is all you will be needing to log in.

2. Launch the API Designer by double-clicking on the API Designer app that you installed previously:

Figure 6.7 – Starting the API Designer

3. Choose the directory/folder where you have downloaded the Swagger files previously. Most likely you already have logged on previously:

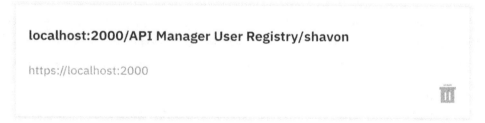

Figure 6.8 – Logged-on environment

4. Click on the localhost:2000 listing. You may get prompted to log in again, so you should use the information that was output when you started the LTE.

5. Select the **Develop APIs and products** tile. You should see the `Patient API` title and perhaps other APIs from prior chapters:

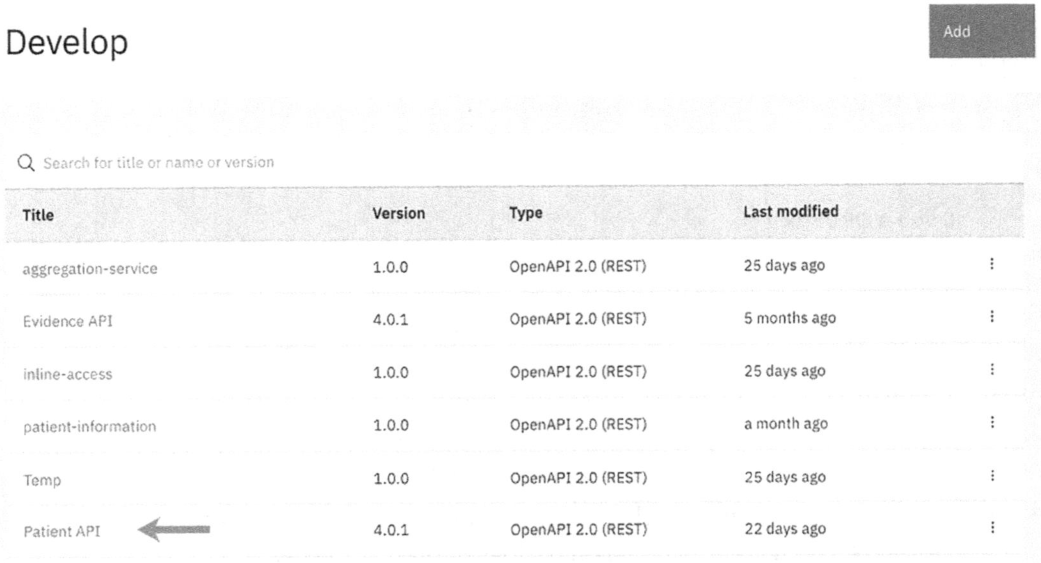

Develop

Add

Q Search for title or name or version

Title	Version	Type	Last modified	
aggregation-service	1.0.0	OpenAPI 2.0 (REST)	25 days ago	⋮
Evidence API	4.0.1	OpenAPI 2.0 (REST)	5 months ago	⋮
inline-access	1.0.0	OpenAPI 2.0 (REST)	25 days ago	⋮
patient-information	1.0.0	OpenAPI 2.0 (REST)	a month ago	⋮
Temp	1.0.0	OpenAPI 2.0 (REST)	25 days ago	⋮
Patient API ⟵	4.0.1	OpenAPI 2.0 (REST)	22 days ago	⋮

Figure 6.9 – Imported Patient API

Notice that the **Version** is **4.0.1**. This represents the version of FHIR that this API supports. You can also see that it supports **OpenAPI 2.0** specifications.

> Tip
> If you do not find the patient API file loaded, please download it from the GitHub site and use the LTE to add a new API using the existing OpenAPI import. You will find these files in GitHub using this URL: `https://github.com/PacktPublishing/Digital-Transformation-and-Modernization-with-IBM-API-Connect`. Once completed, you can continue with these steps.

6. Select **Patient API** to bring up the designer so you can review the FHIR API and make modifications. You should explore the following sections to gain a better understanding of what FHIR requires.

Click on **Design** and then the **General** drop-down menu. Click on **Base Path** to review the base path. This is shown in *Figure 6.10*:

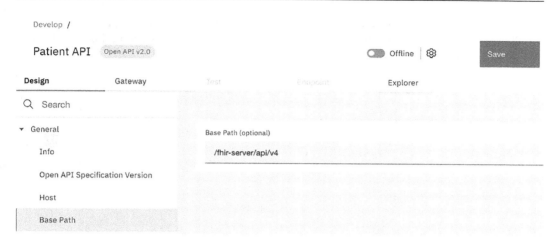

Figure 6.10 – Patient API Base Path

Another important section relating to the URL is the parameters. Click on **Design |
Parameter Definitions**. This is where you will see all the parameters defined within
the API. *Figure 6.11* shows some of the parameters:

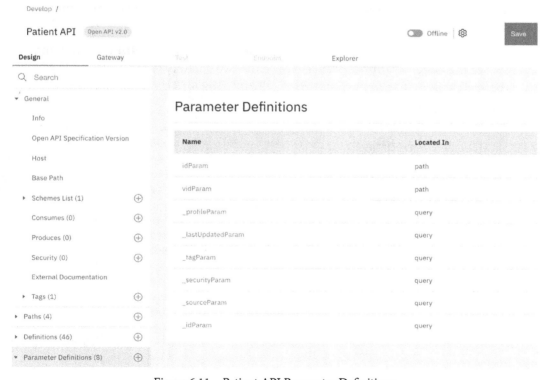

Figure 6.11 – Patient API Parameter Definitions

Another important section is your schema definitions. Click on **Design |**
Definitions to observe what is already defined. These are the elements for the FHIR
resource. This is shown in *Figure 6.12*:

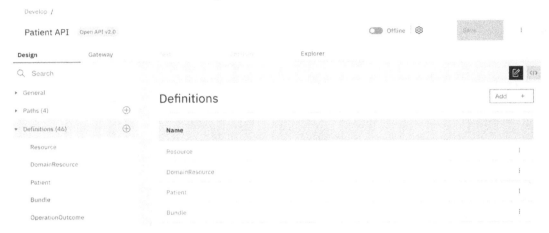

Figure 6.12 – Definitions shows the FHIR data elements

Next, you can review how API Connect provides the ability to enhance the runtime
of the API.

Click on the **Gateway** tab and then select **Gateway and portal settings**. Scroll until
you see the **Gateway (option)** label. Notice the two gateway types (`datapower-`
`api-gateway` and `datapower-gateway`). *Figure 6.13* shows the location of the
Gateway dropdown:

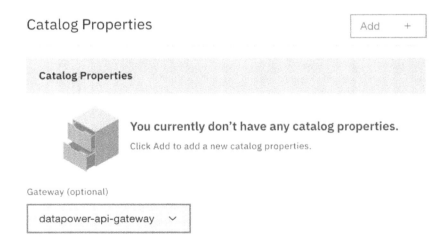

Figure 6.13 – Gateway selections on the Gateway tab

The two gateways allow the API developer to choose which implementation of gateway will satisfy the type of API developed. For APIs migrated from the previous version 5 release of API Connect, `datapower-gateway` provides version 5 compatibility. For newly developed APIs, `datapower-api-gateway` provides additional functionality and an increase in the API performance.

Also on the Gateway tab is the **Properties** section. Notice the default endpoint. It is named `target-url` and its associated value is `https://hapi.fhir.org`:

Figure 6.14 – Properties displays references to variable fields

As you can see, a lot has already been defined for you and there are places where you can add documentation about your API. If you were doing this with an empty OpenAPI Swagger file, you would have had to create all of those details. Luckily for you, it was all handled previously.

7. Locate the **Policy** menu item on the left. Notice how the page just shows the **invoke** policy. When we start applying logic to the API, this screen will be updated to show visually how the flow will be executed:

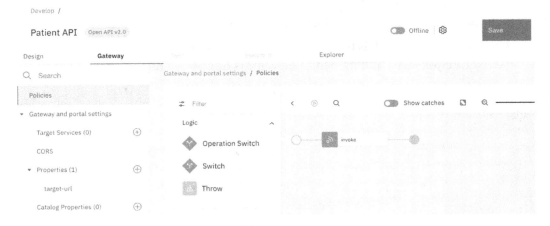

Figure 6.15 – Default policies for FHIR API

Now that you have familiarized yourself with the default implementation, it's time to make some updates so we can test our API. The changes you will make will be adjusting the target URL and the basepath, so it adheres to the endpoint: `http://hapi.fhir.org/baseR4/Patient?_pretty=true`:

1. Navigate to the **Design** tab and select **Base Path**. Change `/fhir-server/api/v4` to `/baseR4` and click on **Save**:

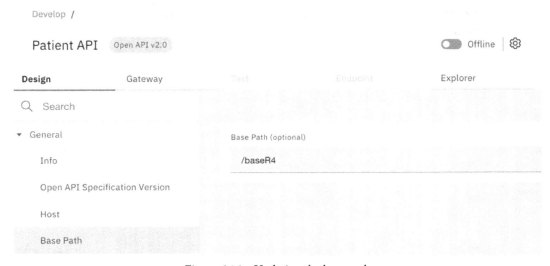

Figure 6.16 – Updating the base path

Figure 6.16 shows the results of your change. What you just did was adjust the base path to match the target endpoint URI to specify `/baseR4` versus the default that was provided in the download. When the API executes, API Connect will then append the new base path to the host endpoint. Next, you'll update the target host within the `targetURL` property.

2. Navigate to the **Gateway** tab and select **Properties**. You will notice `targetURL`. Click on the link to allow for changes.

    Notice that the default value is `https://localhost/fhir-server/api/v4/`. You will be changing this to `http://hapi.fhir.org`. Make that change and click **Update**, and then click **Save** once again.

3. While still on the **Gateway** tab, you need to adjust the **invoke** policy that you saw in the **Policies** section.

    You should see the single **invoke** policy. Click on the policy itself to open the property editor:

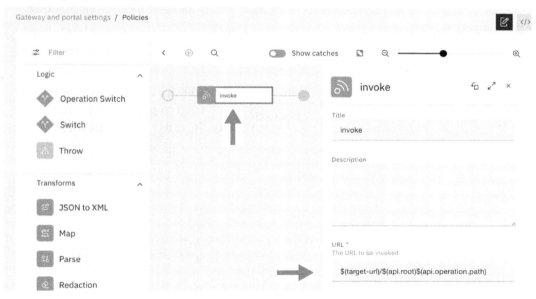

Figure 6.17 – Invoke property editor

As you can see in *Figure 6.17*, the URL property has the following replacement properties:

```
$(targeturl)$(api.operation.path)$(request.search)
```

You are already familiar with `target-url`. The other two are context variables. You were introduced to context variables in *Chapter 4, API Creation*. When you want to utilize the submitted path to an operation, you reference the `api.operation.path` context variable. If you reference `request.search`, you are essentially taking the request details after the question mark. An example would be `http://hapi.fhir.org/baseR4/Patient?_pretty=true`.

You will adjust these to reflect the changes you want to make. In the URL property, change the value of `$(targeturl)$(api.operation.path)$(request.search)` to `$(targeturl)$(api.root)$(api.operation.path)`.

This will now call the backend API using the values you changed. You should test it to verify it is working. Ensure your API is online. If it shows offline, slide the button to online.

Once your API is online, click on the **Test** tab and perform the following steps:

1. On the **Test** panel, use the **Request** dropdown to see GET `https://localhost:9444/localtest/sandbox/baseR4/Patient`. This is shown in *Figure 6.18*.

    You may have noticed that API Connect had all of the HTTP methods support – a real timesaver. Click **Send** to invoke the API.

2. If all goes well, you will be returned the JSON payload from the FHIR server (shown in *Figure 6.18*). If any errors occur, you will need to research the errors. If you make corrections to the API, ensure you save it, and then execute these steps again:

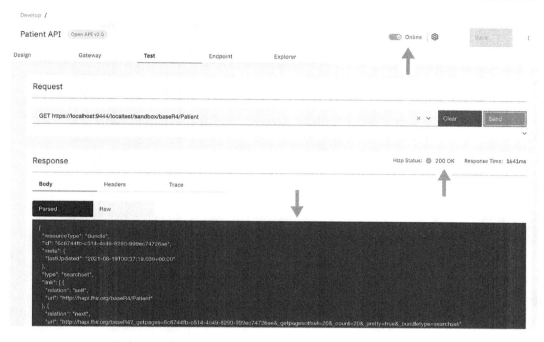

Figure 6.18 – Successful patient results from the test

You have successfully created and tested an FHIR GET request and have seen data returned. You did that with very minimal steps.

Before you learn about adding logic policies, let's first move your API up to the API Manager so you can experience development and testing on that platform.

> **Important Note**
>
> It is an assumption that you have access to the API Manager and belong to a provider organization. If you have yet to get access to the API Manager, please request access from the provider organization owner. To get to the API Manager user interface, use this URL: `https://[api-manager-server-ui]/auth/manager/sign-in/`.

3. Next, log in to the **API Manager** and select **Develop APIs and products**:

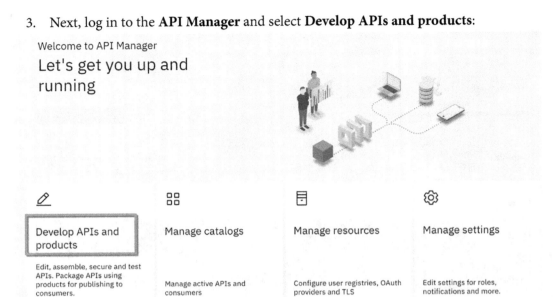

Figure 6.19 – Using the API Manager for development

You will be presented with a list of existing APIs. This is a list of all the APIs within the provider organization.

4. Click on the **Add** button and choose **API** (from **REST**, **GraphQL**, or **SOAP**) so you can import the `Patient API` file you created with the LTE. Drag and drop your API or click to upload the file using a file browser that pops up to select the API.

5. Choose **Existing OpenAPI** and select your `Patient API` file:

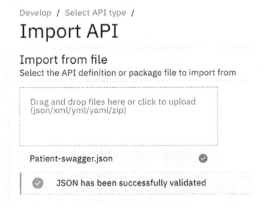

Figure 6.20 – Importing the Patient API file to the API Manager

Your API is now available for additional development in the API Manager.

You might notice that the page display is very similar to the LTE environment:

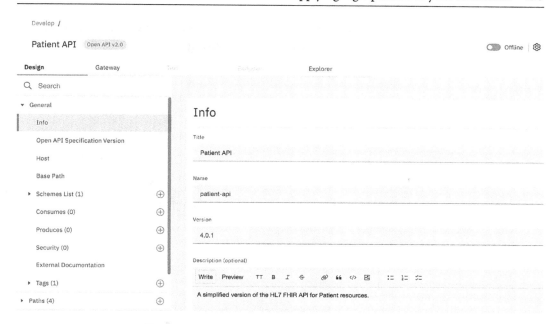

Figure 6.21 – Patient API viewed on the API Manager

Now, it's time to add some logic constructs to your API to allow greater flexibility.

Applying logic policies to your FHIR API

When you were working within the Gateway policies, you may have noticed the logic policies that were listed in the left panel under **Logic**. While the options are specific to conditional operation, the **Throw** policy is also provided for conditional error handling. Depending upon the gateway, the number of options differs. Actually, the logic policies have been simplified between the **DataPower V5 Compatible** Gateway and **DataPower API Gateway**. The **If** logic policy is implemented within the **Switch** logic policy of the **DataPower API Gateway**. *Figure 6.22* shows that change:

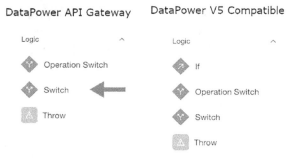

Figure 6.22 – The If policy is replaced with Switch in the new API Gateway

If you are confused about how these options change within the Gateway policies, simply click on the **Gateway** tab and then **Gateway and portal settings**:

Figure 6.23 – Navigating to the Gateway tab

On the Gateway and portal settings, scroll down and you will see the **Gateway (optional)** label, showing what you have selected for this API:

Figure 6.24 – Gateway type selected

Now that you are aware of the types of logic you can add, we'll put that into practice. You'll start with **If** and **Switch**.

The If and Switch logic policy

In the near future, the DataPower Gateway V5 Compatible offering will be used very little, as the DataPower API Gateway is touted to perform 10x faster. That being said, since the DataPower API Gateway supports If logic, with the Switch policy, you'll learn how to do both logics with the Switch policy.

Let's add some simple logic that is actually useful while you are doing development. The scenario would be that whenever you are testing your API, you would like to have the endpoint changed based on the APIm_Debug header value. You only want to debug if you are executing your test case data. So how would you do that?

1. On the API Manager, select **Develop APIs and products** and click on **Patient API**.

2. Select the **Gateway** tab and then **Policies**. From the left side, drag and drop a **Switch** logic policy just before **invoke**. The drop area is a small square on the line before **invoke**:

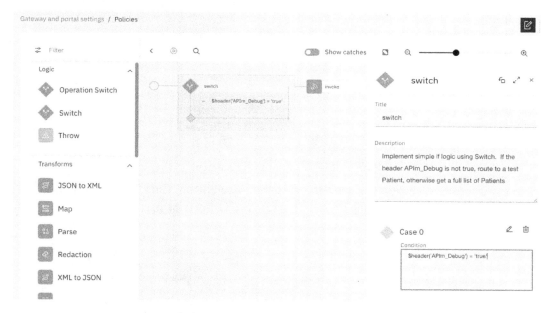

Figure 6.25 – Using Switch logic in the API

3. The **Properties** pane will appear and you can begin providing the relevant information. Type in a description. You can use any description or use the description in *Figure 6.25*.

4. The $header context variable will be referenced in **Case 0** to determine the value of APIm_Debug. If its value is not true, we will take a different path for our endpoint. Click on the *pencil* sign next to **Case 0** and a new panel will display.

5. Using the dropdowns, choose **$header**. Then, enter the header name as `APIm_`
    `Debug`. Choose the equal (=) operator and then type the `true` value. Click **Done**.

    You just specified that if the `APIm_Debug` flag is not on, process as normal.
    This is representing the `If` statement. Now, we can create the **Add otherwise** case,
    as shown here:

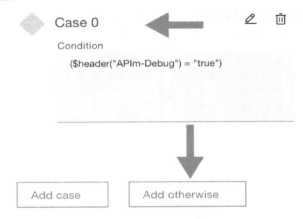

Figure 6.26 – Creating an If statement

6. If you scroll down, you can see the **Add otherwise** button. Click that button and
    you will see the **switch** logic update showing the **Otherwise** branch:

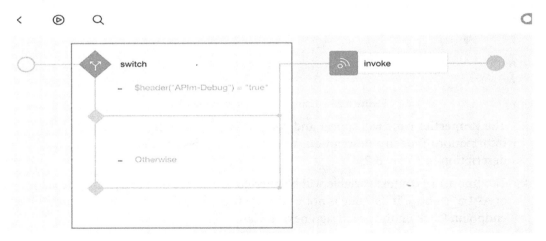

Figure 6.27 – Adding the Otherwise branch

It is within the **Otherwise** branch that you will add a new **invoke** policy, specifying
a patient that you will be allowed to debug.

7.   Drag an **invoke** policy from the left pane and drop it on the line below **Otherwise**:

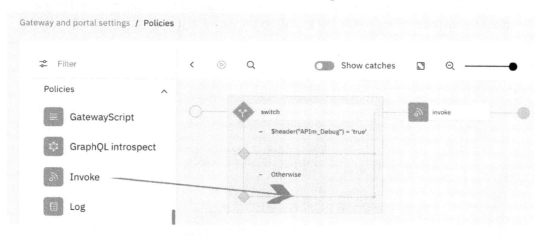

Figure 6.28 – Dragging an invoke policy and dropping on the line

8.   Now, you will update the properties for the **invoke** policy to specify the URL and
patient ID to act as our default test user. In the **Properties** pane, enter this in the
URL field: `http://hapi.fhir.org/baseR4/Patient?_id=1263576`.

Figure 6.29 – Adding values to the URL

9. Lastly, we want to ensure that we don't run the same **invoke** policy for both cases, so we will drag the first **invoke** policy to our case, checking for the header. Drag and drop the **invoke** policy to the left and drop on the **switch** case for our header check:

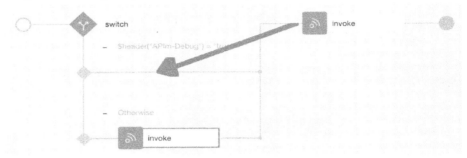

Figure 6.30 – Drag the invoke policy to the first switch case

10. Your final assembly is complete. Click on the **Save** button.

Figure 6.31 – Final switch completed

You have updated your API to apply some logic that will run debugging only if the debug header is set to `true`. You are ready to run a test. You will accomplish that by using the **Test** tab, as shown here:

Figure 6.32 – Navigating to the Test tab

Perform the following steps here:

1. Click on the **Test** tab and use the dropdown to select **Get patient operation**. Change the **Value** of the APIm-Debug parameter to false:

Required	Key	Located In	Type	Value	Delete
Parameters				Body	

Required	Key	Located In	Type	Value	Delete
✓	Accept	header	string	application/json	🗑
	APIm-Debug	header	string	false ⬅	🗑

Figure 6.33 – Running a test using the Test tab

2. Click on the **Send** button and review the response. You should see information about the single patient you set within the **Otherwise** case:

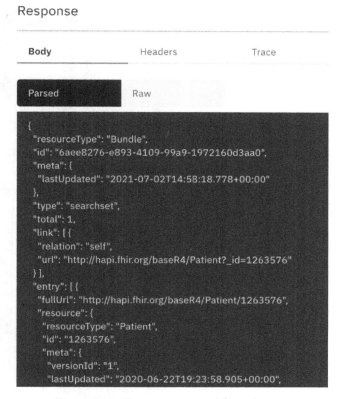

Figure 6.34 – Response returned from the test

Now you have successfully applied If logic to an API. While that was a simple example to introduce you to logic switches, you will probably be more interested in seeing how to apply a switch that routes to the proper operations you have created for your API. You will do that next.

The operation switch logic policy

The **operation switch** policy is used to separate the programming logic by the HTTP operation verb (POST, GET, PUT, DELETE, and so on). By splitting it this way, it provides you with the opportunity to supplement each operation with additional policies, such as security or transformations. Before we implement the operational switch, let's review where the operations are set up by looking at a completed example:

1. Log in to your API Manager.

2. Choose the **Develop APIs and products** title.

3. Click the **Add** button and choose the API.

4. Click on **Import Existing OpenAPI** and select `Evidence-swagger.json`.

    Your file will be uploaded as follows:

Title	Name	Version	Type
Patient API	patient-api	4.0.1	OpenAPI 2.0 (REST)
Evidence API	evidence-api	4.0.1	OpenAPI 2.0 (REST)

    Figure 6.35 – Uploaded Evidence API

5. Now click on **Evidence API** and navigate to **Paths**. It is under **Paths** that you can create a new path and assign HTTP methods:

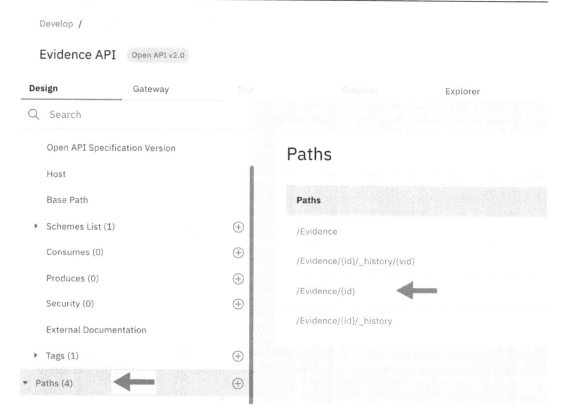

Figure 6.36 – Paths are where you add HTTP methods

You can see the paths already created in this API. These RESTful paths were created using the FHIR specifications on how you interact with an FHIR server.

6. Locate /Evidence{id} in the **Paths** menu list, and expand it and **Operations** to see what operations have been set:

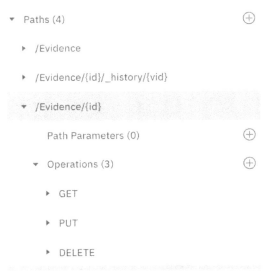

Figure 6.37 – Operations defined for /Evidence{id}

In *Figure 6.37*, you can see the *plus sign* (next to **Operations**) button visible on the right. This is how you will add new operations. Go ahead and click on the button to see how you would accomplish it:

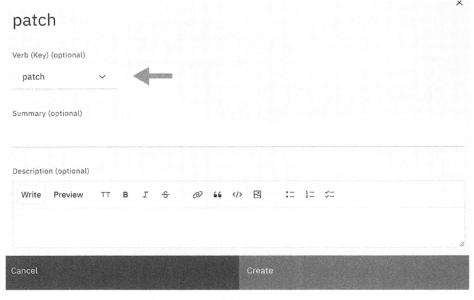

Figure 6.38 – Adding an operation

As you can see, there is a dropdown that has all the operations not currently added. You can choose any of the other operations and provide a description of that operation.

7. Choose `patch` and click **Create**:

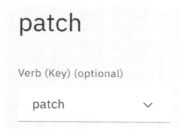

Figure 6.39 – patch added to the operations

You will now see PATCH added on the left menu under **Operations**. You are required to add a response for the PATCH operation. Expand PATCH and you will see **Responses**.

8. Click on the plus sign (next to **Responses**) to add a response.

This pops up a new page to add details about the operation:

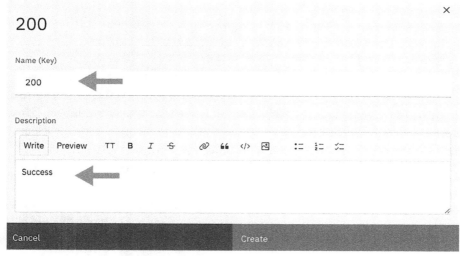

Figure 6.40 – A response code is required for patch

The default is the 200 HTTP response code. You'll accept this default and click **Create**. Ensure you enter a descriptive message, such as Success. Click **Create** to create the response code.

9. Next, you will define the response schema returned with your 200 response code:

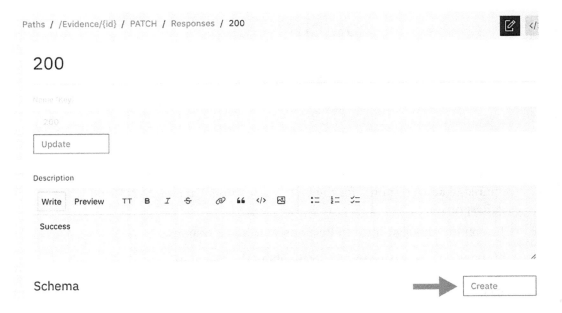

Figure 6.41 – Create a new schema for the 200 HTTP code

In the **Schema** section, click on **Create** to define a new schema. There are two method tabs to choose from. The first is **Definition**. When you choose **Definition**, you create the schema from scratch. You can add a title and a description, but you must select a type. Since the response is a simple text response, you will select string. The other option is **Reference**. Since our imported API already has existing definitions, we want to choose this. Click on **Reference**:

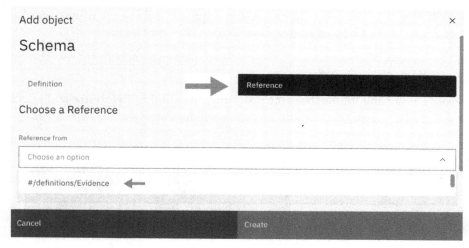

Figure 6.42 – Choose an existing schema definition

10. Click **Create** to create the schema. To go along with your PATCH operation, you'll require some parameters. One parameter (idParam) and definition (Evidence) were already created for you when you uploaded the API. Click the plus sign within the PATCH operation to add a new parameter:

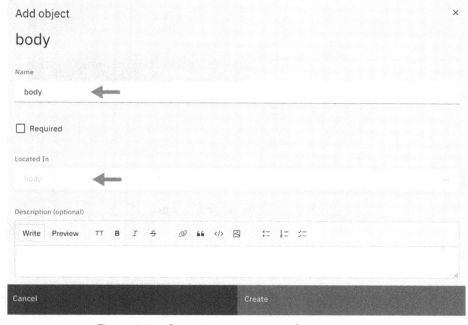

Figure 6.43 – Operation parameters and response setup

11. You will type in `body` as the name, set it as required, and use the dropdowns to select `body` in the **Located In** dropdown. Then click on **Create**.

The screen refreshes and you will notice that the schema shows an error. You can choose **Fix Reference** or **Create**.

12. Choose **Create** and follow the same procedure of selecting **Reference** and selecting the `#/definition/Evidence` choice. It will look exactly like *Figure 6.42*.

You'll notice you still have one error:

Figure 6.44 – Error show need for more parameters

You also need to reference the `{id}` parameter for the PATCH operation. Click on the parameter plus sign again and add the `idParam` reference. *Figure 6.45* shows that you select **Reference** and `#/parameters/idParam`:

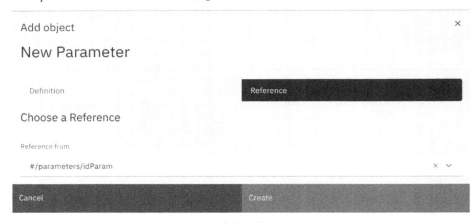

Figure 6.45 – Add the idParam parameter

13. Click **Create** to add your parameter. Your error should disappear, with the results shown in *Figure 6.46*:

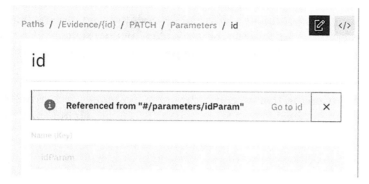

Figure 6.46 – Successful reference of idParam

14. Save your API. Next, navigate to the PUT operation and review **Responses**:

Figure 6.47 – The Evidence PUT operation settings

As you can see in *Figure 6.47*, similar to the PATCH operation we just created, a body of **Evidence** type was required for this API, and it is allowing two status codes that represent what actually happened in the FHIR server. Before we return to the operation-switch example, let's first review the **Evidence** schema:

1. Click on **Definitions** so that we can review how **Evidence** is constructed. Click on **Evidence** and then switch to the **Source** icon. This is shown in *Figure 6.48*:

Figure 6.48 – Definitions will show how data is organized

2. The OpenAPI definition is very lengthy and complex. The benefit of using the *form* icon to navigate to **Definitions** and choose **Evidence** is that when you click on the **Source** icon, it places you on the line number where **Evidence** is defined. As you learned before with updating parameters using the **Source** icon, you can also make updates to your **Definitions** in the same manner.

Now, you should have a good understanding of what is contained within the path and how operations are part of it, as well as references to definitions and how to add items using the **Source** icon. Let's return to operation-switch to show how we implement those paths.

Updating the Evidence API with operation-switch

You have learned that the Evidence API has four operations within the /Evidence/ {id} path. Since each may have different requirements upon execution, it would be good to separate those in the assembly so that you can apply different policies to them. We'll use operation-switch to set that demarcation.

> **Warning**
>
> Nesting an operation-switch component inside an `if` or `switch` construct, or another operation-switch component, is not supported.

To use the operation-switch, perform the following steps:

1. Make sure you are in the **API Manager** and have chosen the **Develop APIs and products** tile. Locate **Evidence API** and click on it:

2. Locate the **Gateway** tab and select it. Ensure you are in the **Policies** section. You will see an **invoke** policy, as shown in *Figure 6.49*:

Figure 6.49 – Evidence invoke policy

3. Drag and drop an **Operation Switch** component just before the **invoke** policy on the canvas.

    From the **Logic** menu, drag and drop the **Operation Switch** component and release it before **invoke**, as shown in the following screenshot:

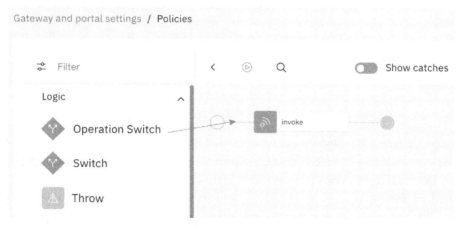

Figure 6.50 – Dragging the Operation Switch component to the canvas

The **operation-switch** component will then show before `invoke`, as shown in *Figure 6.50*. You will notice that there is an **empty case** displayed. You will start providing those cases:

Figure 6.51 – operation-switch

4. You already know that there are operations set for **Evidence API**. They were GET, PUT, DELETE, and PATCH.

5. On the **operation-switch** property panel and scroll down until you see **case 0**. Click on where it says **search operations...** and a list of operations will be displayed:

Figure 6.52 – Operations display for GET, PUT, DELETE, and so on

What you may have noticed is the operators you thought would show up (GET, PUT, DELETE, and so on) are not there. Instead, there are various values, such as readEvidence, updateEvidence, and so on. How did they get there? You'll learn shortly, but you should finish this first.

6. Select readEvidence and click on **Save**.

7. Now, click on the **Design** tab. Then find the `/Evidence/{id}` path.

8. Scroll down to see **Operations** and expand it to choose `GET`, as shown in the following screenshot:

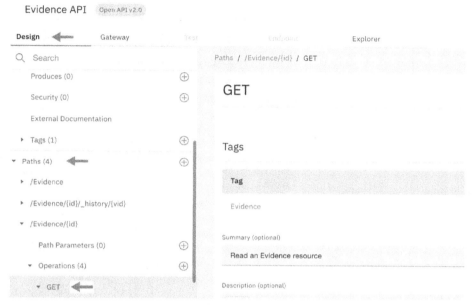

Figure 6.53 – Navigating to review operation details

Continue to scroll down the page until you see the **Operation (optional)** label:

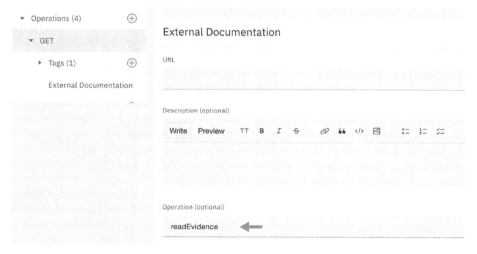

Figure 6.54 – Adding an operation is good for readability and documentation

Notice how the **Operation (optional)** field is showing **readEvidence**. You saw this optional field when you added the PATCH method earlier, but you left it blank. Since you didn't supply any **Operation (optional)** fields to PATCH, it will create a default name and list it on the case properties field on the invoke policy. Now you know how an Operation (optional) field works, you can continue.

9. Click on the **Gateway** tab so that we can finish our **operation-switch** cases.

10. Click on **operation-switch** to bring up the **Properties** pane. Your screen should show your **Case 0** selection of readEvidence.

11. Finally, click on the **Add case** button, as follows:

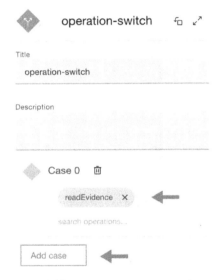

Figure 6.55 – Adding more cases to operation-switch

When you click on **Add case**, it will add a new case field ready for your selection:

12. Then, click on **search operations…**. Notice the difference between an **Operation (optional)** field with data or left blank:

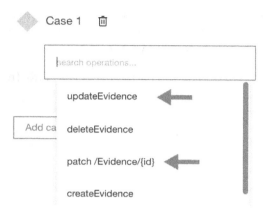

Figure 6.56 – The search operations… dropdown

You will choose `updateEvidence` to assign the `PUT` method to **Case 1**. Notice how `readEvidence` is not shown and the `PATCH` method you added is displayed in the dropdown. It shows your `PATCH` method and the path, instead of a readable **operation (optional)** name.

Tip

Having an **Operation (optional)** name makes your visual programming more readable.

---

13. Repeat *step 11* to add the **deleteEvidence** case. Your updates should reflect on the canvas as three total cases, as follows:

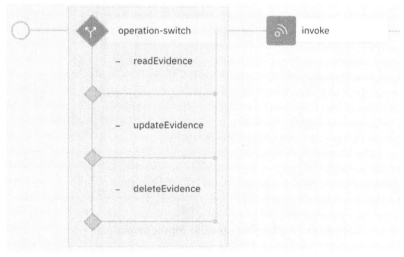

Figure 6.57 – Three operations are set up

As you can see, **operation-switch** shows readable cases so that you can understand what they will be doing. We can make it more readable by changing the **invoke** policy to a more descriptive name. To do so, perform the following steps:

1.   Click on **invoke** to bring up its property pane. Change **Title** from `invoke` to `get Evidence by id`:

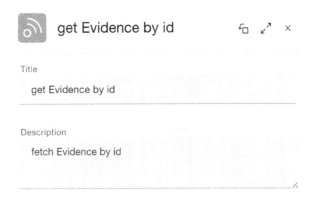

Figure 6.58 – Updating the title from invoke to get Evidence by id

After you make that change, **operation-switch** will be updated:

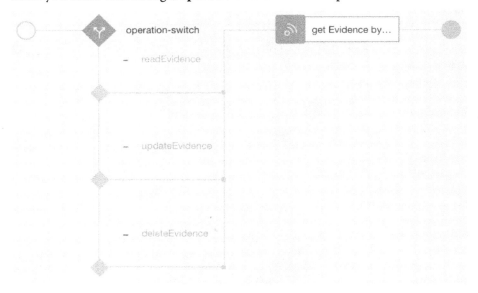

Figure 6.59 – Updating the title for the invoke policy

It is definitely more readable. Now, you will need to place the **get Evidence by id** invocation under the **readEvidence** case. Drag and drop the **get Evidence by id** invoke policy on the line under **readEvidence**:

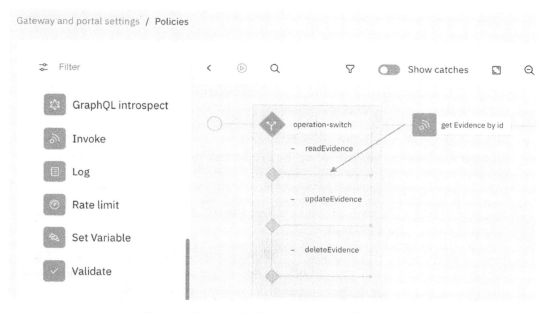

Figure 6.60 – Drag Evidence by id to readEvidence

Now, whenever a **readEvidence** operation (GET) is called, **get Evidence by Id** will be invoked.

You will now need to test the conditional switch, but first, you need to update the base path and invoke URL. Since you are still on the Invoke pane, you should update the invoke properties now.

2. On the invoke pane (now called **get Evidence by id**), find the URL field and change the current value from $(target-url)$(api.operation.path)$(request.search) to $(target-url)$(request.path).

3. Click **Save** to save your assembly work. You'll update the target-url property base path next. Click on the **Design** tab.

4. Scroll down and select **General | Base Path**. Change the base path from /fhir-server/api/v4 to /baseR4:

Figure 6.61 – Update base path to FHIR path

5. With the base path updated, you next need to update the **target-url** endpoint. Make sure to save the API by pressing the **Save** button.

6. Click on the **Gateway** tab and locate the **Properties** menu item. Click on **Properties** and select the target-url link. Change the property value from https:// localhost/fhir-server/api/v4/ to http://hapi.fhir.org/.

7. Save your changes and then click on the **Test** tab. If the **Test** tab is disabled, change the slider from **Offline** to **Online**.

You'll be presented with a page that will allow testing. One of the benefits of using the **Test** facility is it already has all of your operations available for execution – a real timesaver. It also knows the parameters that are required and displays them:

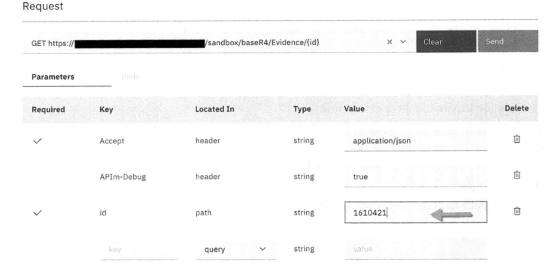

Figure 6.62 – The test tab is used to test the API

8. On the **Test** page, use the dropdown to select the GET request that is calling `https://<host>/sandbox/baseR4/Evidence/{id}`.

9. In the **Parameters** section, update the required id field and enter 1610421 as the value.

10. Click **Send** to initiate the request to the FHIR server. The response will return the Evidence FHIR resource with the ID of 1610421, as shown here:

Figure 6.63 – A successful test

You have successfully executed one of the operation-switch cases. But what about the other two that you set up? If you test them, nothing happens. It's because we need to add invoke policies to the other two. You'll do that next:

1. Return to the **Gateway** tab and drag and drop an invoke policy on the **updateEvidence** case. The properties will display invoke, and you should update the title and URL. Change the invoke title from invoke to put Evidence by id.

2. Replace the URL with $(targeturl)$(request.path). Your screen should look like *Figure 6.64*:

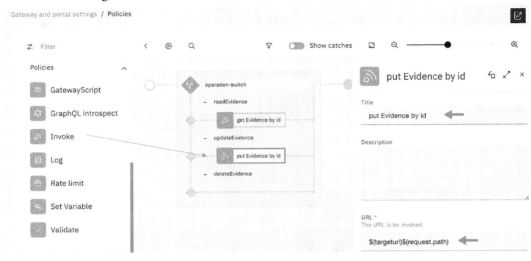

Figure 6.64 – Adding a new invoke policy for updateEvidence

3. Now, perform the same steps you just did but instead drop the invoke policy on the **deleteEvidence** case and change the title of the invoke policy to delete Evidence by id. Finally, you will also update the URL with $(targeturl)$(request.path).

After saving your changes, your final **operation-switch** should look like *Figure 6.65*:

Figure 6.65 – Completed operation-switch with three operations

You have now successfully implemented an **operation-switch**. With this pattern of developing with the **operation-switch**, you can now add additional policies before and after your new invokes (**get Evidence by id**, **put Evidence by id**, and **delete Evidence by id**).

> **Information**
>
> The HAPI FHIR server is available for testing online. If you will be testing using operations that update and delete resources, it is suggested that you create a new resource first, save the `return` identifier, and use that identifier in future operations.

This concludes the steps to creating RESTful services using API Connect. You've learned a lot, so let's review the key skills learned.

Summary

While this chapter was rather lengthy, it was packed with good information. You were introduced to FHIR and how that digital framework is helping healthcare companies successfully exchange information securely. You even learned with actual FHIR resources defined with data definitions.

You learned more about how RESTful services are created in API Connect and utilized the API Manager drafts to perform the visual development. Within the **Develop APIs and products** tile, you learned how to add existing APIs, modify those APIs with the designer, and add logic to flows such as operation-switch. The skills you learned in this chapter will make you very comfortable adding new or existing APIs, and then enhancing the flow of the API using logic constructs.

Finally, you were introduced to the Test facility where you can run a test directly from API Connect and how that facility improved your agility by generating your operation calls.

Your knowledge is building rapidly. The next chapter will build on this one by adding security to your APIs.

7
Securing APIs

In the modern world of technology, security is one of the most essential non-functional characteristics of any platform or solution. It is especially true of an API as an API's outreach is the defining feature of its success. More the number of consumers of an API greater will be its footprint. A secure API supports broader outreach and increases consumers' confidence. A secure API's consumers can fully leverage that API confidence about their data confidentiality and privacy.

In the previous few chapters, you looked at methods of building APIs using various features and functions available within **API Connect** (**APIC**). *Chapter 4, API Creation*, helped you understand and explore multiple API building constructs, and in *Chapter 5, Modernizing SOAP Services*, you learned methods to expose your existing SOAP assets as RESTful APIs. One of the primary goals of both chapters was to make you comfortable with the idea of API development and help you start your modernization journey using APIC. The success of digital modernization efforts hinges upon how the entire security requirements of the enterprise, system, application, API, function/method, or even a **line of code** (**LOC**) are addressed. Security encompasses everything from the enterprise to LOC. The focus of this chapter will be to coach you on techniques for securing the APIs on the APIC platform.

In this chapter, you will be learning about:

- **Out-of-the-box** (**OOTB**) security capabilities of APIC
- Protecting APIs with Basic authentication and Client ID (API key) methods
- Applying the OAuth2 security method to secure APIs

- Implementing **OpenID Connect (OIDC)** to secure APIs
- Using JWT policies
- Adding additional security measures

After you have finished going through each of the preceding topics, you should become comfortable with the following:

- Understanding the full spectrum of API security capabilities of the APIC platform
- Choosing between different methods of API security and understanding the key advantages and disadvantages of each method
- Applying various security methods to secure your APIs
- Developing an understanding of the methods to test API security

Technical requirements

Unlike *Chapter 4*, *API Creation*, and *Chapter 5*, *Modernizing SOAP Services*, which utilized local development tools such as API Designer and **Local Test Environment (LTE)**, this chapter will require you to have access to an existing APIC cloud implementation. This APIC cloud access can be provided by your company's APIC cloud administrator.

The exercises in this chapter will require you to have the following access:

- API Manager UI.
- Access to the mock authentication URL service: `http://httpbin.org/basic-auth/user/passwd`. This service returns `HTTP Response Code 200` for the `username=user` and `password=pass` credentials. For any other username and password values, this service returns `HTTP Response Code 401`.
- Access to `https://stu3.test.pyrohealth.net/fhir/Patient/d75f81b6-66bc-4fc8-b2b4-0d193a1a92e0` from your API Gateway.
- Access to a curl tool for executing tests.

Once you have access to these, you are ready to start exploring various API security features available in APIC.

Out-of-the-box security capabilities of APIC

When securing APIs with APIC, you are provided with three out-of-the-box security capabilities that can be applied. These are as follows:

- **API key**: This method involves configuring **Client ID** and **Client Secret** security definitions as part of defining an API. Once defined in the API's Security definition, a consumer may pass values for these API keys as part of a request's query (`X-IBM-Client-Id`, `X-IBM-Client-Secret`) or header (`client_id`, `client_secret`).

- **Basic authentication**: This option allows you to perform API authentication by validating the supplied credentials against an Authentication URL or an LDAP-based user registry.

- **OAuth**: The OAuth option allows us to secure APIs utilizing the standards set forth for OAuth2 (and OIDC).

These three capabilities should generally solve most of your API security concerns. When choosing between these security capabilities, you utilize a two-step process within an API definition. *Figure 7.1* identifies these two steps:

Figure 7.1 – Out-of-the-box security options

Step 1 is where you either choose or create a new security scheme. In *step 2*, you apply one or more of the security schemes to your API. **The OpenAPI Specifications (OAS)** terminology identifies these steps as **Security Definition** and **Security Enforcement** (or Security).

Of these three OOTB security options, the utilization of the API key method requires implementing only steps 1 and 2 described here. There is no further setup needed for the API key, whereas Basic authentication (**Basic**) and **OAuth2** methods require you to separately create security providers before these can be utilized in the security scheme step. You will learn how to accomplish this next.

Preparing for the APIC security implementation

Basic authentication and OAuth2 require integration with various servers. They require configuring resources (user registries and OAuth providers) in the Cloud Manager or API Manager. As an architect or administrator, you will need to know how these resources are accomplished. As a developer, you will only be concerned about having access to the configured and functioning resources. This section will give a high-level introduction on how resources are configured for utilization by the provider organization's developers. *Figure 7.2* shows you the **Resources** configuration page:

Figure 7.2 – Configuring Resources

This page is available in API Manager and Cloud Manager interfaces under **Home** | **Manage resources**. It allows creation of the user registries (Local, LDAP, Authentication URL, and OIDC), TLS references, OAuth providers, and Billing. The security components described in this chapter revolve around user registries and OAuth providers. You will be creating these resources in the API Manager interface.

While there are many variations on how you can create these resources, here you will learn how to configure a simple user registry and an OAuth provider. You will learn to define a user registry next.

> **Note**
>
> You must have the necessary permissions to the API Manager to perform these tasks. Typically, you should belong to either of the following two roles to be able to manage user registries: **Organization Owner** or **Administrator**.

Creating a user registry

You can create a new user registry by logging in to the API Manager and navigating to **Home | Manage resources | User registries | Create**. You will be presented with the screen shown in *Figure 7.3*:

Figure 7.3 – Selecting a new user registry type

As shown in *Figure 7.3*, there are four types of user registries. **Local user registry (LUR)** is the first user registry created when you install APIC. The LUR type is used to provide simple registries for **Catalogs**. Each catalog (sometimes called an environment) has its own LUR that contains the authorized users for that catalog. Refer to *Figure 7.2* for the default LUR created for the **Sandbox** catalog.

LUR is the most basic type of user registry provided by APIC. From *Figure 7.3*, you can appreciate the range of user registry types that are supported by APIC.

Next, you will learn to provide some of the common user registry types, starting with an **Authentication URL user registry**.

Configuring an Authentication URL user registry

There are instances when a user's credentials are stored outside of an LDAP system, for example, in a user table in a database, a CRM or third-party authentication system, or a simple file-based repository. A simple REST authentication endpoint can be made available that can authenticate a user against such repositories. From APIC's perspective, the HTTP response code from such an authentication endpoint decides whether a user's authentication is successful (HTTP response code 200) or a failure (HTTP response code <> 200). APIC supports a user registry type called an Authentication URL user registry to utilize such an authentication endpoint.

You will now learn to build an Authentication URL user registry that will use a mock authentication endpoint, `http://httpbin.org/basic-auth/user/pass`. Here is how it goes:

1.   Navigate to the **Home | Manage resources | User registries** page. Click **Create** and then choose the **Authentication URL user registry** tile (refer to *Figure 7.3*). Enter the information shown in the following *Table 7.1*:

Field Name	Value
Title	`HttpBin user registry`
Display namae	`HttpBin`
Url	`https://httpbin.org/basic-auth/user/pass`
TLS client profile	`Default TLS client profile: 1.0.0`

Table 7.1 – Creating a new Authentication URL user registry

Your screen should look like the left-hand section 1 of *Figure 7.4*:

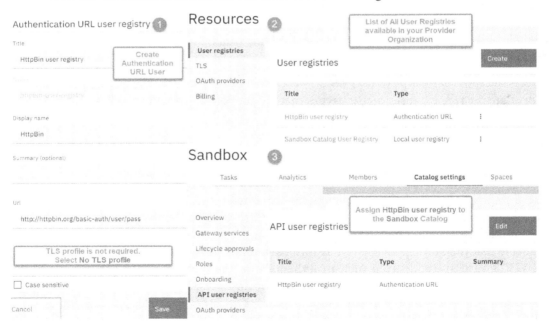

Figure 7.4 – Authentication URL user registry configuration

2. Click **Save** and you are ready to use it. You will certainly appreciate the ease of creating a user registry. You will be able to see your newly created Authentication URL user registry as **HttpBin user registry** in the list of **User registries**. Refer to *section 2* of *Figure 7.4*.

3. The last step is to make the **HttpBin user registry** available to the **Sandbox** catalog. Click on the **Home** screen | **Manage catalogs** tile | **Sandbox** tile | **Catalog settings** tab | **API user registries** section | **Edit** button. In the **Edit API user registries** view, click the checkbox for **HttpBin user registry** and then click **Save**. Refer to *section 3* in *Figure 7.4*.

This completes the creation of an Authentication URL user registry and making it available to the APIs deployed on the Sandbox catalog.

Another scenario you will often encounter is to interface with an LDAP system for authentication purposes. You will next see the method of configuring an LDAP user registry to perform such an authentication.

Introducing the LDAP user registry

Most corporations maintain their system login credentials (employee, applications) in an LDAP repository. They do this for multiple good reasons, such as the systematic organization of records in a hierarchical structure, data security, and platform neutrality.

For such scenarios, APIC supports an LDAP user registry type. Creating an LDAP user registry requires many details and is dependent on your LDAP server configuration. You will most likely need to work with your LDAP administrator in setting up an LDAP user registry. You can still review the sample configuration in the following screenshot to see some of the information that is required to set up an LDAP user registry:

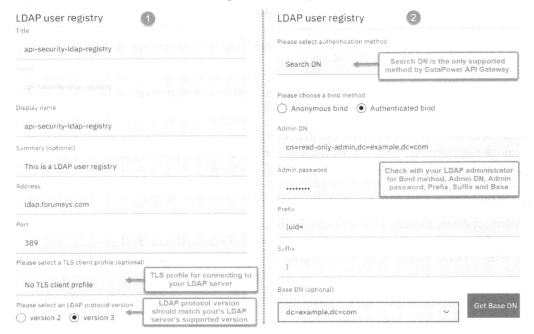

Figure 7.5 – LDAP user registry configuration

As you can see in *Figure 7.5*, setting up such a user registry requires significantly more information, including, the **bind method**, **Prefix**, **Suffix**, **Base DN**, and so forth. You should consult your LDAP administrator before embarking on this configuration.

There is one more user registry type that is supported by APIC, and that is the OIDC user registry. Since this chapter is about API security (instead of user security), OIDC is only briefly covered here.

Introducing OpenID Connect (user registry)

APIC also supports creating a user registry connected to an OIDC **Identity Provider** (**IdP**). It can only be used for onboarding and authenticating Cloud Manager, API Manager, and developer portal users. You cannot use it for securing your APIs, though.

The OIDC user registry will generally be connected to a primary OIDC provider such as Google, Slack, or GitHub (among many others). APIC makes it convenient for you to integrate with the main IdPs, such as Google, by automatically populating many key configuration parameters. You will need to get your client/application registered with your IdP to get a unique set of **Client ID** and **Client secret** credentials that you will need to provide on the **Create OIDC user registry** form. Refer to *Figure 7.6*:

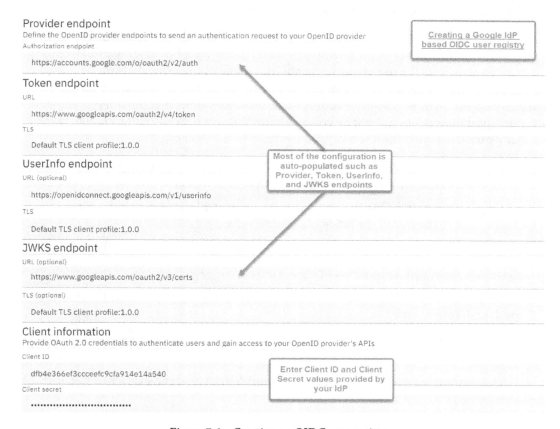

Figure 7.6 – Creating an OIDC user registry

Having covered all the user registry types supported by APIC, your next goal is to learn the OAuth provider configuration. With this OAuth provider configuration, you will complete the main setup required to secure the APIs.

Configuring native OAuth providers

Just what is an **OAuth provider**? An OAuth provider is a service provider that provides authorization services via an **Authorization Server** to the **Resource Owner** (typically the end user) and to the **Client** (typically the applications trying to access the **Resources** on the resource owner's behalf). An OAuth provider is a third party that is trusted by the resource owner and the client alike. To facilitate this OAuth-based communication, the provider provides an **authorization code** and **access token**.

You will use API Manager to create a native OAuth provider. After completing the provider configuration, you will enable this provider in the Sandbox catalog. A native OAuth provider allows the use of native capabilities of APIC to perform OAuth authorization tasks. Of course, APIC allows you to utilize third-party OAuth providers such as Amazon Cognito, Google OAuth 2, Okta, and Facebook Log. The integration with such third-party providers will not be covered here. You will instead create a native OAuth provider by utilizing APIC's OOTB capabilities. Follow these steps:

1. Log in to **API Manager**. Go to **Home | Manage resources | OAuth providers**. Click on the **Add** button and select **Native OAuth provider**.

2. On the **Create native OAuth provider** page and provide the following values. You can leave the rest of the fields as their defaults.

    Title: `api-security-native-oauth-provider`

    Gateway Type: `DataPower API Gateway`

3. Click **Next**. On the **Configuration** page, keep the default values for **Authorization path** and **Token path**.

    Under **Supported grant types**, choose **Access Code** and **Resource owner – password**.

    Under **Supported client types**, keep the **Confidential** checkbox selected. Refer to *section 1* of *Figure 7.7*:

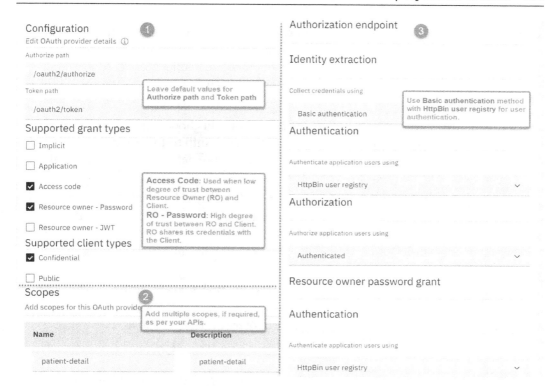

Figure 7.7 – OAuth provider configuration

Here are some important details about the various fields in section 1 of *Figure 7.7*:

- **Authorize path** is the standard OAuth endpoint for authorization. It is a path on the authorization server invoked by the client to retrieve the initial authorization code.

- **Token path** is also a standard OAuth endpoint. It is a path on the authorization server invoked by the client to retrieve the final access token in exchange for the authorization code. The access token is then used by the application to call the protected resource/API.

- **Access code** (also known as an **authorization code grant** type) is a preferred choice when there is a low degree of trust between the resource owner and the client. The resource owner performs an independent login with the authorization server and receives an authorization code. The client then exchanges that authorization code for an access token from the authorization server.

- When you choose the **Resource owner - Password** grant type, you are specifying that the client has access to the resource owner's credentials (username and password) and that there is no need for an extra fetch authorization code step. It is typically used in scenarios where there is a high degree of trust between the resource owner and the client.

- You do not have to choose both grant types. You can choose either of the grant types to adjust how OAuth is executed. For your OAuth testing in this section, you must keep the **Access code** option selected. For more details on the other grant types, visit `https://www.ibm.com/docs/en/api-connect/10.0.1.x?topic=authentication-configuring-native-oauth-provider`.

> **Note**
>
> You can apply the following rules of thumb when selecting one of the preceding grant types:
>
> **Access code**: Use when the resource owner does not want to share the user credentials with the client.
>
> **Resource owner - Password**: Use when the resource owner is willing to share the user credentials with the client.

4. Click **Next**. On the **Scopes** page (*section 2* of *Figure 7.7*), overwrite the default scope, `sample_scope_1`, with `patient-detail`. Do the same for **Description**. A scope is an access control mechanism enforced by the authorization server on the client. It provides the authorization server with a way to notify the resource owner about the resources that are accessed by the client on the resource owner's behalf. After updating the default scope, click **Next**.

5. On the **Authorization endpoint** page, keep the values as shown in *section 3* of *Figure 7.7*. Although you are selecting the default values, the **Identity extraction** and **Authorization** values can also be an HTTP (default or custom) HTML form. This form provides for the ability to customize the resource owner's interaction (generally, the login form and the scope permissions pages) with the OAuth process. This form is not available in the case of a third-party OAuth provider as in that case, the interaction between the resource owner and the authorization server is controlled by that third-party provider itself. Click **Next**.

6. You will be presented with a native OAuth provider summary. Review the summary information. You will notice that the **Base path** value is empty. APIC will automatically generate this value for you after you click on the **Finish** button. The **Base path** value will be used to construct the complete URLs for the **Authorization URL** and the **Token URL**. You will see this later in *Figure 7.15* when you apply the OAuth security definition to your API. Click **Finish**.

With the OAuth provider now created, the next step in the process is to make the OAuth provider available to the catalog where the OAuth protected resource(s) will be deployed. Here is how it is done:

1. Click the **Manage** icon on the left navigation of **API Manager** and select the `Sandbox` catalog.

2. Click the **Catalog Settings** tab | **OAuth providers**. Click on the **Edit** button to add your newly created OAuth provider. Your newly created OAuth provider, `api-security-native-oauth-provider`, will be displayed on the screen. Select the checkbox and click **Save**. Refer to the following screenshot:

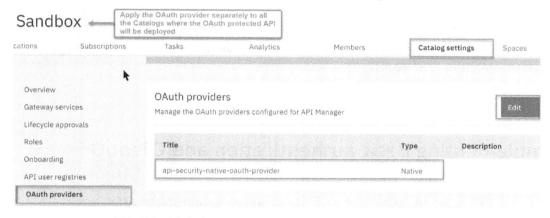

Figure 7.8 – Adding the OAuth provider to the Sandbox catalog

With the preceding configuration, you are now ready to use the OAuth provider to protect the API resources using OAuth2 security.

This concludes the section on User Registry and OAuth provider preparation. The following sections of this chapter will guide you through the process of applying the security to your APIs, using the various resources you just created. You will begin by protecting an API using Basic authentication with the Authentication URL method.

Protecting APIs with Basic authentication and Client ID (API key)

In this section, you will begin developing APIs that use the security features you have just set up. Using Basic authentication with an API key is among the easiest methods of applying authentication security to an API. This method of applying API key security (client ID and client secret) to an API was covered in detail with an example earlier in *Chapter 4, API Creation*, in the *Enabling API Security* section.

To briefly refresh the API key concepts that you have covered earlier, an API key is the most basic level and quickest method of applying authentication security to your APIs. This method involves configuring client ID and client secret (together referred to as an **API key**) security definitions as part of defining an API. Once security schemes are defined, you simply select those definitions in the security section of your API. A consumer can then send these values as part of either a request's query (X-IBM-Client-Id and X-IBM-Client-Secret parameters) or its header (client_id, client_secret). Refer to *Figure 7.1* for details of defining a security scheme and then applying the scheme to the API security.

On the other hand, Basic authentication using the Authentication URL method validates the username and password values passed in the Basic auth headers of the request against the Authentication URL user registry.

You will now develop an API that will be secured by the API key and Basic authentication methods.

Implementing Basic authentication and ClientID in API security

You recently created an Authentication URL user registry entitled HttpBin user registry. It is time to use this registry and a client ID to secure your API:

1. Navigate to **Home | Develop API and Products** and click **Add** to create a new API. Choose a new **From target service** REST proxy and click **Next**. Use the information in *Table 7.2* to complete the wizard entry of your API:

Field Name	Value
Title	`patient-information`
Version	`1.0.0`
Base path	`/patient-info`
Description	`Proxy API from a target service`

Target service URL	`https:stu3.test.pyrohealth.net/fhir/`
	`Patient/d75f81b6-66bc-4fc8-b2b4-`
	`0d1931a92e0`

Security	`Secure using Client ID, CORS`
Path	`/patient`
Operation	GET (Remove all other operations using the 3 dots menu)

Table 7.2 – Proxy for Basic auth and ClientID

2. Your API already has an API key security definition, `clientID`. You will now add a Basic auth security definition. Refer to *Figure 7.9*:

Figure 7.9 – Security definitions for API

3. Go to the **Design** tab | **Security Schemes** | **Add**. Provide the values as per *Table 7.3* and then save the scheme.

Field Name	Value
Name	`http-auth-url`
Description	`Authentication URL based API security`
Type	`Basic`
Authenticate using User Registry	`HttpBin User registry`

Table 7.3 – Basic auth security definition

You should now see two security schemes available for your API. Refer to *Figure 7.10*:

Figure 7.10 – API Security Schemes for Basic auth and API keys

4. To use the `http-auth-url` security scheme, you will need to add this scheme to the security of the API. Go to the **Design** tab | **General** | **Security** | **clientID** in the navigation menu. Select the `http-auth-url` scheme. Refer to *Figure 7.11*. Click on the **Submit** button. Then, click on the **Save** button:

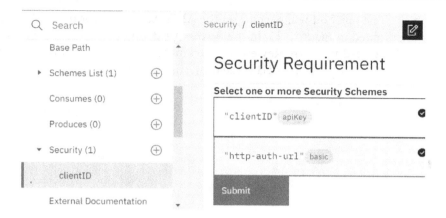

Figure 7.11 – API Security for Basic Auth and API keys

5. Go ahead and publish the API. Click the **Offline** slider to publish the API to the **Sandbox** catalog. Refer to *Figure 7.12*:

Figure 7.12 – Publishing API changes

6. Click on the **Test** tab. You will need to set the API's **Authentication** and **Parameters tabs** as highlighted in *Figure 7.13*. In the **Authentication** tab, set the **Username** to user and **Password** to pass. Review the **Parameters** tab to check how APIC has pre-filled the X-IBM-Client-id value for testing.

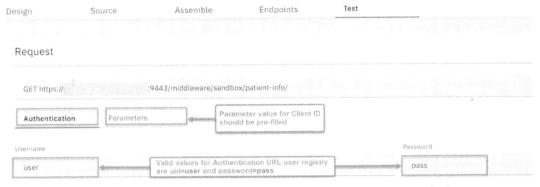

Figure 7.13 – Setting Basic Auth and Client ID parameters in the Test facility

7. Click **Send** to execute the test. You should receive a successful response as shown in the following screenshot:

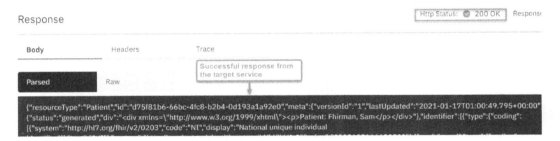

Figure 7.14 – Basic Auth and Client ID successful response

How easy that was! You completed the testing of your API proxy secured using two security methods: API key (client ID) and Basic auth (Authentication URL). As a learning exercise, you can consider modifying your REST proxy by removing one of the security definitions from its security and then re-executing the test (remember to save and publish after any REST proxy changes. Use the toggle facility for quick republishing to the Sandbox catalog). You can also run some negative tests by changing the authentication values of username, password, and X-IBM-Client-Id parameters in the **Test** facility.

One of the goals of this section was to get you familiar with the process of creating and using the security definitions. To that extent, Basic auth and API keys served a great purpose. But as the reach of your APIs expands beyond the realm of your internal organization and its trusted partners, you will certainly need to apply more advanced security modes. OAuth is one such security method, and you will learn this next.

Applying OAuth 2.0

In this section, you will learn about the specifics of OAuth, and later you will learn about OIDC (another similar security standard). Because of similarities between OAuth and OIDC, it might be helpful to know at a high level what is what. It is important to know that while OIDC deals with authentication, OAuth deals with authorization. OIDC gives you a single login for multiple sites. OAuth provides you with the ability to control access to your data on a single site, by multiple sites.

So, what is exactly OAuth?

OAuth 2.0 is a security standard that allows one service (for example, a healthcare provider) to securely access limited data (for example, some lab results) from another service provider (for example, a medical lab) without ever having access to the person's account credentials (username and password) shared with the other (medical lab provider). OAuth solves the authorization problem. Because of OAuth, a patient can now delegate limited account access (lab details) to their healthcare provider.

In the context of APIC, the implementation of OAuth security involves the following main steps:

1. Create an OAuth provider.
2. Make the OAuth provider available to a catalog.
3. Enable OAuth security in an API.
4. Create a test client that will invoke the API.
5. Test the OAuth from the resource owner's perspective.

In the *Configuring native OAuth providers* section, you already took care of the first two steps. We will now build upon the work you have done thus far and complete the remaining steps.

Enabling API with an OAuth security definition

OAuth is one of the security definitions types available to secure an API. You will now secure your patient-information API using the OAuth provider api-security-native-oauth-provider:

1. In **API Manager**, open the patient-information API proxy. Navigate to its **Design** tab | **Security Schemes**. You can keep the existing security schemes.
2. Click the **Add +** button in the **Security Schemes** section.

3. Create an OAuth `oauth-secdef` security definition as per *Table 7.4*. Leave the default values for the **Authorization URL, Token URL,** and **Scope** fields.

Field Name	Value
Name	`oauth-secdef`
Type	`OAuth2`
OAuth provider	`api-security-nativev-oauth-provider`
Flow	`Access code`

Table 7.4 – OAuth security definition

Your completed security definition should resemble *Figure 7.15*:

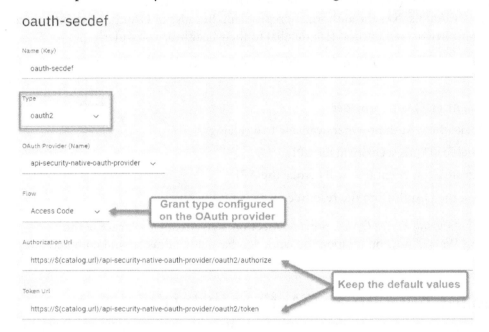

Figure 7.15 – OAuth Security Definition

You will use **Authorization URL** and **Token URL** to test the OAuth flow in the *Testing of OAuth flow* section. Notice that the **Scopes** field of the OAuth provider is pre-populated. In the case of multiple scopes, keep the scope that is relevant for the API on which the security definition is applied. In this example, your OAuth provider only manages the `patient-detail` scope. Therefore, you do not need to make any changes to **Scopes**.

4. Click on the **Save** button to save the `oauth-secdef` security definition.

5. Apply the `oauth-secdef` security definitions to the security of your API. Go to the **Security** section. You will find `clientID`, `oauth-secdef`, and `http-auth-url` security schemes listed under the **Security** view. You should now be able to select `oauth-secdef` as the security scheme and the `patient-details` scope. Unselect the `http-auth-url` definition. Refer to *Figure 7.16*:

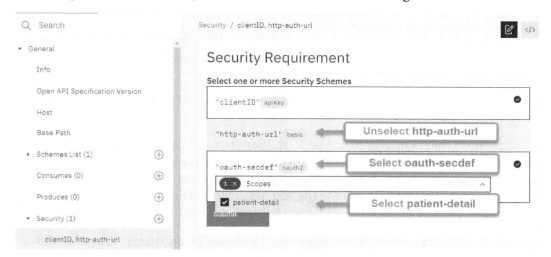

Figure 7.16 – Client ID and OAuth Security

Save your API.

6. Since you removed the `http-auth-url` security from your API, you must republish your changes. Use the toggle facility (refer to *Figure 7.12*) to republish your changes.

From the resource's (API) perspective, what you have done so far applies the OAuth security to your REST proxy. These steps of creating the OAuth provider, assigning the provider to the catalog, and attaching the OAuth security definition to the API make the API OAuth safe. The API provider's role generally finishes after these steps.

There is much that happens from the client's (API's consumer) perspective, though. You should get an understanding of the client's configuration as well. OAuth providers grant type configuration, which you did earlier (refer to *Figure 7.7*), and this directly impacts the client's interaction with the API. Next, you will see the client-side configuration and the steps to register the client with the API.

Creating a client

As previously stated, the client (typically an application) interacts with the resource (API) on the resource owner's (typically the end user) behalf. From the earlier example, the client is the application developed/owned by the healthcare provider. The resource is the service, exposed by the medical lab, that fetches the resource owner's lab results from the lab provider's backend systems. Before a client can attempt to access resource(s) on the resource owner's behalf, the following needs to happen:

1. Registration of the client with the resource server.

2. The client should have a unique client ID and client secret for authenticating with the authorization server.

3. The client should subscribe to the resources that it intends to access on the resource owner's behalf.

You will perform the client's configuration in the **Sandbox** catalog. Typically, a client belongs to a **Consumer Organization**. Client creation is managed by the consumer organization's administrators. Since you do not have a consumer organization set up yet on the developer portal, you will use the API Manager functionality to perform the required configuration:

1. Go to the **Home** screen | **Manage catalogs** | **Sandbox** | **Consumers**. Click the **Add** button. Choose the **Create organization** option.

2. Create a new consumer organization by providing values as per *Table 7.5* and click **Create**:

Field Name	Value
Title	`sandbox-corg`
User registry	**Sandbox Catalog User Registry**
Type of user	`Existing` If your username does not exist in the **Sandbox Catalog User Registry**, select the **New User** option and proceed to create a new user.
Username	`<your-username>`

Table 7.5 – Consumer organization creation

Once an organization is created, you can refresh the **Consumers** tab to view your newly created organization. Refer to *Figure 7.17*:

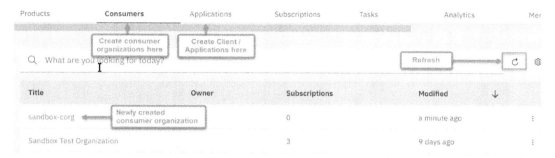

Figure 7.17 – Consumer organizations and applications

3. You will now perform the important step of creating a client (application) and assigning it to the `sandbox-corg` consumer organization. In APIC, a client is also called an application. Creating an application registers that application with the resource server. Go to the **Applications** tab (refer to *Figure 7.18*). Click **Add**. Provide values as per *Table 7.6* and click **Create**:

Field Name	Value
Title	`corg-app`
OAuth redirect URL	`https://example.com/redirect` OAuth redirect URL is a critical component of the overall OAuth flow. It lets the Authorizaion server redirect the Resource Owner to the Client (Application) after successful authentication. Along with the redirect, the Authorization Server sends an Authorizaiton Code or an Access Token to the Client.
Consumer organization	`sandbox-corg`

Table 7.6 – Client (application) creation

The system will provide a unique set of **Client ID** and **Client secret** values (refer to *Figure 7.18*). Ensure that you copy and store them safely. It is not possible to retrieve the client secret value after this. Close the **Credentials** view.

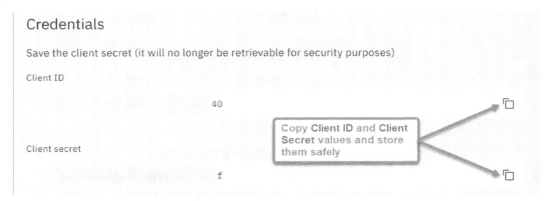

Figure 7.18 – Client/Application credentials

Once the application is created, you can refresh the **Applications** tab to view your newly created application.

You will now need to create a **Subscription** for the corg-app application to the resources that it is going to access.

4. On the **Applications** tab, open the corg-app application menu (the three dots) and select the **Create Subscription** option from the menu. Refer to *item 1* of *Figure 7.19*:

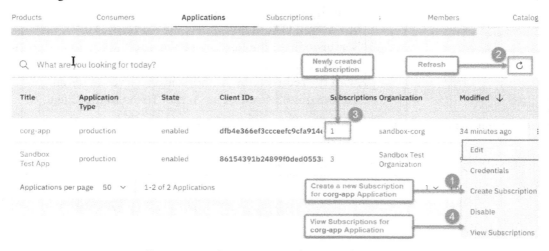

Figure 7.19 – Creating an Application Subscription

5. On the next screen, select the **Plan** that corresponds to your API (`patient-information 1.0.0`) and click **Create Subscription**. Refer to *Figure 7.20*:

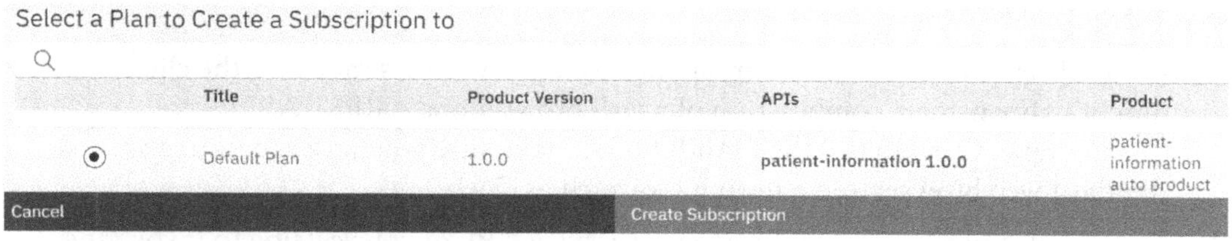

Select a Plan to Create a Subscription to

	Title	Product Version	APIs	Product
◉	Default Plan	1.0.0	patient-information 1.0.0	patient-information auto product

Cancel   Create Subscription

Figure 7.20 – Select a Plan to create an application subscription

Click on the **Refresh** button on the **Application** tab. You can view the `corg-app` application's subscription through the **View Subscriptions** menu option. Refer to *item 4* of *Figure 7.20*.

You have completed all the necessary configurations for securing your API with OAuth security. The configuration involved steps for creating an OAuth provider, assigning the provider to a catalog, enabling API with the OAuth security definition, and finally configuring a client that can access the API. After all these steps, the API is ready to be tested and you are now going do so.

Testing OAuth flow

Like other OAuth configuration steps, testing the OAuth flow is also a multi-step process. This is due to the nature of multiple interactions between all the parties involved in OAuth processing. Apart from the complexity of the multiple parties, OAuth flow also changes based on the configured grant type in the OAuth provider.

You will be performing testing for the access code, sometimes referred to as the **three-legged flow**. You will remember that in the case of the access code or three-legged flow, the client does not have any access to the resource owner's credentials. The resource owner is directly authenticated by the authorization server before the control is returned to the client along with the authorization code. The client then makes a separate call to the authorization server to request the access token and to access the resource on the resource owner's behalf. Let's see how we can begin with testing.

Prerequisites for OAuth testing

The following are these steps that are required to be performed for conducting a successful test of the OAuth flow:

1. In the absence of a testable client/application, you will be simulating the client functionality using a combination of a web browser and a REST API testing tool (for example, Curl) to test the OAuth flows. Make sure you have access to any standard web browser and a testing tool such as Curl.

2. You will also need access to the gateway service URL of your catalog to invoke the authorization path, token path, and API path. To get to the Gateway service URL of your catalog, navigate to the **Home** screen | **Manage catalogs** tile | **Sandbox** tile | **Catalog settings** tab | **Gateway services** section. Refer to *Figure 7.21*:

Figure 7.21 – Gateway services URL

The required URLs for testing the OAuth flow should look like the URLs provided in *Table 7.7*:

URL Name	Path
Authorization Path	`https://$(catalog.url)/api-security-native-oauth-provider/oauth2/authorize`
Token Path	`https://$(catalog.url)/api-security-native-oauth-provider/oauth2/token`
API Path	`https://$(catalog.url)/patient-info/patient`

Table 7.7 – OAuth testing URLs

Replace $ (catalog.url) with the captured Gateway URL value (refer to *Figure 7.22*).

3. You will need the client ID and client secret values that you copied earlier.

4. You will need the scope value of the patient-information API/resource. The scope value is patient-detail.

5. Lastly, you will need valid Basic Auth user credentials: user and pass.

Now that you have gathered all the required prerequisites, it is time for you to test your OAuth flow.

Access Code or three-legged flow

This flow is known by multiple names: access code, authorization code, or a three-legged flow. As was discussed earlier, this flow involves all the participants, that is, the resource owner, client (application), and resource server/authorization server. The basic steps to test this flow are as follows:

1. First, we fetch the authorization code. The client initiates access to a protected resource on behalf of the resource owner. It uses the authorization path for fetching the authorization code. A sample authorization path will look like this: https://$(catalog.url)/api-security-native-oauth-provider/oauth2/authorize?&client_id=dfb4e366ef3ccceefc9cfa914e14a540&response_type=code&scope=patient-detail.

2. Replace the $(catalog.url) and client_id values as per your environment. Notice scope=patient-detail. There could be multiple scopes (each scope could be attached to numerous APIs) that the same client could access from the resource server. Therefore, the client needs to let the resource server know about the scope to which the current access is requested.

3. After replacing the values of $(catalog.url) and client_id, copy the
    authorization path in your browser window and press *Enter*. The resource server's
    OAuth security kicks into action. It utilizes the facilities of an authorization server
    to initiate a resource owner authentication process. Your browser should challenge
    you with an authentication prompt. Fill in the fields in the authentication prompt as
    Username=user and **Password**=pass. Refer to *Figure 7.23*:

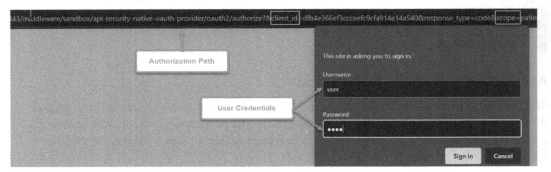

Figure 7.22 – Authorization path and authentication challenge

4. Press *Enter*. Once successfully authenticated, your browser URL should change to
    the redirect URL you supplied while configuring the client. The authorization server
    will present an authorization code. Notice the code query parameter along with the
    long authorization code value that you have received from the authorization server.
    The address in your browser should look similar to https://example.com/
    redirect?code=AAIHooRpu0WL3Zf6EhueWlKEuhT5p8G7I81UhmfoymKw
    S2BcVEfzlMH-O966EvS-1FKY8NVeeLd9uXcwPVGG93r3CGolnXqCkuP8Hd
    ulBD0fqQ.

5. Copy the value of the code query parameter. You will be using this value to fetch
    the access token in the next step.

6. Then, fetch the access token. After successfully getting the authorization code,
    the next step in the OAuth flow is to fetch the access token. The access token is
    a validation of the client by the authorization server. You will use the Curl tool to
    test this part of the flow. The sample curl command should look like the following:

```
curl -v -k
-X POST "$(catalog.url)/api-security-native-oauth-
   provider/oauth2/token"
-u $(client_id):$(client_secret)
-d "grant_type=authorization_code&code=$(Authorization
   Code)"
-H "Content-Type: application/x-www-form-urlencoded"
```

Build your Curl command along similar lines (sample command):

```
curl -v -k
-X POST "$(catalog.url)/api-security-native-oauth-
   provider/oauth2/token"
-u dfb4e366ef3ccceefc9cfa914e14a540:3ce9e2601e7b296115
   d6b60207f707af
-d "grant_type=authorization_code&code=AALU6PpJBTnG
   GC2w7pcY14fiM1A3JewkBHT0a7HzDY5aVtUvzJIMvpFOq8V
   CX41eARGSmy7sw6kykqvn-DKIJVrUz1Uli3pkGN19E9T78GUm4A"
-H "Content-Type: application/x-www-form-urlencoded"
```

7. Replace `$(catalog.url)`, `$(client_id)`, and `$(client_secret)` as per your environment. `$(Authorization Code)` should be replaced by the authorization code value you fetched in the *Fetch Authorization Code* step.

8. Next, send your curl request. The Curl command should return an access token in the `access_token` field. The sample response should resemble the following:

```
{
    "token_type": "Bearer",
    "access_token": "AAIgZGZiNGUzNjZlZjNjY2NlZWZjOW
    NmYTkxNGUxNGE1NDC7571_QNkH9DGjU23GqRx58N
    96SeRDccX9RwbLmt5y5locz0Y-ww4K_DFGEqFqELa_
    nfELawWuJiTmRQ_L8ouqjTQ_bghbz1QClkrlhEtIZg",
    "scope": "patient-detail",
    "expires_in": 3600,
    "consented_on": 1623919415
}
```

The returned `access_token` value can be used by the client to access the resource. This `access_token` is only valid until the duration specified in the `expires_in` response parameter. The default expiry is set to `3600` seconds that is, 60 minutes. The client can make multiple resource calls within this duration using the same `access_token`. This default value of `3600` seconds can be changed while configuring the OAuth provider.

9. Then, make a resource/API call. This is the last step in your testing process. You (acting as the client) will now make the resource/API call using the access token (`access_token`) that was retrieved in the previous step. In the Curl tool, create a new request as per the following Curl request example:

```
Curl command
curl -v -k
-X GET "$(catalog.url)/patient-information/patient"
-H "Authorization: Bearer $(Access Token)"
-H "Accept: application/fhir+json"
-H "X-IBM-Client-Id: $(client_id)"


Sample command
curl -v -k
-X GET "$(catalog.url)/patient-information/patient"
-H "Authorization: Bearer AAIgZGZiNGUzNjZlZjNjY2NlZW
  ZjOWNmYTkxNGUxNGE1NDD_QpPhIKzfdr-vZ791qShm
  4ZGWLXKP2dJBc9YV3OgJguP6d_vz9mZ_nJ_Iy_MdQ
  rDQIr_Q6XJWxD0PuU6_jFz3XtAjgoDc6VFo5KCPuJmnfg"
-H "Accept: application/fhir+json"
-H "X-IBM-Client-Id: dfb4e366ef3ccceefc9cfa914e14a540"
```

Replace $(client_id) as per your environment. $(Access Token) should be replaced by the Access Token value you fetched in the *Fetch Access Token* step.

10. Finally, send the `curl` request. You should receive a valid response back from your `patient-information` API.

This concludes the testing of the access code or three-legged OAuth flow.

In this section, you learned how OAuth can be used to secure the APIs. At its core, OAuth is an authorization mechanism. It still leverages an external authentication framework for carrying out user authentication without any means to fetch meta-information about the user itself. This is where OIDC comes into the picture. It is a layer that sits on top of OAuth and enhances the OAuth flow by closing some of these user information-related gaps. You will now look in detail at how APIC supports the OIDC standard.

Implementing OpenId Connect (OIDC)

OAuth was built for authorization and cared most about the permissions and scopes of the protected resources. These permissions were then assigned to a client on a resource owner's behalf. OAuth's fundamental limitation was that it did not provide any standard way for the client to fetch any meta-information about the logged-in user. The client application was oblivious to information such as the resource owner's email ID, name, account creation date, last logged-in date, and profile picture. The OIDC standard was built on top of OAuth to solve some of these limitations of OAuth.

Most of the steps related to enhancing a resource's security using OIDC remain identical to the OAuth implementation that you just covered in detail. Some of those steps (OAuth) are enhanced to add OIDC capability to an API. This section will cover those changes to the OAuth configuration. These changes to the OAuth flow to accommodate OIDC can be classified into two main categories: **OAuth provider configuration changes** and **OAuth flow changes**. Let's take a quick look at these:

- OAuth provider configuration changes involve the following:

  - Enabling OIDC capability in the provider.

  - Providing a security object to sign the **ID token**. The ID token is the resource owner's profile token. This helps the client in ensuring the integrity of the ID token that is returned by the authorization server.

- OAuth flow changes have the following effect:

  - Adding OIDC scope to the *Fetch Authorization Code* request.

  - Adding OIDC scope to the *Fetch Access Token* call. This call then returns an ID token (JWT format) along with an access token.

The ID token is the main addition to the existing OAuth flow. The client can then use this ID token to fetch more details about the logged-in resource owner/user. You will notice that there are no changes in the resource configuration. The only requirement is to republish the resource/API once the OAuth provider configuration changes are done.

To understand the OIDC functionality and the ways in which it can be used in APIC to further complement the API's security, you will be building upon the work done in the *Applying OAuth 2.0* section of this chapter. You will modify the existing OAuth provider and will make modifications to the testing flow that was covered in the *Testing OAuth flow* section.

OAuth provider changes

There are a few provider changes required to accommodate OIDC. Let's walk through those changes:

1. Log in to **API Manager**. Go to **Home | Manage resources | OAuth providers** section. Click on the **api-security-native-oauth-provider** OAuth provider.

2. Click on **OpenID Connect**. Select **Enable OIDC**. Clear all the selections under **Support hybrid response types**. Select **Auto Generate OIDC API Assembly**. Refer to *Figure 7.23*:

Figure 7.23 – Enable OpenID Connect in an OAuth provider

3. In the **OpenID Connect** view, you will either need to provide an **ID token signing crypto object** or an **ID token signing key** (a **JSON Web Key or JWK**) to sign the ID token generated by APIC. The reason to sign the ID token is so that the receiver of this token (the client) can be sure of the validity of the token generator. For the purposes of this tutorial, you will generate a JWK using `https://mkjwk.org/`. Refer to *Figure 7.24*:

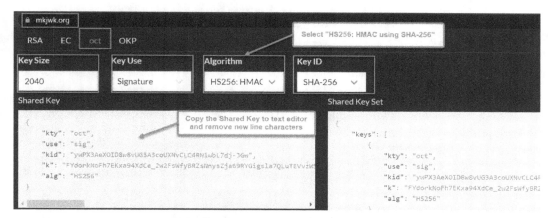

Figure 7.24 – JWK generation

JSON Web Key (JWK)

JWK is a JSON object representing the cryptographic key used for signing. You can generate a JWK using publicly available open source libraries (for example, `https://openid.net/developers/jwt/`), public websites (for example, `https://mkjwk.org/`), or the libraries/methods approved by your company's security department. The use of an online tool such as `https://mkjwk.org/` is for demonstration purposes only. It is not a recommendation. Please consult your organization's security administrator for the approved method of JWK generation and usage.

---

4. Copy the generated JWK to the **ID token signing key** field. You can consider using a text editor to convert the generated JWK into a single line so that it can be copied easily to this field.

5. Select HS256 in the **ID token signing algorithm** field. Refer to the following screenshot. Click **Save**.

Figure 7.25 – ID token signing

> **JWK versus Crypto Object**
>
> ID token can be signed using two mechanisms: Using a JWK (like you just configured), or by using crypto objects available in the DataPower appliance (in which case you will need to supply the crypto object's name in the **ID token signing crypto object** field). Crypto objects take precedence over JWK, in case both are specified. These crypto objects must be in the APIC domain on the DataPower appliance. Crypto objects can either use a shared secret key or a crypto key (private key) to encrypt or sign the ID token. Refer to the IBM documentation for algorithm support for a shared secret key and crypto key: https://www.ibm.com/docs/en/api-connect/10.0.1.x?topic=policies-generate-jwt.

From the OAuth provider configuration perspective, these are the only required changes to enable OIDC security. Make sure that you do the following:

*   Republish your patient-information API to ensure that you are testing against the latest API version. Refer to *Figure 7.26*:

Figure 7.26 – Republishing the API changes

- Recreate the subscription for your `corg-app` application. Refer to *Figure 7.19*.

Next, you will cover the testing of OIDC security.

OAuth flow changes

OIDC security testing follows along the same path as OAuth flow testing, with some minor changes. From the resource owner's standpoint, the overall interaction flow looks identical to OAuth. The notable differences rest in the following:

1. **Fetch Authorization Code**: Instead of a single scope value, the client sends two scope values to the authorization server; for example, in the case of OAuth flow, the `scope=patient-detail` value is used. In the case of OIDC flow, the same authorization code call is made with `scope=patient-detail+openid`. Notice the plus (+) symbol used between the two scope values. A sample authorization path URL will look like this: `https://$(catalog.url)/api-security-native-oauth-provider/oauth2/authorize?&client_id=dfb4e366ef3ccceefc9cfa914e14a540&response_type=code&scope=patient-detail+openid`.

> **OpenID scopes**
>
> Notice the two scope values sent as part of the OIDC call. The primary scope value is `patient-detail`. The other scope value is `openid`. The scope value of `openid` is part of OIDC's list of built-in scopes. These scope identifiers, as they are called, have different purposes. Some of these are briefly explained here:
>
> `openid` (required): This indicates that the client intends to use OIDC to verify the user's identity.
>
> `profile`: The client wants to request access to `name`, `family_name`, `given_name`, `middle_name`, `picture`, and such.
>
> `phone`: The client wants to request access to `phone_number` and other phone number-related information.

Replace the $(catalog.url) and client_id values as per your environment. Copy the authorization path URL to your browser window and press *Enter*. After logging in, you will be redirected to a URL that contains the authorization code.

2. **Fetch Access Token**: There is only one difference between the access token call that you had earlier constructed while testing the OAuth flow and the access token call for OIDC. In the case of OIDC, the client sends two scope values as part of the *Fetch Access Token* call. Notice the space (%20) between the two scope values of openid and patient-detail. It is different from the *Fetch Access/Authorization Code* where the two scope values had a plus (+) between them:

```
Curl command
curl -v -k
-X POST "$(catalog.url)/api-security-native-oauth-
  provider/oauth2/token"
-u $(client_id):$(client_secret)
-d "grant_type=authorization_code&code=$(Authorization
  Code)&scope=openid%20patient-details"
-H "Content-Type: application/x-www-form-urlencoded"
```

```
Sample command
curl -v -k
-X POST "$(catalog.url)/api-security-native-oauth-
  provider/oauth2/token"
-u dfb4e366ef3ccceefc9cfa914e14a540:3ce9e2601e7b2961
  15d6b60207f707af
-d "grant_type=authorization_code&code=AAKwNX_f-
  GuOfdx17N78HYCkyj1p7YiW7A886zOK6uM3yMywG1ZWza37P
  pu7PYI7igKFejXxsgiYwYKnfpKmZuVdQbNy9eF-caNPS2L7vw
  Lh8w&scope=openid%20patient-detail"
-H "Content-Type: application/x-www-form-urlencoded"
```

Replace $(catalog.url), $(client_id), and $(client_secret) as per your environment. $(Authorization Code) should be replaced by the authorization code value you fetched in the *Fetch Authorization Code* step earlier.

3. Send your Curl request. The Curl command should return an access token. The sample response should resemble the following:

```
{
    "token_type": "Bearer",
    "access_token":
        "AAIgZGZiNGUzNjZlZjNjY2NlZWZjOWNmYTkx
        NGUxNGE1NDBxVgVhACO5T17OZg_yS1QbBZ5W18-
        31A_mudnPS3VuY8YP31jQ783oSyPXlFJZrdf3E5
        ucboDL9d139AmEQ4oz7NHjulTWJBMJZ7vrCArhWA",
    "scope": "patient-detail openid",
    "expires_in": 3600,
    "consented_on": 1624007362,
    "id_token": "eyJraWQiOiJ5d1BYM0FlWE9JRDh3OHZVRzNBM
2NvVVhOdkNMQzRSTjF3Ykw3ZGotSkd3IiwiYWxnIjoi
SFMyNTYifQ.eyJqdGkiOiI5ODk2NTcwNS1mNmNmLTQ1NDct
YTYxMS02NTU1NmMxOTg5Y2QiLCJpc3MiOiJJQk0gQVBJQ29
ubmVjdCIsInN1YiI6InVzZXIiLCJhdWQiOiJkZmI0ZTM2Nm
VmM2NjY2VlZmM5Y2ZhOTE0ZTE0YTU0MCIsImV4cCI6MTYy
NDAxMDk2MywiaWF0IjoxNjI0MDA3MzYzLCJhdF9oYXNoIj
oiLWJUVVFJSXBDMXF4QjNUZmY1V0ZWZyJ9.NnTcb92WQp2
RNWMwjk6SNd75ODLwQWRdnxAFqhE0Cxk"
}
```

id_token is a **JSON Web Token (JWT)** format token value. access_token can be used by the client to access the resource (just like you did in the *Access Code or three-legged flow* section, in *step 9, Make a resource/API call*). id_token contains standard user information (email, name, and some other meta-information) from the identity provider and the signature of the authorization server (so that it can be independently verified by the client).

This section was an introduction to OIDC support in APIC. OAuth and OIDC together provide a formidable standards-based authentication and authorization framework. The last few sections demonstrated, with the help of theory and examples, that they both are comprehensively supported in APIC. You should leverage them to secure your APIs as much as possible.

The next section will introduce you, in detail, to the method of applying JWT-based security to APIs.

Using JWT policies

JWT (pronounced *jot*) is one of the methods of defining the identity information of a user/system (a client) in a JSON format. It is primarily used in authorization scenarios where its usage is employed to pass an authenticated client's meta-information (identity and claims) to the server in a secure and verifiable format. JWT removes the burden of storage of an authenticated client's information on the server in such a use case. Generally, without JWT, websites use cookies to pass an authenticated client's access information to the server and rely on the server to store cookies on the server side. The use of cookies is discouraged for the reasons of privacy and security.

In the authorization scenario, the client and server interaction works as follows:

1. The client performs its initial authentication with the authentication system.

2. The authentication system generates a JWT that contains various meta-information, such as the client's identity and claims, signed by the authentication system using a JWK. A limited validity JWT is then returned to the client.

3. The client then passes this JWT, along with its resource access request, to the resource server.

4. The resource server extracts the JWT and performs the verification of the signed information before allowing access to the protected resource.

Now that you have learned about JWT's primary usage in general, you will be eager to know about its support in APIC. Not only does APIC support JWT verification, but it also lets you generate a JWT. As you learned here, JWT's authorization use case requires two parts – JWT generation and JWT validation. You will learn how to perform these steps next.

JWT generation

JWT generation primarily involves three steps: client authentication, token generation, and token signing.

There are two methods for generating a token—using APIC's Generate JWT policy or by using your organization's identity provider. You will learn the method of using the Generate JWT policy to generate a JWT token. Configuration of your organization's identity provider to support JWT token generation is outside the scope of this book. Here are the steps for the JWT policy method:

1. Log in to **API Manager**. Go to **Home | Develop APIs and products | Add | API (from REST, Graph QL or SOAP)**.

2. On the **Select API type** page, ensure that **Open API 2.0** is selected. Select **New OpenAPI**. Then, click **Next**.

3. On the **Create new OpenAPI** page, provide the details as per *Table 7.8* to create a new API that you will enhance shortly for the purpose of generating a JWT:

Field Name	Value
Title	`jwt-generate`
Base path	`/jwt`
CORS	`Enabled`
Security	`API Keys (clientID, clientSecret)` You will need to create a Security definition for Client Secret API Key in the header with a parameter name of `X-IBM-Client-Secret` and then apply that to your API's Security.

Table 7.8 – JWT generation proxy

4. The **Security Schemes** and **Security** sections of your API should look similar to *Figure 7.27*:

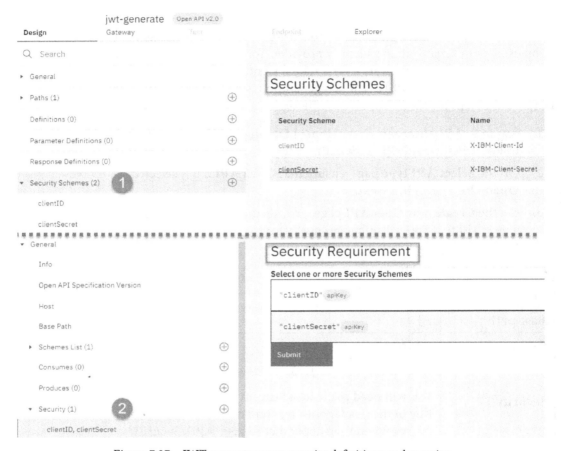

Figure 7.27 – JWT generate proxy security definitions and security

5. Now that your base API proxy's structure is set up, you will now enable this API to generate a JWT. Modify the API's default Path (/):

   - **Path name**: /generate
   - **Operations**: GET (remove all the other operations by using the three vertical dots menu)

   Click the **Save** button after making the modifications.

6. Next, define the parameters for the GET operation. You will send these parameters in the request's headers. Here you will define two important required headers. Navigate to **Paths | /generate | Operations | GET | Path Parameters** in the navigation menu. Click the **Add** button to add the parameters. Add the iss-claim and aud-claim parameters. These parameters are located in header, have a string type, and are required. Refer to *Figure 7.28* for this:

Figure 7.28 – Claim headers for JWT generation

These headers specify the **Issuer** (iss) and **Audience** (aud) claims. The JWT needs to contain verifiable claims. iss-claim should contain the identification information of the server that issued the JWT. aud-claim should contain the identification information of the resource server or the resource that the client/application will access using the generated JWT. You will pass these claim values in the request to the jwt-generate API. The jwt-generate API will then insert these header values in the JWT that it returns to you. This is a method of generating dynamic claims. Save your changes.

7. Then, go to the **Gateway** tab and delete any existing policies from the flow by hovering over the policy and clicking the garbage can. You will now add three new policies to your message processing flow.

8. Drag and drop the **Set Variable, Generate JWT**, and **GatewayScript** policies onto your message processing flow as per *Figure 7.29*. Rename each policy's **Title** property as per *Figure 7.29*:

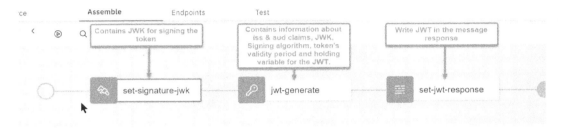

Figure 7.29 – JWT generation message processing flow

9. Set the `set-signature-jwk` policy's property values as per *Table 7.9*. Click the **Add action** button to set these values. Click on the **Save** button after setting the property values.

Property	Value
Action	Set
Set	hs256-key
Type	String
Value	You can either use the same JWK that you generated for the OIDC configuration, or you can generate a new JWK through https:// mkjwk.org/ using the method described earlier.

Table 7.9 – set-signature-jwk property values

10. Set the `jwt-generate` policy's property values as per the value in *Table 7.10*. Click on the **Save** button after setting the property values:

Property	Value
JSON Web Token (JWT)	`generated.jwt` Variable name that contains the generated JWT. This vaeriable is used in the `set-jwt-response` GatewayScript policy.
Issuer Claim	`request.headers.iss-claim`
Audience Claim	`request.headers.aud-claim`
Validity Period	`300` **(in seconds)**
Sign JWK variable name	`hs256-key` Variable name that contains the security key set in the `set-signature-jwk` policy.
Cryptographic Algorithm	`HS256` This should match your JWK's Algorithm

Table 7.10 – jwt-generate property values

11. Lastly, set the generated JWT in the message response body. Copy the code named `setjwtresponse.js` from the GitHub repository into the `set-jwt-response` policy as follows:

```
var jwt = context.get('generated.jwt');
var accessToken =
{
    "jwt": jwt,
};
context.message.header.set('Content-
  Type',"application/json");
context.message.body.write(accessToken);
```

12. Save and publish your API (remember to use the toggle facility). This completes the configuration of your `jwt-generate` API.

You will now run a quick test to test the configuration. Go to the **Test** tab of your API. You will see that the platform has already pre-filled many header values. Enter the `iss-claim` and `aud-claim` values as per *Figure 7.30*. **Send** the request. You should receive a JWT in the **Response** section.

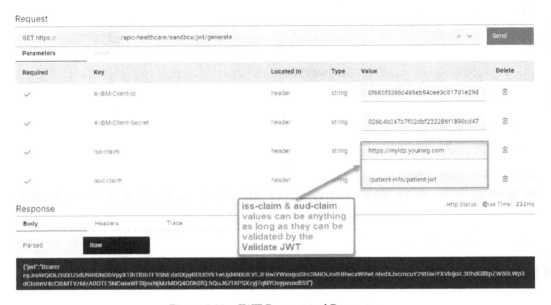

Figure 7.30 – JWT Request and Response

The *JWT Generation* section covered details about using the APIC framework for JWT generation. You must consult your organization's security administrator to confirm the usage of the APIC framework for JWT generation. Many organizations use external identity providers to generate JWT. If that is the case, then please use the approved method for JWT generation.

Now that you are familiar with the process of JWT generation, next, you will learn the method of performing JWT verification to protect your API resources.

JWT verification

Verification of a JWT is an authorization process. It assumes that the JWT presented has been issued to an authenticated client (within the confines of the token's validity period). Thus, JWT verification concerns itself with ensuring that the client is authorized to access the protected resource and that the JWT has been signed by a verifiable and approved identity provider.

You will now configure a new API to use the Validate JWT policy to verify a presented JWT before allowing access to the backend service, as follows:

1. Create a new `patient-jwt-information` REST API proxy using the information provided in *Table 7.11*:

Field Name	Value
Type	`From target service`
Title	`patient-jwt-information`
Version	`1.0.0`
Base path	`/patient-info`
Target service URL	`https://stu3.test.pyrohealth.net/fhir/Patient/d75f81b666bc-4fc8-b2b4-0d193a1a92e0`
Security	`API Keys (Client ID), CORS`
Path	`/patient-jwt`
Operation	GET (Remove all the other operations by using the three vertical dots menu)
Parameters	**Required:** `Yes` **Name:** `authorization` **Located in:** `header` **Type:** `string`

Table 7.11 – JWT Verification Proxy configuration

Click the **Save** button after making the modifications.

2. Open the **Gateway** tab. Drag and drop **Set Variable** and **Validate JWT** policies in front of the **Invoke** policy as per the following screenshot. Also, implement a `Default` catch flow to handle any JWT verification failures. Drag and drop a **GatewayScript** policy on the `Default` catch block. Rename each policy's **Title** property as per *Figure 7.31*:

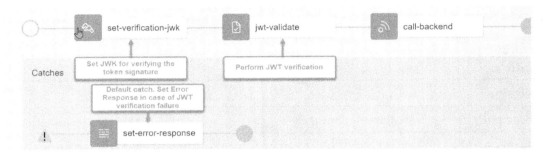

Figure 7.31 – JWT verification message processing flow

3. Set the `set-verification-jwk` policy's property values as per *Table 7.12*. Click the **Add action** button to set these values.

Property	Value
Action	`Set I`
Set	`hs256-key`
Type	`String`
Value	Use the same exact key that you used in the `set-signature-jwk` policy of the *JWT Generation* section.

Table 7.12 – set-verification-jwk property values

4. Set the `jwt-validate` policy's property values as per *Table 7.13*:

Property	Value
JSON Web Token (JWT)	`request.headers.authorization` Request header that contains the JWT.
Issuer Claim	`^https:\/\/myidp.ibm.com` This is a **Perl Compatible Regular Expressions** (**PCRE**) string to match the iss claim value in the JWT. This example string matches any value that start with `https://myidp.ibm.com`. You can change this value to as per your environment. Do not include the PCRE string with any quotes/apostrophes in the Assemble tab.
Audience Claim	`/\patient-info\/patient-jwt` This is a PCRE string to match the aud claim value in the JWT. This will typically be the resource path of the resource that your client is requesting to access.
Verify Crypto JWK variable name	`hs256-key` The variable name that contains the security key you have set in the `set-verification-jwk` policy.

Table 7.13 – jwt-validate property values

5. Lastly, set up a default `catch` block to handle any JWT verification errors. Set the JWT verification error in the message response body. Copy the code named `jwtverification-errorresponse.js` from the GitHub repository into the `set-error-response` policy:

```
var errorName = 'JWT Validation Error';
var errorMessage = context.get('jwt-validate.error-
    message');
var errorResponse = {
    "name": errorName,
    "message": errorMessage};
context.message.header.set('Content-
    Type',"application/json");
context.message.body.write(errorResponse);
```

6. Save and publish your API. (Use the toggle facility for publishing to the Sandbox catalog.)

This completes the configuration of your `patient-jwt-information` API.

Having completed this proxy configuration, let's test the JWT verification process. You generated your JWT in the last part of the *JWT Generation* section, which was a while ago (remember that in your JWT generation configuration, you have set the validity period as `300 seconds`). Hence, before testing the verification of the JWT, you will want to generate a new JWT. Refer to *Figure 7.31* for the steps to generate a new JWT. Once you have generated the new JWT, use it in the **Test** tab of your `patient-jwt-information` proxy to send a verification request. Refer to *Figure 7.32*. You should get a successful response.

Figure 7.32 – JWT verification successful request

Wait for around 5 minutes and execute the same test again. You should receive a JWT validation error, as shown in *Figure 7.33*:

"{"name":"JWT Validation Error","message":"JWT validation failed, because the JWT expired at Mon Jun 14 2021 06:19:00 GMT-0400 (EDT)."}"

Figure 7.33 – JWT verification failure

You just learned about the comprehensive support for JWT in the APIC framework. The APIC framework provides the ability for JWT generation and JWT verification. These facilities can be used independently of each other. There might be cases where you will want to use the JWT generation capabilities of your organization's central identity provider. That is a perfectly acceptable use case. In such scenarios, you will simply use the Validate JWT policy in your API proxy to validate the JWT.

In the last few sections, you learned in detail about the techniques for Basic authentication (LDAP and custom repository), Client ID (API key), OAuth, OIDC, and JWT policies to secure your APIs. The following section will briefly describe the many capabilities of APIC that you can employ to secure your APIs through additional methods.

Adding additional security measures

The last few sections provided a comprehensive overview of APIC's OOTB security features to secure APIs. But by no means are these the only security features that you can use. You can build almost any security mechanism using a combination of user-defined policies and GatewayScript policy. You can further secure your services using **Transport Layer Security** (**TLS**) profiles, a User Security policy (for authentication and authorization), a Client Security policy, and even apply a SAML-based authentication to your OAuth provider. SAML and OAuth integration can be achieved by modifying your OAuth provider implementation to include a GatewayScript policy (to extract the SAML assertion from a request), a Client Security policy (to pass the SAML assertion to an authentication registry), and a custom authentication registry endpoint to validate the SAML assertion. You can build your authentication endpoint (for SAML assertion validation) within DataPower and expose it to APIC. Refer to *Figure 7.34*:

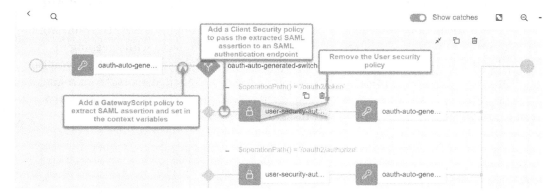

Figure 7.34 – Overview of SAML and OAuth integration

Again, the possibilities of expanding upon OOTB features are limitless because of the built-in flexibility of the APIC components. As the saying goes, *"The sky's the limit"*.

Summary

This chapter covered multiple methods provided by APIC to secure the APIs (or resources, as they are often called). Security is a vast and multi-layered subject, one of the most critical layers being authentication of parties seeking access to these Resources. These parties could be a typical user, a Resource Owner, or a Client/Application that intends to access these Resources. A comprehensive platform, such as APIC, provides many techniques to apply security to its protected Resources, and authenticate various parties that are trying to access these Resources.

This chapter was about exploring details of many such API security methods provided by APIC. All the security methods, except for the API Keys method, depend upon the setup of a user registry. If the security mechanism is OAuth and OIDC, then the setup also requires setting up an OAuth provider resource first. You covered these setups in the early part of this chapter. In the *Preparing for the APIC security implementation* section, you reviewed the LUR and performed the detailed steps for setting up an Authentication URL registry. You were also introduced to the LDAP user registry and OIDC user registry (this is only used for APIC user authentication).

The rest of the chapter was dedicated to exploring various security methods to secure the APIs through examples. You built examples to secure the APIs using the API key (client ID and client secret) method, Basic authentication using the Authentication URL user registry, OAuth/OIDC security, and by using JWT policies.

In each of these examples, you also learned the testing methodology to test secured APIs.

With all this information, you can appreciate the lengths and depths the APIC framework goes to in order to fulfill the API security needs of various audiences while keeping up with the latest industry security standards. Hopefully, by applying this knowledge, you can now secure your APIs and protect them from any kind of unsolicited access from rogue actors out there. Let's all of us do our part to "keep it safe" out there! You will next learn one of the other critical aspects of APIs – message transformations.

8
Message Transformations

You have now learned how **API Connect** (**APIC**) can quickly and easily be configured to organize, socialize, analyze, and provide a secure gateway to your existing APIs. Hopefully, you have seen the tremendous value in using it to expose your existing APIs where the request and response messages are simply passed to/from the backend API that is being proxied. There may be times, however, that the request and/or response messages will not be what the consumer is passing and the backend API is expecting. Or perhaps you wish to modernize an existing backend service by exposing it as a RESTful API. Take, for example, an existing SOAP Web service that you wish to expose as a RESTful API to your consumers. Or maybe you will just need to enrich or redact certain data within the request/response messages. There are many reasons why you might to alter the request and response data as it passes through your API. This is where API Connect provides many options depending on the type of data you are transforming and the complexity of the transformation itself.

In this chapter, we will cover several ways and use cases that can help you implement data transformations within your API definition. These topics will include basic transformations using the built-in "drag and drop" features as well as more complex transformations that can be accomplished programmatically. When you finish this chapter, you should have a good understanding of all of the possible transformation capabilities available, how to implement them, and how to choose the best policy for each of your use cases. To accomplish this, we will cover the following topics:

- Introduction to API Connect pre-built transformation policies
- Using a Map policy
- Redacting fields
- Applying JSON to XML or XML to JSON policies
- Implementing advanced transformations with XSLT and GatewayScript

Technical requirements

The examples presented in this chapter will require an installation of API Connect or API Connect LTE to configure and execute them. A working knowledge of XSLT and JavaScript will also be extremely beneficial in the later sections of this chapter.

Introduction to API Connect pre-built transformation policies

In *Chapter 4, API Creation*, you were introduced to pre-configured, or built-in, policies within the policies of an API flow. Of the many to choose from, some can be used to transform your data either on the request or response flow. As you are configuring the Gateway policies of your API, you will notice that all of the pre-defined policies on the left of the page are logically grouped based on the type of functions that they perform. You will notice a **Transforms** grouping, which will contain five different policies that you can use within your Gateway policy flow to transform your data. Each one will perform a specific type of transformation or pertain to a specific data format to transform to and from. In addition to these five policies listed under the **Transforms** heading, the GatewayScript policy can also be very useful in transforming data. This policy is listed under the **Policies** heading. *Figure 8.1* shows these different policies that we will be working with in this chapter to perform data transformations highlighted in yellow:

Figure 8.1 – Pre-defined policies for transformations

Just as you saw in *Chapter 4, API Creation*, you will simply drag the desired policy onto the Gateway policy flow and configure it appropriately. We will cover each of these policies in this chapter, describing when you might use them and how. We will start with the more simple use cases and build up to the more complex. Let's get started with the Map policy.

Using a Map policy

The Map policy is perhaps one of the most user-friendly methods provided to transform request and response data. This policy provides a way to define your data structure or invoke the schema from your Definitions configuration within the API configuration itself. Once the to and from data definitions are defined, this policy provides a convenient drag and drop feature to simply connect the fields to be mapped. The Map policy also provides the ability to add conditional logic, data formatting, calculations, and more. Although this policy can provide all of these features, you should use some restraint when being tempted to get too complex with it as the simplicity may come at a cost. That cost could be the performance or your API.

To explain and demonstrate how to configure the Map policy, let's take a simple order service where we are exposing the API as a RESTful service that expects JSON as the request. The backend, or target service, however, has not been modernized and expects the request to be in XML format. This is a relatively small request and there is a simple 1:1 conversion from the JSON fields to the XML fields. This would be a good opportunity for us to take advantage of the built-in Map policy provided by API Connect.

Our example API that is configured to receive JSON in the request message is expecting the following format:

```
{
    "createOrder":{
        "customerId":"187562X",
        "itemNumber":"12876",
        "itemQuantity":2,
        "itemDescription": "APIC widget",
        "orderTotal":12.75
    }
}
```

The target service that our API will forward requests to is expecting XML as the request message format in the following request message format:

```
<createOrder>
    <customerId>187562X</customerId>
    <itemNumber>12876</itemNumber>
    <itemQuantity>2</itemQuantity>
    <itemDescription>APIC widget</itemDescription>
    <orderTotal>12.75</orderTotal>
</createOrder>
```

You can see the relationship between these two messages and mapping should be evident. We can accomplish this within our API flow using the Map policy. Let's take a look at how we do this:

1. To begin, we must be on **Gateway** tab within the **Policies** section of the API. Here we can simply drag the **Map** icon onto our flow before the **invoke** policy so that the message is transformed before invoking the target service, as shown in *Figure 8.2*:

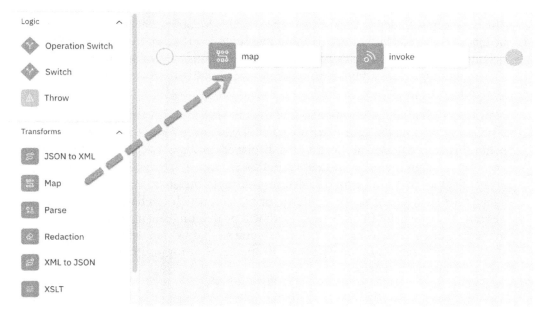

Figure 8.2 – Dragging the Map policy onto the policies flow

Once the map policy is within our flow, the configuration panel should become visible. If you do not see the configuration panel for the map policy, clicking the policy on the main panel will bring it up.

2. From the configuration screen, we can begin to define our input message by clicking on the pencil icon next to the **Input** label, as shown in *Figure 8.3*:

Figure 8.3 – Map policy configuration screen

3. From this screen, clicking the **Add input** button will render the screen where you can now define your input message format. You will notice several fields that you can configure, the first being the **Context** variable. This is where you can specify which variable contains the message that we are defining. For our example, we will be working with the request message as it came into the API that is stored in the `request.body` variable, and which just so happens to be the default, so we will leave that as is. We can also leave the **Name** field as the default value of `input`.

4. The next step is to define the Content type for this incoming message. Since we will be expecting a JSON request, this should be set to `application/json` from the dropdown. You will notice that the dropdown contains a list of several values for the content type that can be set to reflect your incoming message and content-type header as your use cases see fit.

5. Now that we have defined where to find the message to transform and the content type, we can define the data structure itself. If you haven't guessed it already, we will begin this configuration at the last field displayed on this screen, labeled **Definition**. Looking through the values presented in this dropdown, you will notice a wide array of values that can be used to define your input message. Many will simply be a data type such as `integer`, `string`, or `float`, which can be used if you were defining a single field. In our use case, we will be defining an entire input JSON message. For this, we will select **Inline schema** so that we can define our incoming message. Once selected, a new popup will appear, allowing you multiple options on how to define your schema.

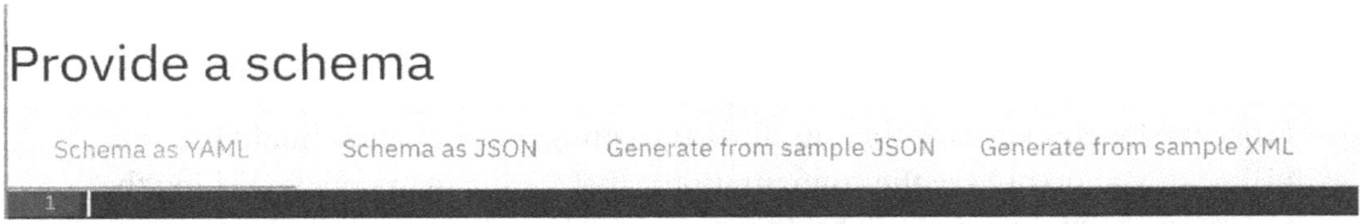

Provide a schema

Schema as YAML   Schema as JSON   Generate from sample JSON   Generate from sample XML

1 |

Figure 8.4 – Provide a schema popup for inline schemas

As shown in *Figure 8.4*, you can define your schema in YAML or JSON format. Alternatively, if you already have a sample JSON or XML file, you can generate the schema based on the sample. To make things simple, you will use **Generate from sample JSON**. Click on **Generate from sample JSON** and paste in the JSON input shown earlier in this section. *Figure 8.5* shows an example:

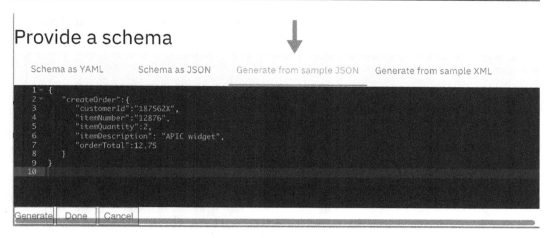

Figure 8.5 – Generating schemas using JSON

After pasting the JSON, click the **Generate** button to create the schema.

Your schema should now look as follows:

```
description: ''
type: object
properties:
  createOrder:
    type: object
    properties:
      customerId:
        type: string
      itemNumber:
        type: string
      itemQuantity:
        type: number
      itemDescription:
        type: string
      orderTotal:
        type: number
example: >-
  {"createOrder":{"customerId":"187562X","itemNumber":"1
  2876","itemQuantity":2,"itemDescription":"APIC
  widget","orderTotal":12.75}}
```

6. At this point, we have defined the basic configuration for our Map policy, as shown in *Figure 8.6*:

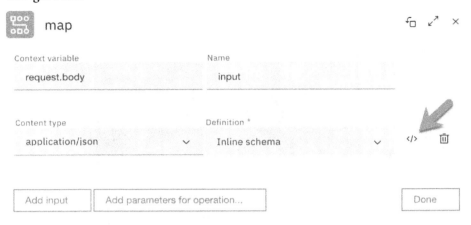

Figure 8.6 – Basic Map policy configuration

7. To review your inline schema, click the **Done** button on your input definition screen shown in *Figure 8.6*. This will now bring you back to the main configuration screen for your Map policy, which will show your input defined, as shown in *Figure 8.7*:

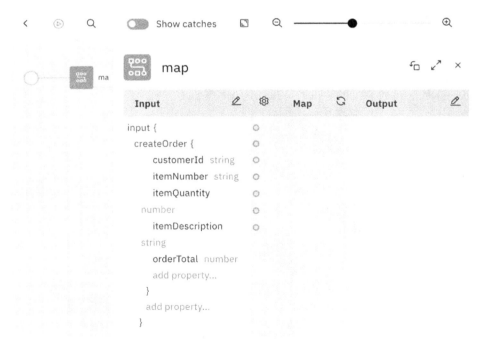

Figure 8.7 – Generated Input configuration

8. Since you used the **Generate** option, you are ready to go, but what if you didn't have a sample? You can directly add your schema by clicking the **add property** link within your **Input** field and continue to add your JSON fields, or you can click the </> button in the top-right corner to edit the source itself and provide the schema there. We will show the latter by adding the JSON schema directly in the API source. For this, we will describe how to perform this manual edit, but you don't need to do this since the schema was already generated previously. But *step 9* describes how this can be done manually so you understand how it can be accomplished.

9. Click the </> button. Once the </> button is clicked, you will see the source for the entire API definition. From within the source, you will see the `assembly` section. This is where you will find the definitions for any and all of your policies. Within this section, you will see the map configuration for the Map policy you just added. Since you didn't add any metadata defining your input or output messages, this configuration will be more of a skeleton at this point. It is within the `inputs` section of your map configuration that you will add your inline JSON schema in YAML format. *Figure 8.8* shows our schema definition for our `createOrder` input JSON message added to the map configuration source. As you can see, we also provide an example message within the source:

```
16    assembly:
17      execute:
18        - map:
19            version: 2.0.0
20            title: map
21            inputs:
22              input:
23                schema:
24                  description: ''
25                  type: object
26                  properties:
27                    createOrder:
28                      type: object
29                      properties:
30                        customerId:
31                          type: string
32                          name: customerId
33                        itemNumber:
34                          type: string
35                          name: itemNumber
36                        itemQuantity:
37                          type: number
38                          name: itemQuantity
39                        itemDescription:
40                          type: string
41                          name: itemDescription
42                        orderTotal:
43                          type: number
44                          name: orderTotal
45                      name: createOrder
46                example: >-
47                  {"createOrder":{"customerId":"187562X","itemNumber":"12876","itemQuantity":1,"itemDescription":"APIC
48                  widget","orderTotal":12.75}}
49            variable: request.body
```

Figure 8.8 – Map policy input definition source

10. Now that you have configured your input message, you can save your API and click the **Form** button to bring you back to the graphical representation of your policy flow. This button is right next to the </> button you clicked to view the source.

11. Now you are back at the policies configuration screen where you should see your map policy.

12. It is now time to configure your output message. You can accomplish this by following the same procedure as you did when defining your input message, but only this time you will configure the output message on the right side of the configuration screen. For our example, we will be transforming our JSON message to an XML message, so when configuring our output message, we will choose application/XML as the Content type.

13. As you did when defining our input message, we will click the </> button to define the output message format. We will again utilize the option to generate our schema from a sample message, as shown in *Figure 8.9*:

Provide a schema

Schema as YAML	Schema as JSON	Generate from sample JSON	Generate from sample XML

```
1   <createOrder>
2       <customerId>187562X</customerId>
3       <itemNumber>12876</itemNumber>
4       <itemQuantity>2</itemQuantity>
5       <itemDescription>APIC widget</itemDescription>
6       <orderTotal>12.75</orderTotal>
7   </createOrder>
```

Figure 8.9 – Pasting in sample XML

14. Clicking the **Generate** button now generates a schema defining your sample XML request message that you provided, as shown in *Figure 8.10*:

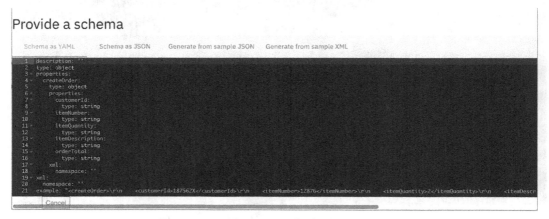

Figure 8.10 – The generated schema is created

15. Once you have provided your XML schema within the output section of your map policy source, you can click the **Done** button and save your API to return to the policy configurations screen. Double-clicking the map policy on your editor will again bring up the map policy configuration screen. At this time, you should see both the input and output messages defined within this screen, as shown in *Figure 8.11*:

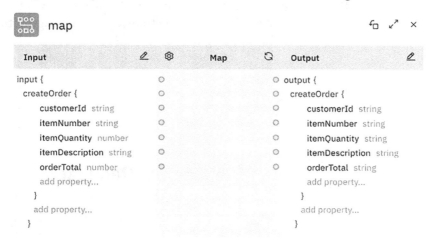

Figure 8.11 – Map policy configuration with input and Output messages defined

16. From this screen, you can now simply connect the dots from your input message fields to your output message fields as you would like them to be mapped. This is accomplished by dragging your cursor from one field of your input to the corresponding mapped field of your output message. *Figure 8.12* shows our completed example mapping:

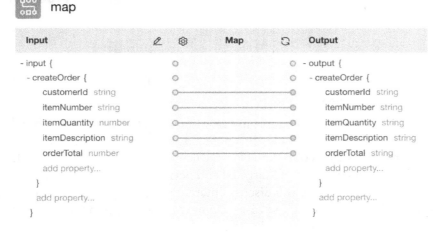

Figure 8.12 – Complete Input to Output mapping

17. Now that you have completed the map policy configuration, you can save and publish the API. When the API is invoked, the request message will now be transformed from the JSON format defined in the map policy to the XML format defined. Our sample API has a target that simply returns the message that it received. So, in our example, if we send the expected JSON request, our API will transform it to XML and send it to our target service, which will simply return the same XML request it received. We will then see the transformed message in the response back to the client, as shown in *Figure 8.13*:

Figure 8.13 – Request and response to/from our API

Although we demonstrated a very simple use case in our example, this might be all you need for your API and is a perfect fit for this policy. You can certainly implement more complex transformations with conditional logic, calculations, arrays, and so on. However, they might be better suited for a more targeted and performant implementation. As we progress through this chapter, we will discuss different policies that are targeted and better suited for specific data formats and use cases. Up next is a very specific case for redacting fields within your request, response, or log data.

Redacting fields

When we speak about transforming data within our APIs, we often think about modifying the data structure itself; perhaps converting from one format to another, such as from JSON to XML. The truth is that data transformation means modifying your data in any way and you might have many different reasons to do so. In today's age of technology, security is at the forefront, as it should be, and protecting consumers' sensitive data should be your top priority. There are standards and requirements, such as **Payment Card Industry (PCI)** standards, that govern how we transmit, share, and store data. This is all done in an effort to protect sensitive data in transit or at rest. One way to protect this data is to redact it. This could mean to strip out the sensitive data altogether or simply mask it with some special characters. API Connect provides a built-in policy to do this for us with minimal effort.

The redaction policy within API Connect is a built-in policy with minimal configuration required to redact all of your sensitive data. By simply dragging this policy onto the Policies palette and providing three pieces of information, your policy is all set.

Let's now take a look at this configuration and how to tailor it to your needs:

1. To begin, let's look at the JSON representation of our `createOrder` XML request message and add a `CCNum` field for the credit card number. This is the field that we will need to redact. This sample JSON request message is as follows:

```json
{
    "createOrder":{
        "customerId":"187562X",
        "itemNumber":"12876",
        "itemQuantity":1,
        "itemDescription": "APIC widget",
        "orderTotal":12.75,
        "CCNum": 12356789

    }
}
```

2. Like all other policies, we start by dragging the **Redaction** policy onto the Policies flow before our invoke action, which will automatically bring up the configuration screen for this policy.

3.   The first thing we will want to configure is the `Root` field within the Redaction policy configuration. This field tells your Redaction policy which data source will contain the data to redact or remove. You will provide a JSONata expression here to indicate this source data. To specify the request or response message, you can use the value `message.body` here. Based on where the Redaction policy is positioned on your flow, this value will represent the request or response message. For example, if your Redaction policy comes before the Invoke policy on your flow, this will indicate the request message. If it is after the Invoke policy, it will represent the response message. If no value is provided here, this policy will act on the entire API context. For our example, we want to redact a field on the request message, so we will specify `message.body` with our Redaction policy positioned before the Invoke policy.

4.   The next field we will want to configure is **Path**. This identifies what data elements we want to act on. We do this via a JSONata expression. Although we provide a JSONata expression, if this were an XML document, we could use the `$xpath()` JSONata extension. For more information on the JSONata format, please refer to the IBM documentation center at `https://www.ibm.com/docs/en/ api-connect/10.0.1.x?topic=gateway-constructing-jsonata- expressions-redact-fields`.

5.   Since our example is JSON, we will add the JSONata expression for the path for our request message `CCNum` field. This expression would be `$.createOrder.CCNum`. The `$` value in this expression indicates the root, which we have already configured as the message body. If no root was provided in the `Root` field, the intended root could also be provided in this JSONata expression. For example, we could have used `message.body.createOrder.CCNum` as our JSONata expression.

6.   Now that we've defined what field we want to act on, we will need to define what we want to do with the data identified. This is defined in the **Action** field. You will see that you have two choices here. The first, and the default, is **Redact**. This option will replace the value of the field identified in **Path** with the "*" characters. Your second option here is **Remove**, which will remove the data altogether. For our example, we will choose **Redact**.

7.   We have not yet completed the Redact policy configuration. Our final step to complete the redaction functionality is to drag a Parse policy just before the Redact policy so that APIC will parse the incoming request message for the Redact policy to handle it.

We have now completed our redaction policy to redact the CCNum value from our request message. This configuration is shown in *Figure 8.14*. We must now save and publish our API in order to test it.

Figure 8.14 – Redaction policy configuration

To see our new redaction policy in action, we can send a request to our API that we configured to redact the **CCNum** field. This API is configured to return the transformed message to the client so that we can see how the data was transformed. *Figure 8.15* shows this request and response where this field has been redacted as expected:

```
 1   {
 2       "createOrder":{
 3           "customerId":"187562X",
 4           "itemNumber":"12876",          <----- Request
 5           "itemQuantity":1,
 6           "itemDescription": "APIC widget",
 7           "orderTotal":12.75,
 8           "CCNum": 12356789
 9
10       }
11   }
```

Body   Cookies   Headers (13)   Test Results

Pretty   Raw   Preview   Visualize   JSON ▼   ⇆

```
 1   {
 2       "createOrder": {
 3           "customerId": "187562X",
 4           "itemNumber": "12876",          <----- Response
 5           "itemQuantity": 1,
 6           "itemDescription": "APIC widget",
 7           "orderTotal": 12.75,
 8           "CCNum": "*******"
 9       }
10   }
```

Figure 8.15 – Request and response with a redacted field

As you can see, the redaction policy is very specific and effective for redacting particular fields in your request, response, or log messages. With very little configuration, you can fulfill this requirement of removing or redacting sensitive data. With some creative XPath expressions, you may be able to fulfill more complex requirements; however, this policy is best suited for basic redaction or the removal of targeted fields.

Applying JSON to XML or XML to JSON policies

As you have seen throughout this chapter, there are two data formats that we have been focusing on – XML and JSON. They are the most prominent in today's API landscape. We have demonstrated the use of the map policy and how it can provide the need for custom transformations from one format to another. There may be times, however, when you just need to convert your JSON to an XML representation, or your XML to some standard JSON representation. Perhaps you are consuming JSON and want to log the payload, but your logging system requires XML, or vice versa. Or perhaps you have some XSLT that processes XML data, but the request is in JSON format. Whatever the case might be, there may be instances where you just need to transform XML to JSON or JSON to XML. For this purpose, API Connect provides two built-in policies, JSON to XML and XML to JSON. As you may have guessed, one transforms your JSON to an XML representation, and one transforms your XML data to a JSON representation.

Let's first take a look at the **JSON to XML** policy. As its name implies, this policy expects a JSON input and will return XML. To use this policy, you will drag it onto the policy flow within the **Gateway** tab at the location in the flow where the JSON message is expected. In our example, we will want to transform our request JSON data to XML, so we drag it as the first policy just before our **invoke** action. Once this is done, you will see the configuration screen for this policy appear. You will notice that there aren't many parameters here to define and you may just leave them all as the defaults. The configurable parameters are as follows:

- **Title**: The display title for this policy.

- **Description**: (optional): You can provide a description for this policy.

- **Input**: The APIC variable that will contain the JSON message to convert. As you have seen with the Redact policy example we discussed, `message.body` is the request message as it came into the API. This is also the default value for this field.

- **Output**: The APIC variable for storing the resulting output message. If the default input message is used, the default for this output variable will also be `message.body`.

- **Conversion type**: Defines how to format the output converted message. Your options here are Badgerfish or Relaxed Badgerfish. Badgerfish is the convention for translating an XML document into JSON. The Relaxed Badgerfish option can be used for more unconventional JSON elements as well as JSON to XML transformation. For our example, we will select the Relaxed Badgerfish option.

- **Root XML Element Name**: The root element of the XML document to be generated. This is only required if the incoming JSON has more than one top-level property.

- **Always output the root element**: A checkbox that would force the previously defined root element to always be outputted in the resultant XML. This is unchecked by default.

- **Element name for JSON array elements**: The XML element name to be used for each JSON array element.

For our example, we want to simply transform our input XML message to JSON. To accomplish this, we will leave the **Input** and **Output** fields blank so that the input request message sent to the API is used as the input to the transformation. For **Conversion type**, we selected **Relaxed Badgerfish** and checked the **Always output the root element** option, as shown in *Figure 8.16*:

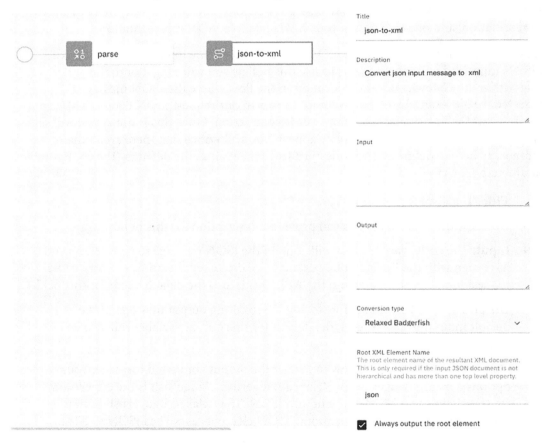

Figure 8.16 – JSON to XML policy configuration

As we did in the Redact example, your final step will be to drag a parse policy before your **json-to-xml** policy, as you can see we did in *Figure 8.13*. Once you have finished your policy configuration, clicking **Save** and publishing your API will make it available to test. For our example, we will alter our JSON request for createOrder by making an array of items to demonstrate how the transformation converts arrays to XML. *Figure 8.17* shows this request JSON and the converted response to XML, as performed by our JSON to XML policy:

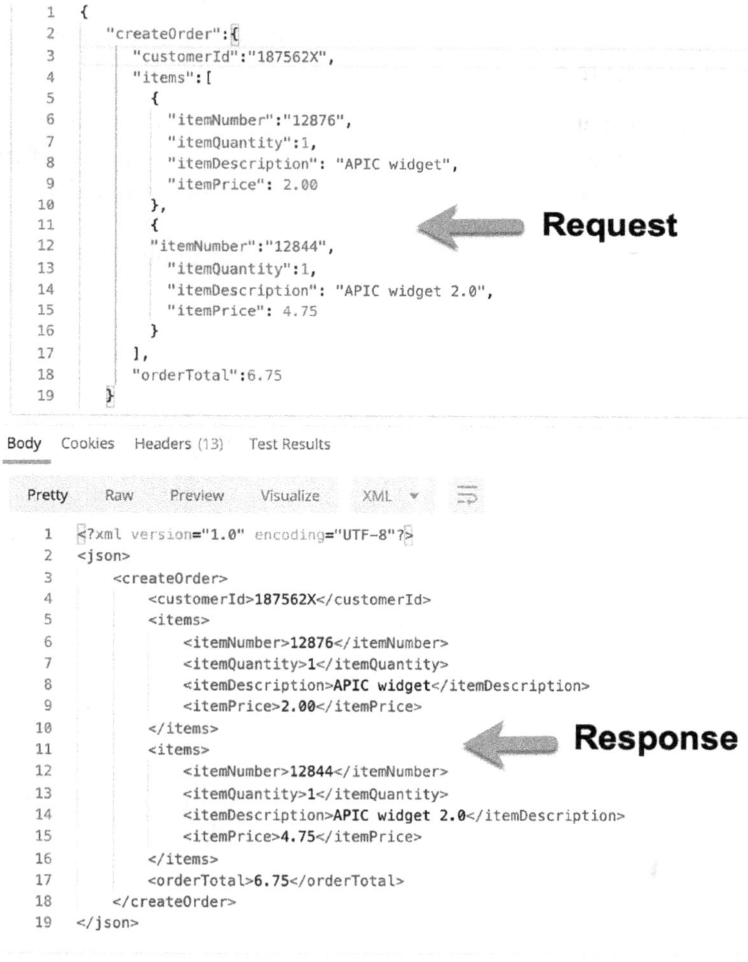

```
 1  {
 2      "createOrder":{
 3          "customerId":"187562X",
 4          "items":[
 5              {
 6                  "itemNumber":"12876",
 7                  "itemQuantity":1,
 8                  "itemDescription": "APIC widget",
 9                  "itemPrice": 2.00
10              },
11              {
12                  "itemNumber":"12844",
13                  "itemQuantity":1,
14                  "itemDescription": "APIC widget 2.0",
15                  "itemPrice": 4.75
16              }
17          ],
18          "orderTotal":6.75
19      }
```

⟵ **Request**

Body   Cookies   Headers (13)   Test Results

Pretty   Raw   Preview   Visualize   XML ▾   ⇥

```
 1  <?xml version="1.0" encoding="UTF-8"?>
 2  <json>
 3      <createOrder>
 4          <customerId>187562X</customerId>
 5          <items>
 6              <itemNumber>12876</itemNumber>
 7              <itemQuantity>1</itemQuantity>
 8              <itemDescription>APIC widget</itemDescription>
 9              <itemPrice>2.00</itemPrice>
10          </items>
11          <items>
12              <itemNumber>12844</itemNumber>
13              <itemQuantity>1</itemQuantity>
14              <itemDescription>APIC widget 2.0</itemDescription>
15              <itemPrice>4.75</itemPrice>
16          </items>
17          <orderTotal>6.75</orderTotal>
18      </createOrder>
19  </json>
```

⟵ **Response**

Figure 8.17 – JSON to XML transformation

Notice in our example transformation message that the resulting JSON contains a top-level element named **json**. This is the result of selecting the **Always output the root element** checkbox in our policy configuration. Also notice that the transformation used the name **json** for this element, as we also specified in our configuration.

You can see how simple it is to transform your JSON to XML using the **JSON to XML** policy. If you thought it couldn't get any simpler, you would be wrong! Let's now take a look at **XML to JSON**. This policy configuration is very similar to the JSON to XML policy, with two small differences in the configuration. The first difference you will notice is that the XML to JSON policy has fewer parameters available. For this configuration, you only have the **Title**, **Description**, **Input**, **Output**, and **Conversion type** fields. The second small difference is the values available for the **Conversion type** field. For this configuration, you can choose **badgerfish** or **apicv5**. Selecting the **badgerFish** option will format your resulting JSON using the `badgerfish` convention, where selecting `apicv5` will convert your XML to JSONx.

To accomplish this transformation, you first need to drag the policy onto **Policy flow** and complete the policy configuration. In our example, we will again be transforming the input message, so we will leave the **Input** and **Output** fields empty. We will then select the **badgerFish** option for our conversion type. Finally, we drag a Parse policy before our **xml-to-json** policy and we are finished. You can see our example configuration in *Figure 8.18*:

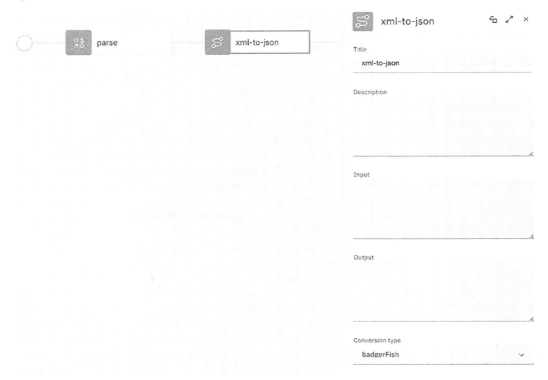

Figure 8.18 – XML to JSON policy

Once we have this configured, we can save and publish our API so that we can test it. *Figure 8.19* shows our XML request and the transformed JSON response for this API:

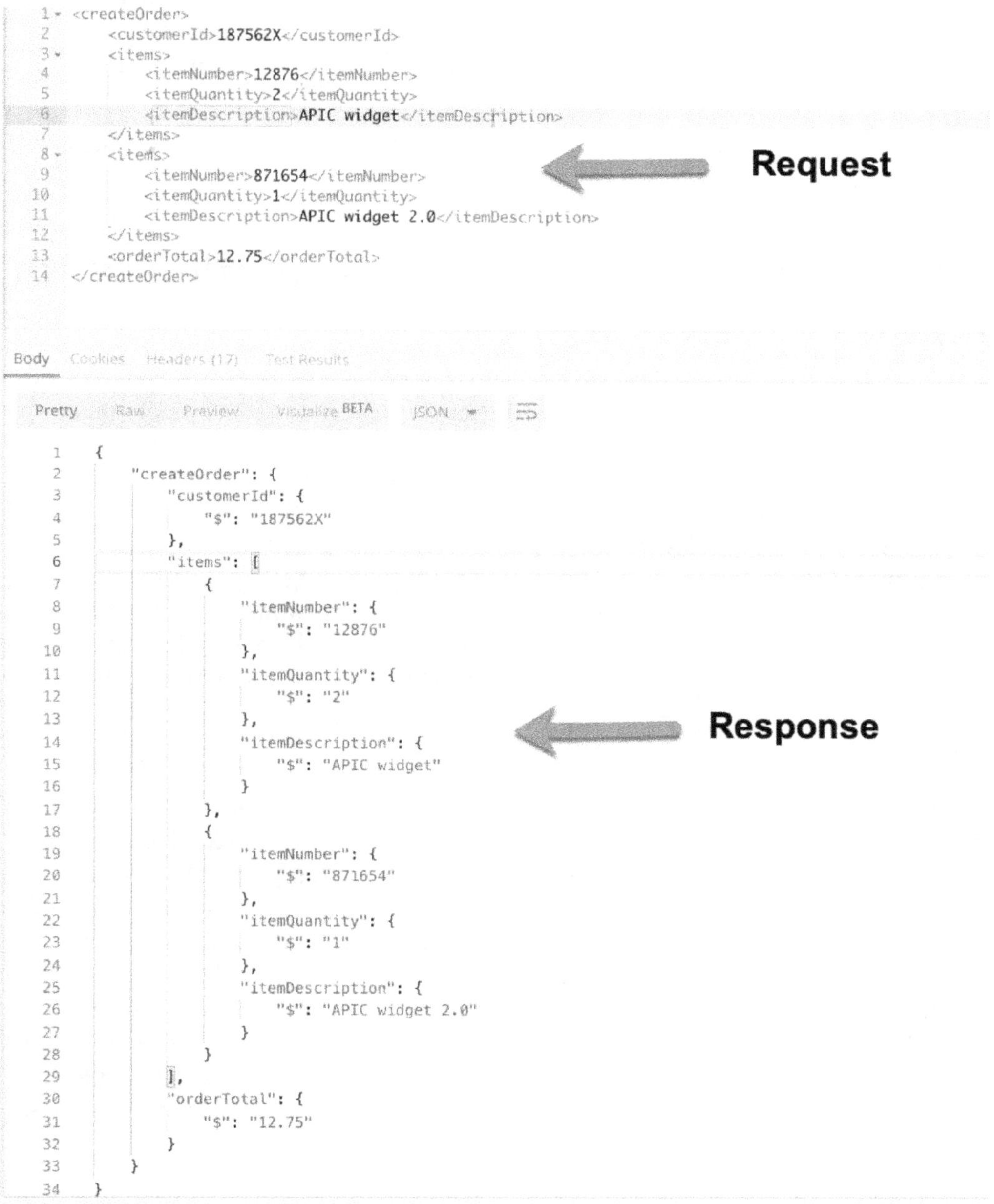

```
1 ▾ <createOrder>
2       <customerId>187562X</customerId>
3 ▾     <items>
4           <itemNumber>12876</itemNumber>
5           <itemQuantity>2</itemQuantity>
6           <itemDescription>APIC widget</itemDescription>
7       </items>
8 ▾     <items>
9           <itemNumber>871654</itemNumber>
10          <itemQuantity>1</itemQuantity>
11          <itemDescription>APIC widget 2.0</itemDescription>
12      </items>
13      <orderTotal>12.75</orderTotal>
14  </createOrder>
```

Request ⟵

Body   Cookies   Headers (17)   Test Results

Pretty   Raw   Preview   Visualize BETA   JSON ▾   ⇌

```
1   {
2       "createOrder": {
3           "customerId": {
4               "$": "187562X"
5           },
6           "items": [
7               {
8                   "itemNumber": {
9                       "$": "12876"
10                  },
11                  "itemQuantity": {
12                      "$": "2"
13                  },
14                  "itemDescription": {
15                      "$": "APIC widget"
16                  }
17              },
18              {
19                  "itemNumber": {
20                      "$": "871654"
21                  },
22                  "itemQuantity": {
23                      "$": "1"
24                  },
25                  "itemDescription": {
26                      "$": "APIC widget 2.0"
27                  }
28              }
29          ],
30          "orderTotal": {
31              "$": "12.75"
32          }
33      }
34  }
```

Response ⟵

Figure 8.19 – XML to JSON request and response

You have now seen how API Connect provides a simple drag and drop method for converting an XML message to a JSON representation, as well as for converting a JSON message to an XML representation using the built-in policies. This is yet another example of convenient solutions for very specific use cases. However, you will most likely find yourself in a situation where you need similar transformations to those discussed, but with more complexity or flexibility.

In the final section of this chapter, we will discuss how this can be accomplished using the two supported programming languages within API Connect.

Implementing advanced transformations with XSLT and GatewayScript

So, you find yourself needing to perform a transformation on your messages, but none of the built-in transformation policies will provide exactly what you need. Perhaps you don't want to transform your messages to a different structure, but you wanted to add to, or enrich, your messages. Or maybe you want to encrypt a specific field instead of redacting it. Countless transformation use cases may present themselves with each one being similar, but different enough that they will require some custom coding. This is where API Connect provides the ability to utilize two powerful programming languages to give you the flexibility to customize your transformations in any way you would like. In this section, we will discuss a few scenarios where this would be helpful, but the possibilities are endless when you have the ability to get back to the power and control of basic programming!

XSLT

Extensible Stylesheet Transformation Language (**XSLT**) literally has "transformation" in the name! XSLT is a language that is specifically designed to transform XML documents into other XML documents. So, if you are looking for a more complex XML to XML transformation where you have more programmatic control, this one is your go-to. XSLT is also considered a Turing-complete programming language, which basically means that it can perform any calculations and computations that can be performed by modern programming languages. Hence, in a nutshell, you have here a complete programming language that is also specifically designed to transform XML documents. If that was good enough, your Gateway service that will be executing the XSLT is most likely going to be DataPower, which, from the beginning, was optimized to execute XSLT transformations.

If you have worked with the DataPower product line, you are most likely very familiar with XSLT and how to write a basic XML to XML transformation. If you are not familiar with XSLT, it would behoove you to become familiar with it so that you can take advantage of this capability with API Connect. You can find plenty of tutorials and information on the web, as well as several books available on the topic. Instead of covering the fundamentals of the XSLT language, we will focus on some more advanced and API Connect-specific features, such as accessing API Connect variables and EXSLT extension functions, which are also supported.

To demonstrate some of the features available to us in our XSLT policy, let's take our simple createOrder request XML message. Perhaps we have a requirement to enrich this XML with a timestamp, the host header that came in on the request, and we require a total count of the items sent. This can easily be accomplished within the XSLT policy by utilizing simple XSLT, an EXSLT extension function, and an API Connect function to access the custom variables.

Before writing your XSLT, you will need to drag the XSLT onto the Gateway policy pane, as you did with the other policies. Once your policy is in place, a configuration screen will appear with an editor where you can write your XSLT code. You will also notice a checkbox to indicate whether the input context for this policy should be used as the input to the XSLT. Checking this box indicates that it will be. For our example, we will be using the input context to create our output XML document, so we will check this box. The editor provided is not all that big and would be difficult to code in in the current state. Clicking the double arrows at the top right of the configuration will expand the configuration screen, making the editor much bigger. Other than the one checkbox, there aren't any other configuration options for this policy other than the code itself, so let's dive into that:

```
<xsl:stylesheet version="1.0"
  xmlns:xsl="http://www.w3.org/1999/XSL/Transform"
   xmlns:date="http://exslt.org/dates-and-times"
    xmlns:apim="http://www.ibm.com/apimanagement">
  <!--Use  for V5 compatible gateway -->
  <xsl:import href="local:/isp/policy/apim.custom.xsl"/>
  <!-- Use for API  gateway
  <xsl:include href="store:///dp/apim.custom.xsl" />
  -->
  <xsl:template match="/">
```

```
<createOrder>
  <requestHost>
    <xsl:value-of select=
      "apim:getVariable('request.headers.host')"/>
  </requestHost>
    <timestamp><xsl:value-of select="date:date-
      time()"/></timestamp>
    <customerId><xsl:value-of select=
      "createOrder/customerId"/></customerId>
    <xsl:for-each select="createOrder/items">
      <items>
        <itemNumber><xsl:value-of select=
          "itemNumber"/></itemNumber>
        <itemQuantity><xsl:value-of select=
          "itemQuantity"/></itemQuantity>
        <itemDescription><xsl:value-of select=
          "itemDescription"/></itemDescription>
      </items>
    </xsl:for-each>
    <orderTotal><xsl:value-of select=
      "createOrder/orderTotal"/></orderTotal>
    <itemCount><xsl:value-of select=
      "count(createOrder/items)"/></itemCount>
  </createOrder>
 </xsl:template>
</xsl:stylesheet>
```

Let's take a closer look at this stylesheet and the parts that are not basic XSLT.

To start with, take note of the namespace declarations and you will notice that in addition to the standard `xsl` namespace declarations, we also declare the `date` and `apim` namespaces. The `date` namespace is declared so we can use the EXSLT dates and times' elements and functions. The `apim` namespace is declared, so we can use the API Connect-specific elements and functions.

To use the API Connect built-in functions, we must use the `xsl:import` directive to import the XSLT provided with the product. Notice that this will be different if your gateway is configured for V5 compatibility versus an API gateway.

As we proceed through the XSLT, we have our standard template definition and we start our output XML document. You can see that our first element after the root element is named `<requestHost>`. This is where we want to use the host header that came in on the request to populate this element. For this, we use a built-in API Connect function, `getVariable`. You can identify this as such by the `apim` namespace. This is a very powerful function that provides you with access to all of the API Connect context variables as well as the API property variables. For more information on the specific variables available, please refer to the IBM Info Center at `https://www.ibm.com/support/knowledgecenter/SSMNED_v10/com.ibm.apic.toolkit.doc/capim_context_references.html`.

In our example, we are retrieving the `request.headers.host` context variable, which will contain the value of a request header sent in the request with the name `host`.

The next element in our transformed request XML will be `timestamp`. As its name implies, this will be the current date and time that the XSLT is executed. This is retrieved using the EXSLT function, `date-time()`.

The remainder of our XSLT is standard XSLT functionality, where you can see that we loop through each `items` element using the `for-each` statement. Finally, we use the XSLT `count()` function to return the number of `items` elements in the request.

This completes our simple, yet effective XSLT. You can see a sample request and response in *Figure 8.20*:

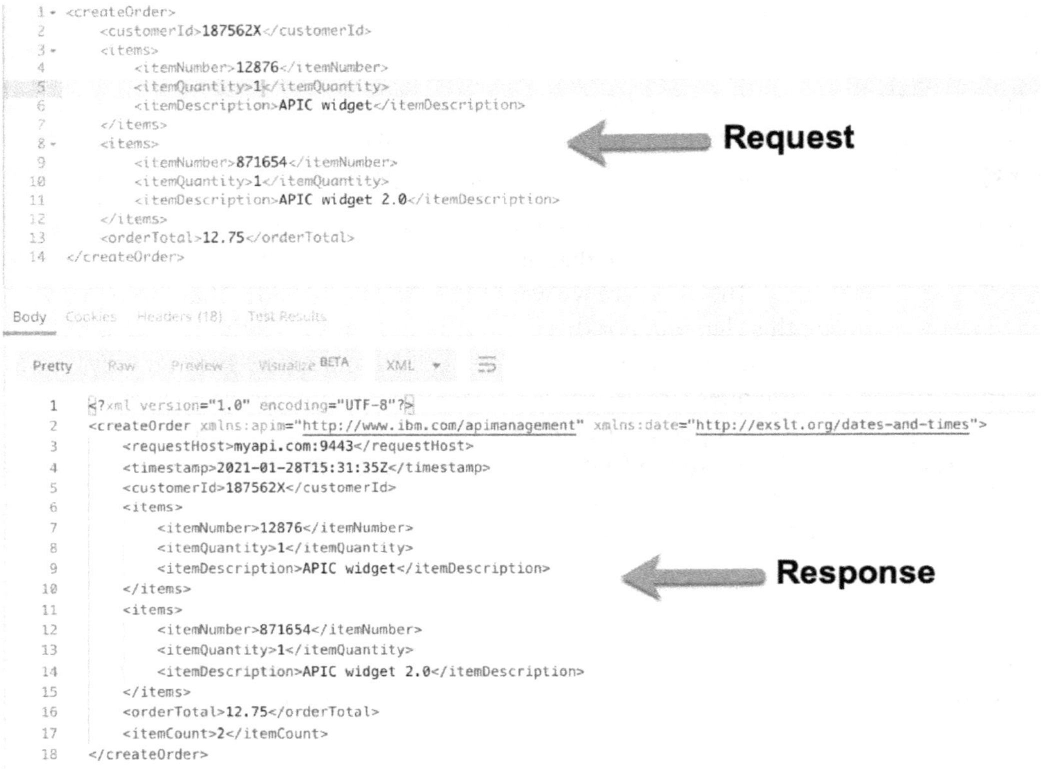

```
 1 ▾ <createOrder>
 2      <customerId>187562X</customerId>
 3 ▾    <items>
 4        <itemNumber>12876</itemNumber>
 5        <itemQuantity>1</itemQuantity>
 6        <itemDescription>APIC widget</itemDescription>
 7      </items>
 8 ▾    <items>
 9        <itemNumber>871654</itemNumber>
10        <itemQuantity>1</itemQuantity>
11        <itemDescription>APIC widget 2.0</itemDescription>
12      </items>
13      <orderTotal>12.75</orderTotal>
14   </createOrder>
```

Request ⬅

Body　Cookies　Headers (18)　Test Results

Pretty　Raw　Preview　Visualize BETA　XML ▾

```
 1   <?xml version="1.0" encoding="UTF-8"?>
 2   <createOrder xmlns:apim="http://www.ibm.com/apimanagement" xmlns:date="http://exslt.org/dates-and-times">
 3        <requestHost>myapi.com:9443</requestHost>
 4        <timestamp>2021-01-28T15:31:35Z</timestamp>
 5        <customerId>187562X</customerId>
 6        <items>
 7          <itemNumber>12876</itemNumber>
 8          <itemQuantity>1</itemQuantity>
 9          <itemDescription>APIC widget</itemDescription>
10        </items>
11        <items>
12          <itemNumber>871654</itemNumber>
13          <itemQuantity>1</itemQuantity>
14          <itemDescription>APIC widget 2.0</itemDescription>
15        </items>
16        <orderTotal>12.75</orderTotal>
17        <itemCount>2</itemCount>
18   </createOrder>
```

Response ⬅

Figure 8.20 – XML request transformed via XSLT

Although we are only scratching the surface here with the power of XSLT, EXSLT, and API Connect functions, we demonstrated in one simple stylesheet at least one useful component of each. Since you have the power of a complete programming language, the possibilities are endless. It is important, however, to keep in mind that XSLT is designed and optimized to process XML. So, when your transformation consists of only XML, this should be your go-to transformation language. Once you start introducing other formats such as JSON, you might then want to consider another language that is both supported on API Connect and better optimized for other message formats. That would be GatewayScript, which will cover in the final section of this chapter.

GatewayScript

We saw the power of having a complete programming language for working with XML messages to not only transform the XML but to access API Connect-specific variables and other specialized functions. But what if we are not working with XML? What if we are dealing with a more modernized RESTful service receiving JSON requests and responses? This is where GatewayScript comes in. GatewayScript is nothing more than DataPower's implementation of JavaScript. Since DataPower is your gateway service, it is only natural that API Connect provides a pre-built policy for implementing your own GatewayScript code. Much like the XSLT policy, GatewayScript also provides the ability to access context and API property variables and is also a complete programming language. If you are not familiar with JavaScript, or just need a refresher, there are many books and resources available on the internet. Much like the previous section on XSLT, we will not cover the basics of this language, just the API Connect-specific details regarding implementation.

To implement your own GatewayScript code, you would begin the same as you did for every other built-in policy by dragging the GatewayScript policy onto the Gateway palette. Once this policy is in place, the configuration screen will become visible, presenting a text editor to enter your GatewayScript code. Again, it would be advisable to maximize this screen to provide an expanded view of this text editor. If your code becomes too long or complex, it may be beneficial for you to work in a text editor or **Integrated Development Environment (IDE)** and then copy/paste your code into the policy editor when finished.

Let's again take our JSON version of the `createOrder` request message and perform the same apply transformation requirements that we had in the XML version. Again, we will want to produce an XML output while adding three additional fields – `requestHost`, `timestamp`, and `itemCount`, as we did in the XSLT example. The big difference here is that we are receiving a JSON message as the request message. This example will demonstrate how we can deal with different message formats, access API Connect-specific variables, and also utilize a built-in JavaScript function. Let's take a look at our example GatewayScript. Line numbers have been added for ease of reference:

```
1.   var apim = require('./apim.custom.js');
     //  var apim = require('apim');
2.   apim.readInputAsJSON(function (error, json) {
3.   if (error)
4.   {
5.     apim.error('Parse Error', 500, 'Internal Error',
           'Failed to parse JSON input');
```

```
6.   }
7.      else
8.   {
9.      var d = new Date();
10.     var itemCount = json.createOrder.items.length;
11.     var xml = "<createOrder>";
12.     xml += "<customerId>" + json.createOrder.customerId
           + "</customerId>";
13.     xml += "<timestamp>" + d.toISOString() +
           "</timestamp>";
14.     xml += "<requestHost>" +
           apim.getvariable('request.headers.host') +
            "</requestHost>";
15.     xml += "<itemCount>" + itemCount + "</itemCount>";
16.     for (var i = 0; i < json.createOrder.items.length;
           i++){
17.        xml += "<items>";
18.        xml += "<itemNumber>" + json.createOrder.items
              [i].itemNumber + "</itemNumber>";
19.        xml += "<itemQuantity>" + json.createOrder.items
              [i].itemQuantity + "</itemQuantity>";
20.        xml += "<itemDescription>" +
              json.createOrder.items[i].itemDescription +
               "</itemDescription>";
21.        xml += "</items>";
22.     }
23.     xml += "</createOrder>";
24.     session.output.write(XML.parse(xml));
25.     apim.output('application/xml');
26.     }});
```

Just as we had to provide an import statement with our XSLT so that we can reference built-in API Connect functions, we will need to do the same for our GatewayScript. Line 1 in our code shows how we set an `apim` variable and added the `require` statement. The parameter passed to the `require()` function will be different if you have your gateway services set to be V5 compatible or not. For V5-compatible gateways, you will set this as `./apim.custom.js`, whereas API gateways will require `apim`.

Since we know that we should be expecting a JSON request message as input to the GatewayScript, we must specify in our GatewayScript to read in the data as JSON, which is what we are doing in line 2 in our sample code. We then proceed to some error handling in case there is an issue parsing the incoming JSON.

Once all of our housekeeping is done upfront, we can now begin to get the values for our new XML request. You can see a JavaScript function being used in line 9 to obtain the current date and time, which will be converted to a string in line 13 where we use it. Line 10 is a good example where we obtain the total number of items in the JSON array to be used for our `itemCount` element. It is also interesting to see here how you can navigate through the incoming JSON message using a dot notation. This will be more apparent as you look through how we populate the output XML elements.

We build our output XML message by simply appending to one variable named `xml` throughout the code, being certain to include the proper XML begin and end tags. Again, we can populate these XML elements with the incoming JSON data by referencing the JSON elements via the dotted notation. On line 16, you can see that we begin a `for` loop, iterating through the incoming JSON array and populating our XML document.

Finally, we have completed our XML document and are ready to send it to the output stream. To do this, we use the `session.output.write()` function, as you can see in line 24. As you can see, we do not just pass in our string that we have been using to build our XML document. We have to be sure to parse this as XML, using the `XML.parse()` function. Once we have set the output, we must then use the API Connect function, `apim.output()`, to set the output to `application/xml`.

Once we have coded our GatewayScript in our policy, we can save the API and publish it so that it can be invoked. *Figure 8.21* shows a sample request and response to our API that converted our JSON request into an XML document while adding a few new fields:

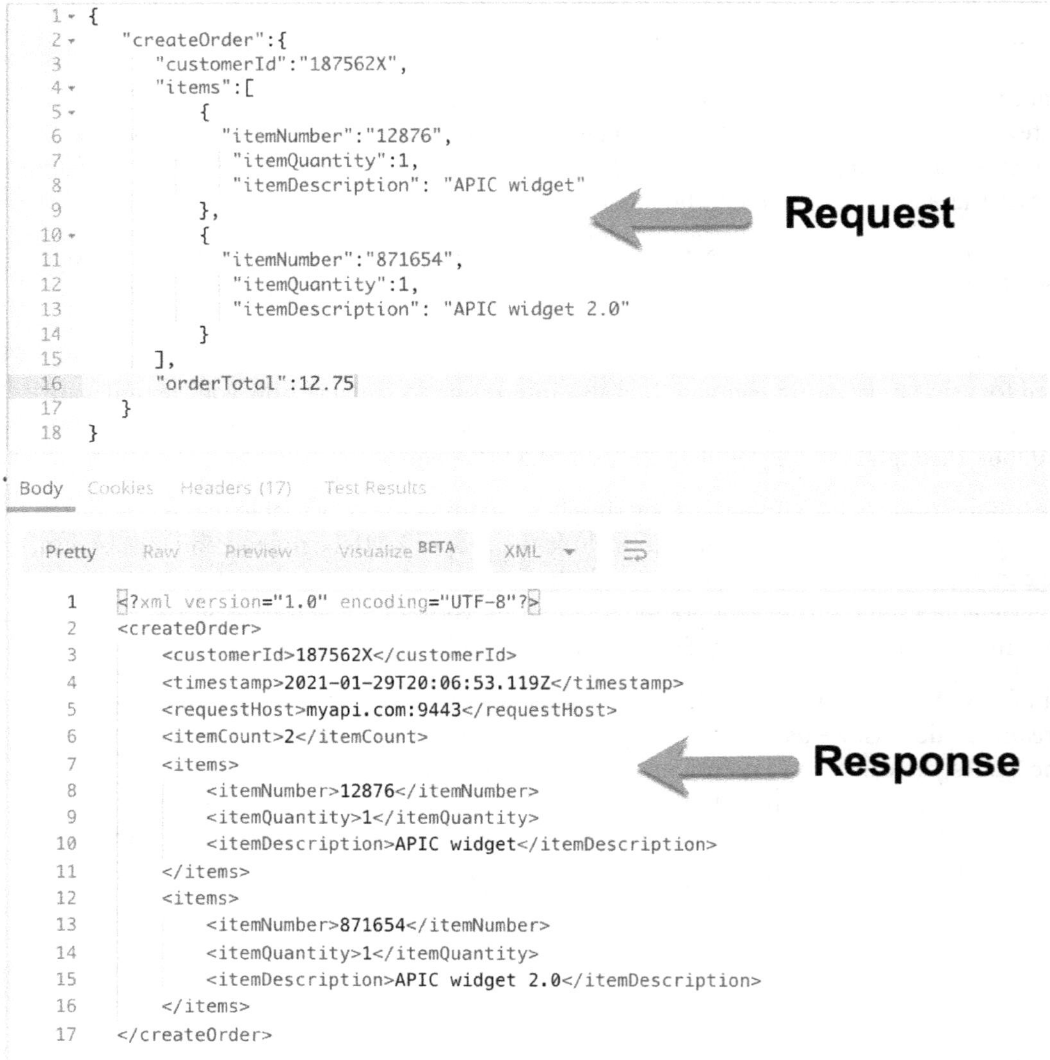

```
 1 ▾ {
 2 ▾     "createOrder":{
 3           "customerId":"187562X",
 4 ▾         "items":[
 5 ▾             {
 6                   "itemNumber":"12876",
 7                   "itemQuantity":1,
 8                   "itemDescription": "APIC widget"
 9               },
10 ▾             {
11                   "itemNumber":"871654",
12                   "itemQuantity":1,
13                   "itemDescription": "APIC widget 2.0"
14               }
15           ],
16           "orderTotal":12.75
17       }
18   }
```

Request

Body   Cookies   Headers (17)   Test Results

Pretty   Raw   Preview   Visualize BETA   XML ▾   ⇥

```
 1   <?xml version="1.0" encoding="UTF-8"?>
 2   <createOrder>
 3       <customerId>187562X</customerId>
 4       <timestamp>2021-01-29T20:06:53.119Z</timestamp>
 5       <requestHost>myapi.com:9443</requestHost>
 6       <itemCount>2</itemCount>
 7       <items>
 8           <itemNumber>12876</itemNumber>
 9           <itemQuantity>1</itemQuantity>
10           <itemDescription>APIC widget</itemDescription>
11       </items>
12       <items>
13           <itemNumber>871654</itemNumber>
14           <itemQuantity>1</itemQuantity>
15           <itemDescription>APIC widget 2.0</itemDescription>
16       </items>
17   </createOrder>
```

Response

Figure 8.21 – JSON to XML transformation using GatewayScript

You have now been introduced to the GatewayScript policy and only scratched the surface with the capabilities of the language itself. We have demonstrated a few of the capabilities to get you started; however, you are encouraged to experiment and discover all of the capabilities that it has to offer.

Summary

As our technology landscape continues to evolve, so will the latest and greatest formats for communicating and sharing data. It feels like just yesterday that service-oriented architecture was going to change our world using SOAP messages and WSDLs. Then, just as we all got our service converted to facilitate the use of SOAP messages, it became RESTful services and JSON was all the rage. It would be extremely difficult to keep up with these technology and architecture shifts with all of our applications and systems; however, we will need to expose our legacy service to our modernized consumers using the latest and greatest. This is where the power of message transformation comes into play.

You have seen how we can easily transform our message from one format to another, or enrich the data as it comes in. You have also seen how API Connect provides built-in policies for some very specific use cases and transformations, making it a simple drag and drop configuration. Of course, there will be more complex scenarios, as we demonstrated with the two programming languages, that will give you total control and flexibility. Choosing the proper built-in policy for your transformations will allow you to quickly configure these policies and provide the most performant and efficient solution when it comes to runtime.

Throughout this chapter, you have learned about the built-in, drag and drop features within APIC to perform your message transformations. For cases where these do not quite fit the bill, you have also learned how to customize your own message transformations using two powerful programming languages. Armed with this knowledge, you are now equipped to implement almost any type of message transformation you might have.

If the custom programming within this chapter piqued your interest, be sure to continue on to the next chapter, where you will be introduced to the GraphQL features within APIC, providing even more flexibility and customization for your APIs.

9
Building a GraphQL API

So far, you have been learning about how to create APIs that provide defined payloads. You have learned about how to utilize existing backend services to deliver data to consumers through REST and SOAP APIs. If your services have been in existence for a while, you may have had situations where you need to update those services with either new business requirements or additional data elements. Perhaps you've even had situations where you have shared your APIs with different business partners and each required the same API but with different requirements. In those cases, you may have created new APIs or updated the APIs to support the additional business requirements for multiple consumers. In either of those cases, versioning would have been required.

Managing multiple versions of APIs can complicate your API strategy. That is the price you pay when you develop and support many versions of APIs. In this chapter, you will be introduced to GraphQL, which provides an alternative method that allows more flexibility regarding how API consumers consume your APIs. You will learn how to integrate GraphQL with API Connect with the appropriate safeguards.

In this chapter, we will cover the following main topics:

- Why GraphQL?
- Learning how a GraphQL API addresses over-fetching

- Creating a GraphQL API
- Setting weights, costs, and rate limits for your GraphQL API
- Removing fields from GraphQL

By the end of this chapter, you should have the skills and knowledge to build flexible and secure GraphQL APIs with API Connect.

Technical requirements

In this chapter, you will be referencing the use of an Express.js GraphQL server to assist you with learning about GraphQL. You will find this file in this chapter's GitHub repository at `https://github.com/PacktPublishing/Digital-Transformation-and-Modernization-with-IBM-API-Connect/tree/main/Chapter09`.

You should copy the `tar` file from there to your local environment. You will be utilizing the API Manager to perform the development tasks in this chapter, which may require you to download additional software from the web to install the Express components. In addition, you may need to update your `/etc/hosts` file to reach the downloaded GraphQL Express web server implementation. For example, you can add a line similar to `192.168.168.205 mygraphql.com` to `/etc/hosts` that identifies the IP address and hostname of your GraphQL server implementation.

Now, it's time to learn more about GraphQL and the motivation to utilize it in certain circumstances.

Why GraphQL?

Before you learn about the benefits of GraphQL, you might be wondering, what is GraphQL? GraphQL is a query language but not for querying databases. It's a query language for APIs, but in many cases, it may eventually interface with databases. From an API consumer's perspective, it's a query language for the client to specify which data fields it needs.

The motivation for the consumer to use GraphQL surrounds collecting payload data that's returned from APIs. In some cases, you receive more data than what is necessary and that impacts bandwidth. In other cases, you don't get enough data and need to create additional APIs to fetch additional data. Those two cases are referred to as **over-fetching** and **under-fetching**. Let's understand each of them:

- **Over-fetching** happens when an API returns data elements you do not need nor want. This extraneous information can be just ignored but has a cost, especially if there are lots of additional fields. It's a waste and affects latency.

- **Under-fetching** is observed when the endpoint returns too little data and requires an API's client to execute another endpoint (API) to capture the information it needs to interact with the user. Another situation where under-fetching is apparent is when you require a list of elements. For example, you may only want to display 10 elements but there are many more elements you may require if the user doesn't see the element in the first 10. In a traditional API method, you would make another API request to fetch additional elements that are required. When this scenario is present in your APIs, this is referred to as the **{n + 1} request problem**.

- Although not generally included in discussions on over-fetching and under-fetching, another implementation that exacerbates the data access problem is what we call **replication fetching**. This is when you implement similar APIs to support business partners in a B2B situation by replicating the implementation but removing data elements to support the business agreement. This leads to multiple versions and more complex modifications because of the existence of multiple copies.

In all such cases, they are potential performance bottlenecks and worse yet, a versioning nightmare. If you can specify how you want to receive your data, wouldn't you prefer it in a single API? That is the value of GraphQL. Does that mean that REST is going away? No. GraphQL is just an alternative for data-intensive services where the client specifies the required elements.

How someone builds your GraphQL server will require you to learn about a few tenants of GraphQL.

> **Information**
>
> To learn more about the basics of GraphQL, visit `https://graphql.org/`.

Although you will not be learning how to build a GraphQL server from the ground up, it will be beneficial to understand the basics. You will learn about those next.

GraphQL anatomy

Solving the aforementioned fetching problems with GraphQL can help you with your API. However, GraphQL provides more capabilities that are additional benefits. You should be familiar with the following three things:

- Types
- Queries
- Resolvers

You will be introduced to each of these so that you can develop a basic understanding of what the designers of the GraphQL server are doing.

Types

When you build out GraphQL, you will observe that it is strongly typed. As you build your schema, you define various data types that will be used in the GraphQL application. When you create a schema, the syntax you use is based on the **Schema Definition Language** (**SDL**). In SDL, you have various types to utilize. Among them are scalar types of String, Int, Floats, Booleans, and ID. You should be aware that every type is nullable. The exception is if the field uses an exclamation point (!) to specify that the field must contain a value. A field that is not nullable is defined as follows:

```
id: ID!
```

As an example, if you wanted to build a schema based on an FHIR resource, you could create a schema based on the FHIR specification, as follows:

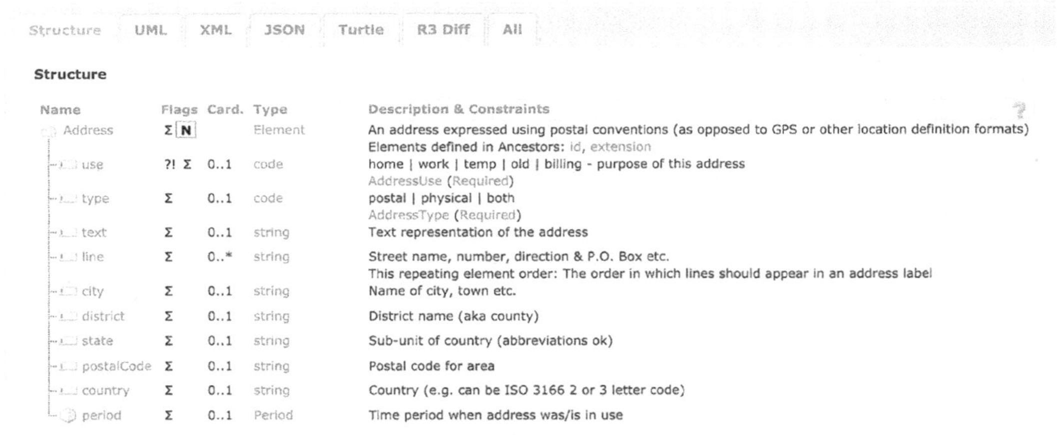

Structure	UML	XML	JSON	Turtle	R3 Diff	All

Structure

Name	Flags	Card.	Type	Description & Constraints
Address	Σ N		Element	An address expressed using postal conventions (as opposed to GPS or other location definition formats) Elements defined in Ancestors: id, extension
use	?! Σ	0..1	code	home | work | temp | old | billing - purpose of this address AddressUse (Required)
type	Σ	0..1	code	postal | physical | both AddressType (Required)
text	Σ	0..1	string	Text representation of the address
line	Σ	0..*	string	Street name, number, direction & P.O. Box etc. This repeating element order: The order in which lines should appear in an address label
city	Σ	0..1	string	Name of city, town etc.
district	Σ	0..1	string	District name (aka county)
state	Σ	0..1	string	Sub-unit of country (abbreviations ok)
postalCode	Σ	0..1	string	Postal code for area
country	Σ	0..1	string	Country (e.g. can be ISO 3166 2 or 3 letter code)
period	Σ	0..1	Period	Time period when address was/is in use

Figure 9.1 – FHIR specification for Address

Using the FHIR specification, you could build a schema that supports the following example:

```
"address": [ {
  "use": "home",
  "type": "physical",
  "line": [ "2125 Triple House Lane"],
  "city": "Arcola",
  "district": "Arcola",
  "state": "Missouri",
  "postalCode": "65603"
```

The FHIR specification would then guide you to build a schema that follows this:

```
type AddressSchema {
  use: addressEnumType
  type: addressTypeEnum
  text: String
  line: String
  city: String
  district: String
  state: String
  postalCode: String
  country: String
}
```

This is just an example of part of the schema. With GraphQL, you will be wrapping this definition into the Query type. The hierarchy would be a Patient schema that has an address that references AddressSchema:

```
type Patient {
  resourceType: String
  id: ID!                 .
  active: Boolean
  name: [NameSchema]
  gender: String
  address: [AddressSchema]
}
```

All of these structures are referenced in the Query type, which is what the API will be using to specify the elements it requires. The Query schema is defined as follows:

```
type Query {
  patient: [Patient]
  patientById(id: ID!): Patient
}
```

At this point, you probably have an understanding of how schemas are created. You have seen how the Query type references the Patient type, which contains the AddressSchema type. You will now learn more about queries, mutations, and how data is accessed using resolvers.

Queries, mutations, and resolvers

As you learned in the previous section, the Query type defines the object you will be interacting with. A GraphQL query is used to read or fetch values. In the definitions you saw previously, Query is referencing patient information. When you execute a query, you are searching for data. In cases where you need to write data, you utilize a type called a **mutation**. Mutations follow the same SDL format as queries and other types.

So, you have learned that queries fetch data and mutations, as well as update data, but what you haven't learned is how the data access is initiated. That is accomplished with a resolver.

A resolver is a function that's responsible for populating the data for a single field in your schema. Refer to the following code:

```
const resolvers = {
  Query: {
    patient: () => patient,
    patientById: (parent, args, context, info) => {
      return patient.filter(item=> item.id === args.id)[0];
    }
  }
};
```

Each field on each type is supported by a resolver. The resolver traverses through the fields, returning all the objects until it completes all the resolvers. Once completed, your graph is fully populated with the data you requested from the backend.

> **Note – v10.0.1.5**
>
> Subscriptions are GraphQL read operations that can update their results whenever a particular server-side event occurs. Results are pushed from the GraphQL server to subscribing clients. In earlier versions of API Connect, GraphQL subscriptions were not supported but now, in v10.0.1.5, GraphQL subscriptions are supported in the payload and query requests. See what's new in the latest release to learn more about GraphQL enhancements: `https://www.ibm.com/docs/en/api-connect/10.0.x?topic=overview-whats-new-in-latest-release-version-10030#overview_whatsnew_revised__section_apiconsumers`.

Now that you have developed an understanding of what is being defined and implemented on the GraphQL server, it's time to build a GraphQL API in API Connect. But first, you will need to have a working GraphQL server.

Installing a GraphQL Express server

To try out the GraphQL API proxy in API Connect, you will need a running instance of a GraphQL server. One has been provided for in this book's GitHub repository:

`https://github.com/PacktPublishing/Digital-Transformation-and-Modernization-with-IBM-API-Connect/tree/main/Chapter09`.

The `tar` file named `chap09.tar` contains an Express Node.js server that contains GraphQL. It can be deployed on Mac and Linux.

Perform the following steps:

1. Untar `chap09.tar` to find a `graphql-express-demo.zip` file.
2. Unzip `graphql-express-demo.zip`.
3. In a Terminal/command-line window, change to the `graphql-express-demo` directory.

   Install Node.js on your Linux or Windows environment:

   ```
   $sudo install node.js
   ```

   You can also visit the following website for installation instructions:

   `https://docs.npmjs.com/downloading-and-installing-node-js-and-npm`.

4. Install Express on your Linux environment, as follows:

    ```
    $sudo dnf install express
    ```

5. The installation will be looking for an environment variable set, which specifies which set of code to execute. In this example, please set an environment variable for NODE_ENV=production:

    ```
    $sudo export NODE_ENV=production
    ```

    You can also refer to the following website for more details on how to install Express.js:

    https://expressjs.com/en/starter/installing.html.

6. Update your /etc/hosts file with your IP graphql.apicisoa.com.

7. If you haven't allowed port 443 in your firewall, you should run the following firewall-cmd commands:

    ```
    firewall-cmd --permanent --add-port=443/tcp
    --zone=external
    firewall-cmd --reload
    ```

8. Start your Express server:

    ```
    node index.js
    🚀 Server ready at https://graphql.apicisoa.com:443/
    graphql
    ```

If everything works properly, you should be able to begin using the GraphQL server.

Next, you will walk through how to create a GraphQL API with API Connect. Along the way, you will be introduced to some features that allow you to run GraphQL APIs successfully while avoiding pitfalls.

In the next section, you will be implementing GraphQL in API Connect using an FHIR Patient resource example and learning about the GraphQL capabilities within API Connect.

Creating a GraphQL API

So far, you have created various types of API proxies using API Connect, so creating a new API should be easy. When you create a GraphQL API, there are a few different checkpoints that you will become familiar with. In this section, you will be working with a GraphQL server that is providing FHIR Patient resource data. A few of the things you will be learning about are as follows:

- How nested calls are triggered by a single API

- How throttling prevents usage spikes

- How to estimate the cost of running a complex query

- Where to establish rate limits since GraphQL APIs are not like traditional APIs

The following diagram shows how the client interacts with API Connect's GraphQL implementation and with the GraphQL server:

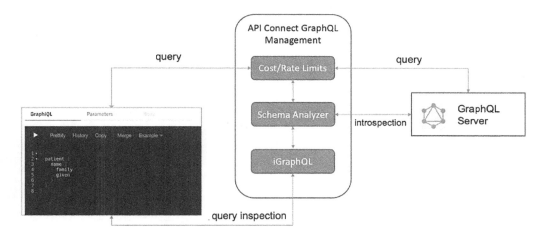

Figure 9.2 – High-level GraphQL flow

The preceding diagram also shows you how a query is passed in and goes through a series of checks before API Connect forwards the query to the **GraphQL server**. In addition, you can see how the **GraphQL user interface** allows you to inspect how the data will be returned based on the schema provided from the **GraphQL server**.

With that in mind, it's time to build your GraphQL API.

Adding the GraphQL proxy

As with previous chapters, you can utilize the API Manager or **Local Test Environment** (**LTE**) to create your GraphQL API. The following steps will walk you through how to create a GraphQL proxy using API Connect:

1. Bring up **Designer** or **API Management Drafts** and click **Add** to create a new API. This will bring up a new page with various options to choose from. This is shown in the following screenshot:

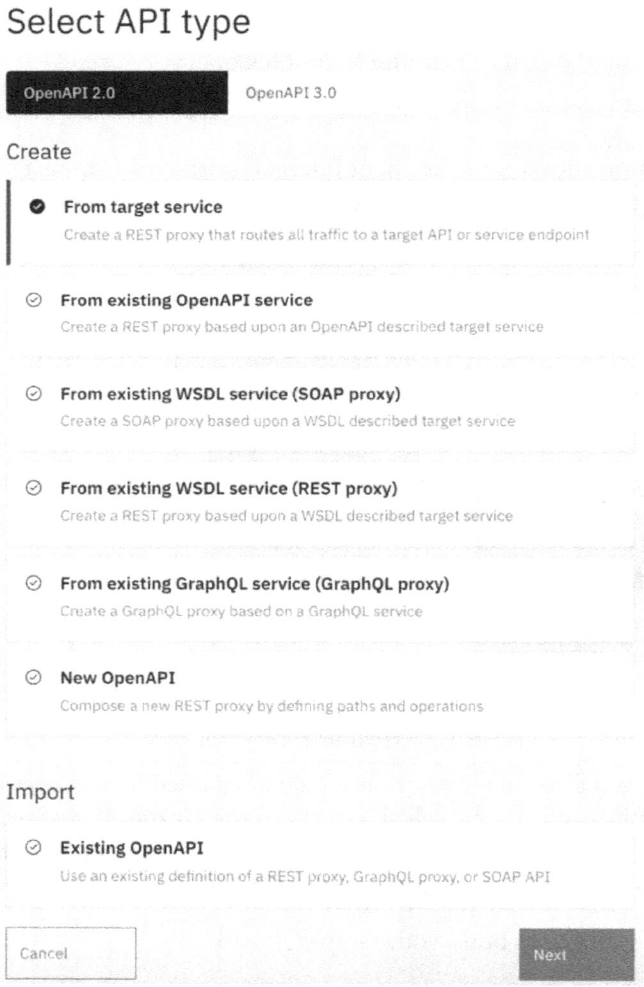

Select API type

| OpenAPI 2.0 | OpenAPI 3.0 |

Create

- ● **From target service**
  Create a REST proxy that routes all traffic to a target API or service endpoint

- ⊘ **From existing OpenAPI service**
  Create a REST proxy based upon an OpenAPI described target service

- ⊘ **From existing WSDL service (SOAP proxy)**
  Create a SOAP proxy based upon a WSDL described target service

- ⊘ **From existing WSDL service (REST proxy)**
  Create a REST proxy based upon a WSDL described target service

- ⊘ **From existing GraphQL service (GraphQL proxy)**
  Create a GraphQL proxy based on a GraphQL service

- ⊘ **New OpenAPI**
  Compose a new REST proxy by defining paths and operations

Import

- ⊘ **Existing OpenAPI**
  Use an existing definition of a REST proxy, GraphQL proxy, or SOAP API

| Cancel | | Next |

Figure 9.3 – Choosing an API type

As you can see, there are various types of APIs you can create.

2. Choose **From existing GraphQL service (GraphQL proxy)**:

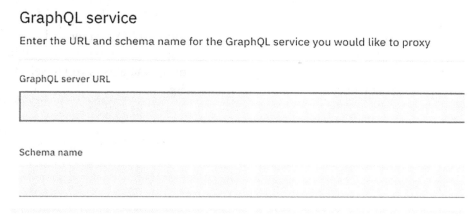

⊘ **From existing GraphQL service (GraphQL proxy)**

Create a GraphQL proxy based on a GraphQL service

Figure 9.4 – Creating a GraphQL proxy

This begins the wizard for creating the GraphQL proxy. Click **Next**.

3. On the next screen, provide a **Title** for your API. At the bottom of the screen, provide a **GraphQL service URL** and **Schema name**:

GraphQL service

Enter the URL and schema name for the GraphQL service you would like to proxy

GraphQL server URL

Schema name

Figure 9.5 – Specifying the location of the GraphQL server

4. You must put in a fully qualified hostname with a path of /graphql.

You must specify that using SSL. You can also enter the schema name that will be saved in the **Definition** section of your **OpenAPI** document. You should enter fhir-patient as the schema name. If you leave the field blank, API Connect will generate a name for you. When you click **Next**, if API Connect cannot reach the server to pull the GraphQL schema, you will get the error shown in the following screenshot:

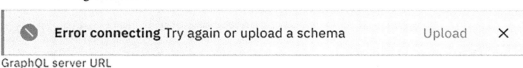

🚫 **Error connecting** Try again or upload a schema   Upload   ✕

GraphQL server URL

Figure 9.6 – Connection error when fetching the schema

A successful connection will look as follows:

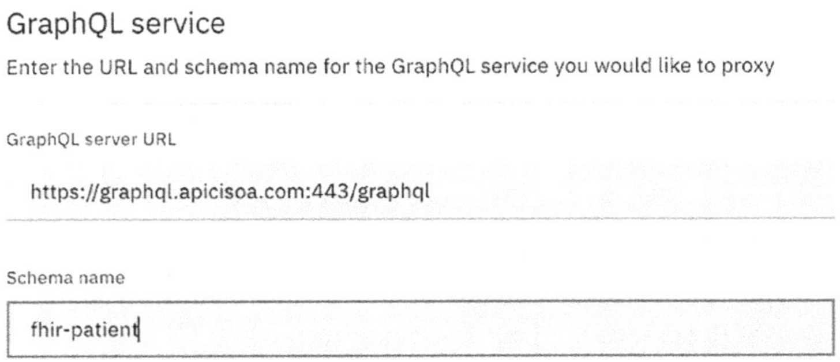

GraphQL service

Enter the URL and schema name for the GraphQL service you would like to proxy

GraphQL server URL

https://graphql.apicisoa.com:443/graphql

Schema name

fhir-patient|

Figure 9.7 – Successfully connecting to a GraphQL schema

This means that API Connect was able to contact the GraphQL server and successfully locate and load the defined schema. After you click **Next**, you will be presented with another screen that highlights warnings and allows you to select some options that will come in handy, as shown here:

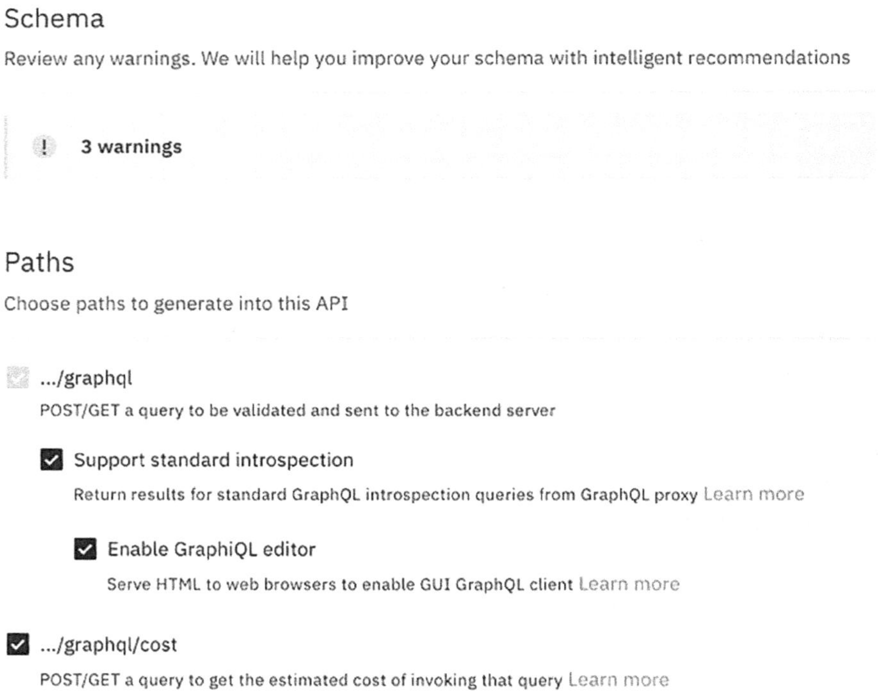

Schema

Review any warnings. We will help you improve your schema with intelligent recommendations

(!)   **3 warnings**

Paths

Choose paths to generate into this API

☑ .../graphql

POST/GET a query to be validated and sent to the backend server

☑ Support standard introspection

Return results for standard GraphQL introspection queries from GraphQL proxy Learn more

☑ Enable GraphiQL editor

Serve HTML to web browsers to enable GUI GraphQL client Learn more

☑ .../graphql/cost

POST/GET a query to get the estimated cost of invoking that query Learn more

Figure 9.8 – Checking the warnings and paths

The first thing you will notice is the **three warnings**. You should click on these warnings to see the concerns that API Connect is highlighting. The following screenshot highlights these warnings:

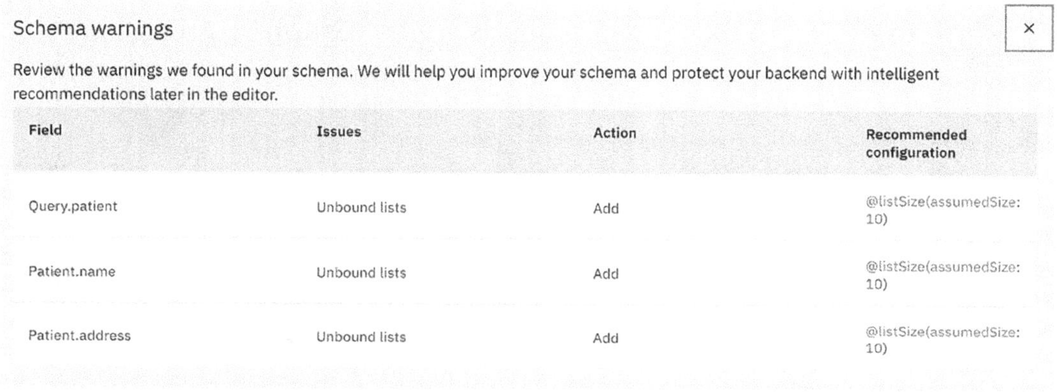

Schema warnings   ☒

Review the warnings we found in your schema. We will help you improve your schema and protect your backend with intelligent recommendations later in the editor.

Field	Issues	Action	Recommended configuration
Query.patient	Unbound lists	Add	@listSize(assumedSize: 10)
Patient.name	Unbound lists	Add	@listSize(assumedSize: 10)
Patient.address	Unbound lists	Add	@listSize(assumedSize: 10)

Figure 9.9 – Schema warnings

As you may recall, each field has a resolver on the backend and the warnings are letting you know that the unbounded lists could return vast amounts of data unless you specify a limitation. Already API Connect is helping you prevent issues by limiting the query. By defaulting to 10, it is keeping the overall size small, but you do have the option to change it.

Returning to *Figure 9.8*, there are four other checkboxes that you should understand. A brief description of each of them is warranted:

- **…/graphql**: This is an automatic setting that lets you know that it will be performing GraphQL-style GETs and POSTs.

- **Support standard introspection**: With this checkbox selected, API Connect will allow the client to update the request data elements. Then, API Connect will validate that it is allowed, based on the schema.

- **Enable GraphiQL Edito…/graphql/costs**: This particular option will allow you to query the cost of making the call. I'm sure you are interested in the cost of fetching your query. This may be an option that you do not want to be made available to consumers in a production environment. It will be up to you to decide.

- **…/graphql/costs**: This is a graphical interactive editor that will allow you to change your query and graphically see the results. It also supports standard introspection.

You can now click **Next** to continue with the wizard.

5. The next page displays some additional choices you need to make about security and whether you want to immediately publish your API to the Sandbox catalog. The following screenshot shows you those options:

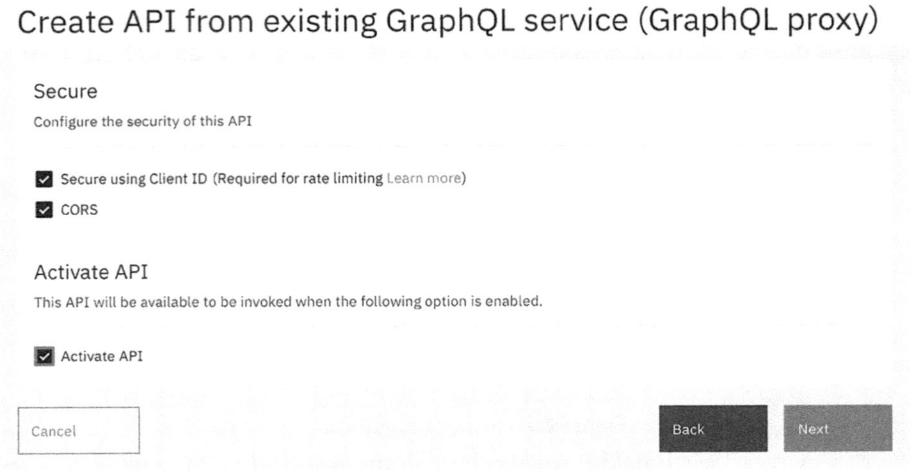

Create API from existing GraphQL service (GraphQL proxy)

Secure
Configure the security of this API

☑ Secure using Client ID (Required for rate limiting Learn more)
☑ CORS

Activate API
This API will be available to be invoked when the following option is enabled.

☑ Activate API

| Cancel | | Back | Next |

Figure 9.10 – Setting security options for GraphQL

Due to *Chapter 7*, *Securing APIs*, you should already be familiar with the fact that enabling CORS checking is valuable but the option to secure using a Client ID has some additional meaning.

In REST, the number of transactions is limited by setting rating limits in the Plan. These are based on requests. With GraphQL, you have different factors that can be applied with regard to rating limits. GraphQL supports a type called **Subscriptions**. With **Subscriptions**, the GraphQL server will periodically send you information about updates to the schema. As the client, you may want to know about changes, but how many times do you want to check? Depending on your Plan, you may be exceeding your rating limits if you are receiving subscriptions consistently. The other consideration that can apply to rate limiting is the cost of the query. Since this calculation can be determined within API Connect and can be added to the Plan, this method of setting the appropriate cost limitations will require some thought. To learn more about this subject, you should review *Securing a GraphQL API using a client ID* at https://www.ibm.com/docs/en/api-connect/10.0.x?topic=api-securing-graphql-by-using-client-id.

Two final notes about *Figure 9.10*. First, you have the option to immediately publish your API with the **Activate API** checkbox. The other thing to note is all of the rating criteria that we discussed earlier are implemented within the Gateway policies. If you uncheck items and want to re-enable them, you need to manually go into the **Gateway** tab and add those policies to the appropriate areas.

Click **Next**.

6. Assuming you select **Activate API**, you will be presented with the **Summary** page, as follows:

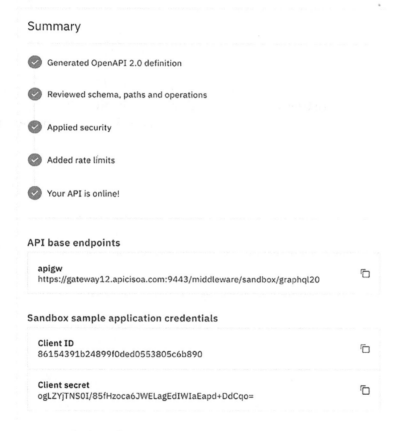

Figure 9.11 – Successfully creating a GraphQL API

The **Summary** page will show you the status of what was applied and some relevant information, such as **API base endpoints** and credentials. Note that your values will be different from those shown in the preceding screenshot.

7. After reviewing the **Summary** page, you should click **Edit API** to review what you just created. Navigate to the **Gateway** tab and review what was generated for you under **Policies**:

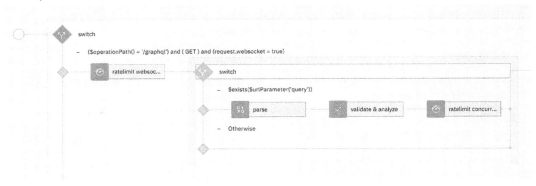

Figure 9.12 – Gateway policies for your GraphQL API

The policies that have been applied to the GraphQL API are rather large. You should scroll to the right to see more policies, as well as scroll down to see more of the switch cases:

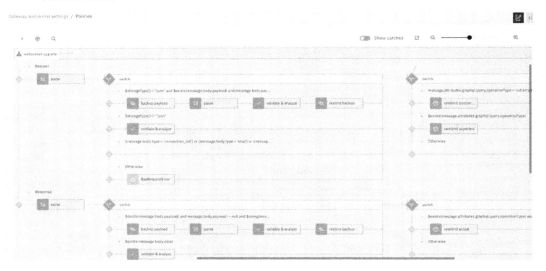

Figure 9.13 – Scrolling right to see the GraphQL policies showing ratelimit

As you will see, the switch handles the various operations and within those operations, you will also see the `ratelimit design` policies. By taking all of the defaults, you now have a GraphQL API you can begin working with.

One concern, given that GraphQL is slightly different than typical REST transactions, is how to measure the rating plans. Before you learn about rating plans, you need to address the warnings you saw. Those warnings have an impact on the rating plans and you will learn about addressing them next.

Addressing warnings in GraphQL

API Connect provides a warning based on your GraphQL schema (SDL) when it has difficulty determining how certain fields will be handled. You saw those warnings when you created your first GraphQL API. Here, you'll learn how to resolve those warnings so that your API executes properly and data is delivered as you expect.

Addressing the warnings

When you initially created your GraphQL API, you noticed there were three warnings. These warnings were there because you had not limited the number of elements that would populate the lists returned by GraphQL. You'll do that now.

1. Edit your GraphQL API and click on the **GraphQL schema** tab. *Figure 9.14* shows the details.

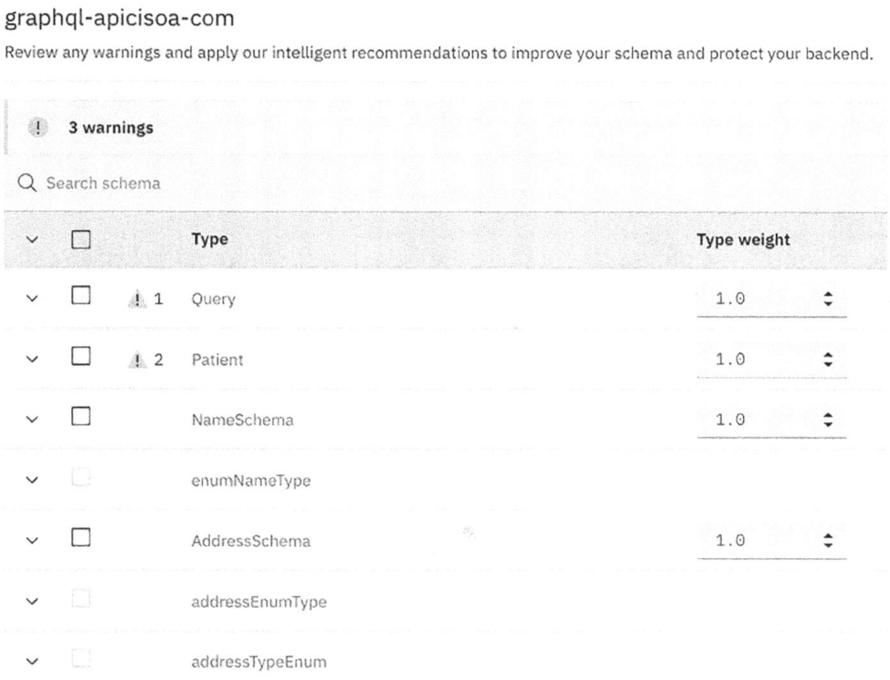

graphql-apicisoa-com

Review any warnings and apply our intelligent recommendations to improve your schema and protect your backend.

⚠ 3 warnings		
🔍 Search schema		

	Type	Type weight
⚠ 1 Query		1.0 ⇅
⚠ 2 Patient		1.0 ⇅
NameSchema		1.0 ⇅
enumNameType		
AddressSchema		1.0 ⇅
addressEnumType		
addressTypeEnum		

Figure 9.14 – Warnings and schema retrieved

Notice the warnings and the list of schema types found. Each type is associated with a weight. The effect of the weight will play a part in the cost of processing your GraphQL queries. You'll learn more about weights in the *Setting weights, cost, and rating limits* section, later in this chapter.

If you click on the yield warning icons, a popup will show you the point of concern:

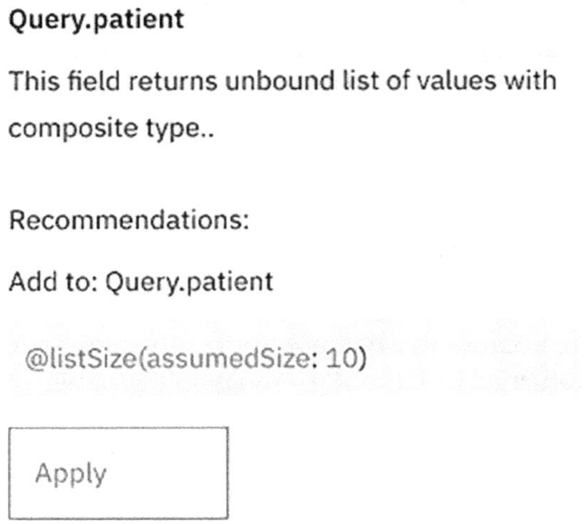

Query.patient

This field returns unbound list of values with composite type..

Recommendations:

Add to: Query.patient

@listSize(assumedSize: 10)

Apply

Figure 9.15 – Warning of unbounded size

This warning specifies that you have an unbounded list of values that has a recommended size. To clear the warning, you can update the size or take the recommendation. Assuming you like the recommendation, click the **Apply** button and the warning will be removed.

Since all of the warnings refer to the same issue but for different fields, you should once again click on the warning and choose **Apply to All**. It will show you the fields as a secondary verification. Go ahead and click **Apply**.

In our scenario, we accepted the warning because we were limiting the size. There are also considerations for **Slicing arguments** from GraphQL. In those cases, you might be specifying exactly where within the list you would like to access the data. Refer to the following screenshot:

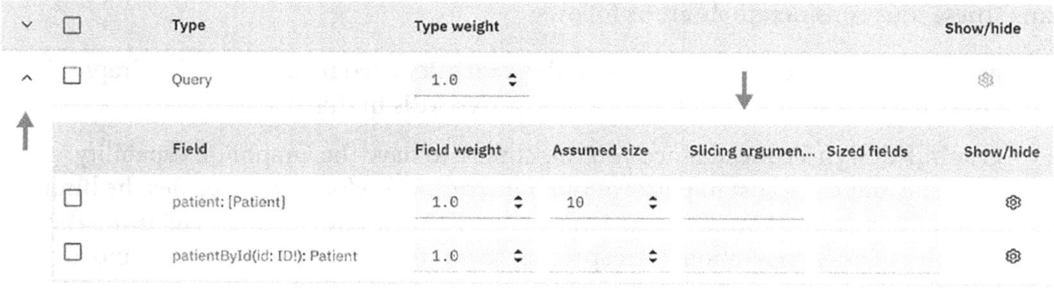

		Type	Type weight			Show/hide
˅	☐					
˄	☐	Query	1.0 ↕		↓	⚙

		Field	Field weight	Assumed size	Slicing argumen...	Sized fields	Show/hide
	☐	patient: [Patient]	1.0 ↕	10 ↕			⚙
	☐	patientById(id: ID!): Patient	1.0 ↕	↕			⚙

Figure 9.16 – Slicing arguments

Now that you have cleared up the warnings, we are almost ready to execute some tests. One consideration you will need to prepare for is how much data might be returned with your query. You will learn about this in the next section.

Setting weights, costs, and rate limits

Recall that when your GraphQL schema was established, every field had a resolver. The resolver would fetch the type and populate the field. The resolver could call a database query, a Kafka event, or maybe even another API. How these fetch actions perform should be a concern of yours. For instance, what if the backend resolver requires an extensive query to a database and performance lags? Also, since your graph may have many fields, you could be calling multiple databases and possibly yours or third-party APIs. These are all good reasons why you need to consider how you establish rate limits, as well as determine the cost of your GraphQL queries. Everything has a cost. Luckily for you, the API Connect gateway parses the GraphQL query and determines a cost based on the weights you establish. Let's check these out in the following sections.

Considerations on performance

Understanding what is happening on the backend GraphQL server is critical when you are looking at how your API will perform and how to establish Plans to offer the consumers.

There are four cost factors to consider. These factors are then established in your Product Plans. These four considerations are as follows:

- **graphql-field-cost**: Applies a limit to the total calculated field cost of the GraphQL query for the calls made to the GraphQL proxy APIs in this Plan.

- **graphql-design-request**: Since you can choose to have the GraphiQL capability available, there is a cost to perform those queries. Therefore, this specifies the limit of how many times you can utilize that feature. This would include a call using the `graphql/cost` operation, introspection requests using just `/graphql`, and any use of the web browser GraphiQL calls.

- **graphql-input-type-cost**: With GraphQL, you can pass mutations (updates) to the GraphQL server, and these also come with a cost. `graphql-input-type-cost` must be set within a rating limit to limit the number of calls of this type.

> Note – v10.0.1.5
>
> `graphql-input-type-cost` from previous versions has been deprecated, so utilization of this cost is avoided. If you previously used this cost factor, you should remove any references to it.

- **graphql-type-cost**: You might like to think of this cost as the *cost of total bytes transferred* in the response to your query. This impacts the provider and consumer because a large payload impacts latency. If you want to set a limit on the total amount of bytes in the response, this is the factor to consider.

To gain more insight on how to set these values, you can use the functionality of API Connect to run some queries with the GraphiQL pane within the API Test facility. You should enter the following query to run a test:

```
query RandomQuery {
  patient {
    resourceType
    active
  }
}
```

As shown in the following screenshot, when a query is run, you can click on the queries listed in the **History** section and see the cost of the query:

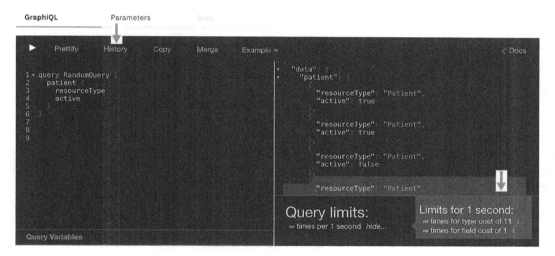

Figure 9.17 – Obtaining the query's cost

These values provide you with valuable input regarding how you will set up rate limits within your Plan. A **type cost of 11** is the cost estimate for running the query and the returned payload. The **field cost** is the cost of including fields within the query. There is a difference between a field that is reading from cache versus the same field executing a third-party service that adds a path length.

Refining the weights

Now that you can observe the costs after a query, you should refine the weights of the types and fields to represent your knowledge of the GraphQL resolvers. This can be accomplished by bringing up the GraphQL Schema editor and making the necessary adjustments. Click on the **GraphQL Schema** tab:

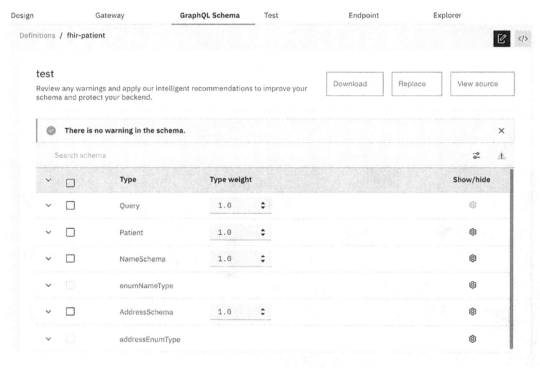

Figure 9.18 – Changing the weights of types and fields

Within the GraphQL Schema, you'll note that the weights have a default value of **1.0**. Here is where you can make adjustments to signify more difficult resolver costs based on types. If you know of a field that calls a resolver that has a more complicated implementation, you can set weights on that field.

To demonstrate this, navigate back to the **GraphQL schema** tab and locate the **Patient** schema. You can assume that the **resourceType** resolver is calling an external web service that fetches **Patient** data from a service. The service is a bit slow, so you want to reflect that within the weight. Change the weight of **resourceType** to 5 . 0. Refer to the following screenshot:

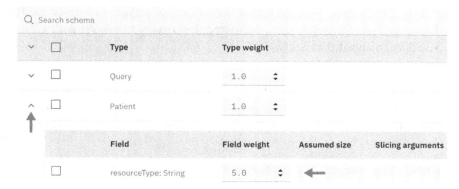

Figure 9.19 – Modifying the weight of a field

Once changed, you can save your API and immediately test it with GraphiQL. Click on the **Test** tab and run your query again using the same values:

```
query RandomQuery {
  patient {
    resourceType
    active
  }
}
```

After it runs, you may have to click on **History** and **Random Query** to see the costs. Your costs should change.

That is how you refine the costing:

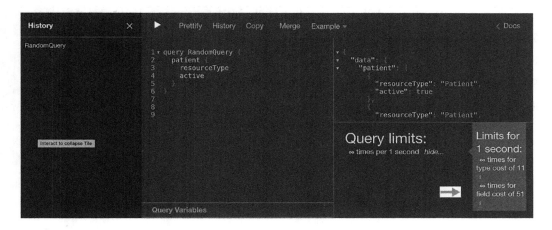

Figure 9.20 – Cost limits changed to 51 for field cost

By understanding these values, you can adjust the weights of certain objects or fields to determine the actual cost. Once you have estimated the costs, you can set your GraphQL rate limits. You'll learn about that next.

Setting rating limits

Previously, you determined the cost of running your query. As the provider of the GraphQL API, you can now create a new Product and a Plan based on cost and GraphQL rating limits.

You are now ready to create the Product for your GraphQL API and specify new Plans based on cost and rating limits. Let's get started:

1. Click on the **Develop** link in the top-left corner and get ready to **Add** a new Product:

Figure 9.21 – Using the Develop link to return to the main menu

2. Click **Add** and select **Product**.

3. Provide the new **Product** with a name, choose your GraphQL API, and click **Next**.

4. Choose your API and click **Next**.

5. Click **Add** to add a new **Plan**. A default Plan name will be generated. Provide a new name for the Plan and continue until you're finished.

6. Choose the **Publish, Visibility,** and **Subscribability** options and click **Next**.

7. On the **Summary** page, click **Done**.

8. Open your new Product and click on **Plans** to choose your newly created **Plan**.

9. Click on the three dots next to your Plan and select **Edit**.

10. Scroll down until you find **GraphQL rate limits**. This is where you will choose the cost amount by unit by using various cost factors:

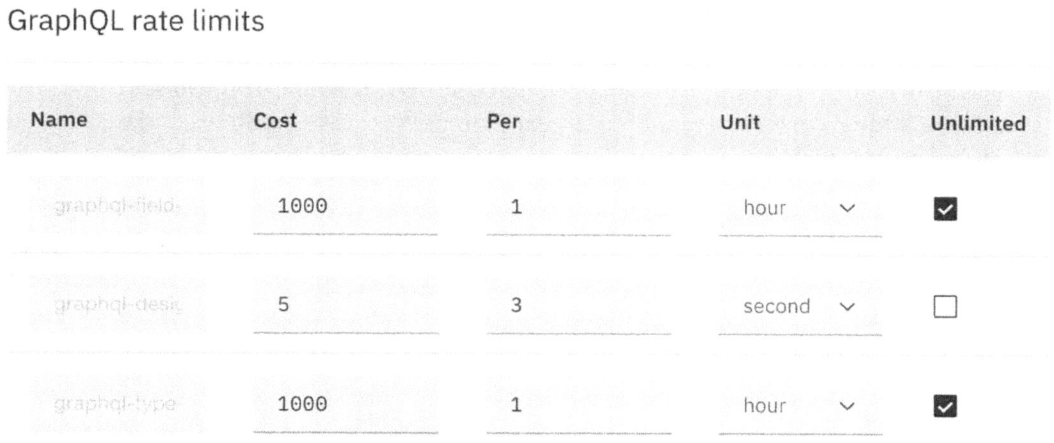

GraphQL rate limits

Name	Cost	Per	Unit	Unlimited
graphql-field	1000	1	hour ⌄	☑
graphql-desc	5	3	second ⌄	☐
graphql-type	1000	1	hour ⌄	☑

Figure 9.21 – Setting rate limits for your GraphQL API

This panel shows the default values. Unless you want unlimited requests, you should uncheck **Unlimited** and apply the costs per unit. You can adjust the values as appropriate based on the Query costs you observed previously.

Another way to reduce costs and reduce security risks is to remove fields from **GraphQL schema**. Next, you will learn how simple it is to do that.

Removing fields from GraphQL

Since you know that the GraphQL fields are developed on backend systems by developers, you might find out that some fields that are presented may not be allowed due to compliance and security reasons. An example would be social security numbers or certain patient history information. This poses a challenge for API developers because they know they need to obfuscate the fields in some way. API Connect provides you with a simple way to ensure that access to fields aren't allowed to the consumers. You can do this by removing the fields in **GraphQL Schema**.

To see how this can be accomplished, perform the following steps:

1. Navigate to any GraphQL API and click on **GraphQL Schema**. You will see a **Show/hide** column where there is a gear icon next to the field. Refer to the following screenshot:

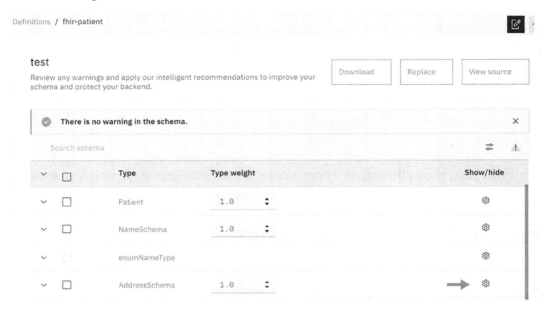

Figure 9.23 – Removing a field using Show/hide

2. By clicking on the gear icon, you can remove the field from the fields presented to the API developer:

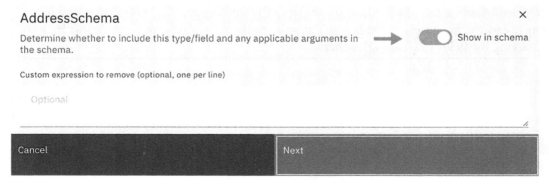

Figure 9.24 – The Show/hide popup

You can toggle this on or off to show the field/object in the schema. When you choose to hide, the field/object will be grayed out, as shown in the following screenshot:

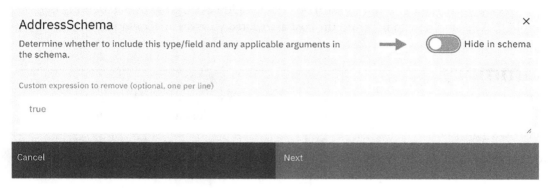

Figure 9.25 – Hiding the field/object in GraphQL

3. After you click **Next**, you are allowed to change your mind as you are presented with a **Summary** panel that shows what you have changed. You can cancel or go back to make modifications or you can click **Done** to complete the removal process.

    The results of a removal show a grayed-out object. In the following screenshot, you can see that **AddressSchema** was hidden:

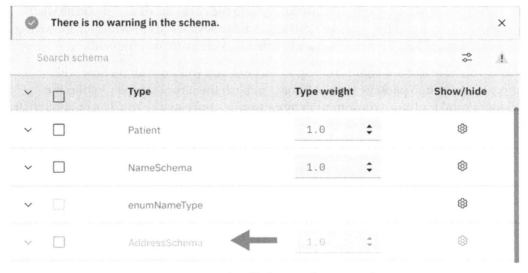

Figure 9.26 – Results of hiding an object in a schema

Of course, you can reverse your decision by clicking on the gear icon again and turning **Show** back on. This is a very simple and easy way to show and hide objects/fields within your GraphQL schema.

That was a considerable amount to learn but you should have a good understanding of GraphQL APIs and how to best find the cost associated with them. Congratulations!

Summary

In this chapter, you were introduced to GraphQL and how to look at it holistically. You learned how the graph is created and what comprises the GraphQL server from a provider viewpoint.

You learned about queries, resolvers, and mutations, as well as how each type field can have an impact on costs and performance.

Developing a GraphQL API follows similar practices to other REST APIs but there were a few wrinkles in how you look at developing a Product plan due to cost factors. You learned that those cost factors provide you with details on how to set up the rating limits in your plan. Adjusting weights at the type level or field level will help determine the cost of your GraphQL API. You also learned how to remove fields from the schema to remove information that you do not wish your consumers to access.

GraphQL is not for every implementation and it certainly doesn't replace your REST APIs, but it does provide benefits in situations where the client needs to determine what information is important for their applications. It also alleviates the providers from having multiple versions of REST services to handle differing consumer requirements.

Now that you have learned about creating APIs, next, you will learn about how API Connect allows you to package those APIs and publish them for discovery within the Developer Portal for future consumption by consumer applications. You'll learn about that in *Chapter 10, Publishing Options*.

10
Publishing Options

At this point, we have covered many different topics around developing your APIs. You should now have a good working knowledge of how to create and secure your APIs in **API Connect** (**APIC**), as well as the different tools, features, and options within the Product itself. Hopefully, you have gained a good understanding of the concepts and implementation of the topics we've covered thus far. Although creating working and secure APIs is critical, it is just as important to think about the entire consumer experience. This includes how your APIs are packaged, published, and discovered, as well as providing any applicable access levels. In this chapter, we will cover how to publish APIs that are easy to discover, well-documented, and mindful of overages.

Specifically, we will cover the following topics:

- Working with Products and Plans
- Creating Rate Limits
- Publishing and configuring Catalogs
- Consumer interaction

By the end of this chapter, you will have a solid understanding of how to take your APIs and Products from a developed state to a published and discoverable state. You will gain a working knowledge of how Plans are configured and how they relate to Products. Finally, you will see how the end consumer will discover and interact with your Products and APIs.

Technical requirements

The examples presented in this chapter will require that you install APIC with the proper access to create and publish APIs, Products, and Plans. For some of the features in this chapter, working knowledge of DataPower is required as they include configuration for the DataPower Gateway itself. The configuration files that will be created within this chapter can be found in this book's GitHub repository: `https://github.com/PacktPublishing/Digital-Transformation-and-Modernization-with-IBM-API-Connect/tree/main/Chapter10`.

Working with Products and Plans

As you may recall from *Chapter 3*, *Setting Up and Getting Organized*, we discussed how to organize APIs, Products, and Catalogs. We intentionally left out Plans in that discussion as they will be covered in detail within this chapter.

As we discussed in *Chapter 3*, *Setting Up and Getting Organized*, your APIs must be grouped within a Product. A Product will be the offering to your end consumers, so they should be logically grouped as one offering. It is the Product that is ultimately published for the consumers to discover.

Each Product must contain at least one Plan. A Plan is configured within a Product to provide different offerings to different consumers for the APIs within it. These offerings could include API access, Rate Limits, and monetization parameters. For example, you could have a Bronze Plan, a Silver Plan, and a Gold Plan that your consumers can be assigned to. Perhaps your lowest Plan level is the Bronze Plan, which provides a very limited of transactions per minute, and any transaction over that limit will be charged to the consumer. Here, your Silver Plan would increase that limit and reduce the charge for the coverage. Finally, your Gold Plan might provide unlimited access to the APIs contained within it.

Although a Product can have multiple Plans, more than one Plan can reference the same API. In addition, a Plan can access APIs within other Products. Let's take a look at how all of this fits together. The following diagram shows the hierarchy of Plans and Products within your API configuration:

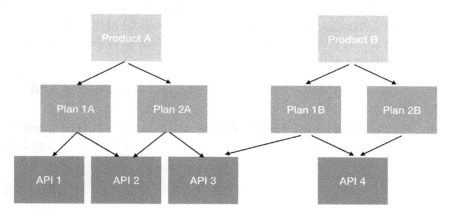

Figure 10.1 – Plan, Product, and API hierarchy

As you can see, each Product contains at least one Plan. This diagram shows two Plans within each Product, though there could be one or more per Product. In this diagram, you can also see that each Plan references at least one API. This is not a one-to-one ratio as multiple Plans can reference a single API. You will also notice that a Plan within one Product can reference an API within another. Let's take a closer look at how to configure Products and Plans.

Configuring Products and Plans

As mentioned earlier in this chapter, a Product is simply a collection of related APIs that make up an offering for a consumer. It is the entire Plan that is published in your Catalog and as you will see in *Chapter 11, API Management and Governance*, organizing your Plans appropriately will play a big part in your versioning scheme. Creating a new Plan is a very simple task. As Plans are created within a specific provider organization, you will start by logging into the API Management Web **user interface** (**UI**) for your provider organization and performing the following steps:

1. From the home screen within the API Manager UI, click on **Develop APIs and Products**.

2. Clicking the **Products** tab on the current screen will display all of the currently available Plans. To add a new Plan, click the **Add** button at the top right and then click **Product**, as shown here:

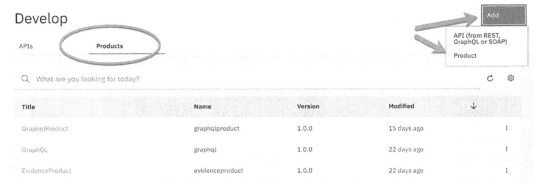

Figure 10.2 – Adding a Product

3. Now, you can choose to create a new Product from scratch or import an existing Product definition from a configuration file. Since we are creating a new Product, we will want to select **New Product**. Since this is the default, this option should already be selected, as shown in the following screenshot. Clicking the **Next** button will bring us to the Product configuration screen:

Select product type

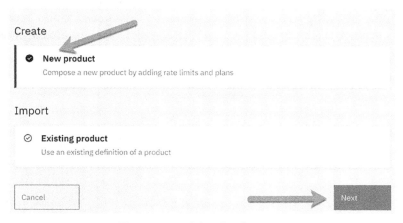

Figure 10.3 – Select Product type

4. You should now see a basic configuration screen for your new Product. From this screen, you must provide a **Title**, **Version**, and an optional **Summary** for your Product, as shown in the following screenshot. You will see that the **Name** field is automatically populated with the *slugified* representation of the **Title** field:

Create new product

Info
Enter details of the product

Title

Member

Name

member

Version

1.0.0

Summary (optional)

Product for APIs pertaining to members

Cancel   Next

Figure 10.4 – Configuring the Plan information

5. Clicking the **Next** button will complete the basic configuration of your new Product and bring you to a screen where you can add existing APIs to your Product, as shown in the following screenshot. From here, you can choose to add existing APIs or simply leave your new Product with no APIs, which can be added later as you create them:

Create new product

APIs

Select APIs to add to this product

	Title	Version	Description
☐	Condition API	4.0.1	A simplified version of the HL7 FHIR API for Condition resources.
☐	Member API	1.0.2	API used to set and get member information

Figure 10.5 – Choosing APIs to include in the new Product

6. Clicking the **Next** button will bring you to the **Plans** screen. Since this is a new Product, it will only contain a **Default Plan**, which is what every new Product gets by default. If the default Plan does not suit your needs for this Product, you can alter it or create one or more new Plans. To create a new Plan, click the **Add** button and a new Plan (**Plan 1**) will be added to the screen. You should provide a **Name**, **Description**, and new **Rate Limit**, as shown in the following screenshot. We will discuss Rate Limits in more detail in the next section since many new Plans can be added as needed in the same manner:

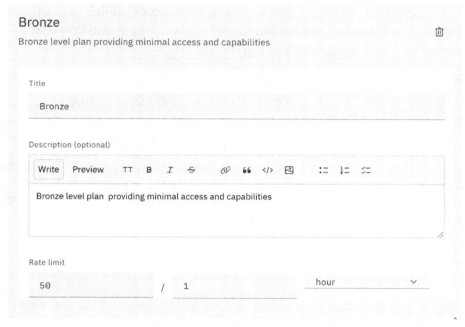

Figure 10.6 – Creating and adding a new Plan to the Product

7. Once you have configured the new Plan parameters, clicking the **Next** button will bring you to the final configuration page, as shown in the following screenshot:

Figure 10.7 – Configuring the Visibility and Subscribability options for Product

This configuration will dictate who can see this Product (**Visibility**) and who can subscribe to this Product (**Subscribability**). Your options for visibility are as follows:

- **Public**: Everyone can see this Product.

- **Authenticated**: Only authenticated users can see this Product.

- **Custom**: This lets you add specific consumer organizations or groups who can see this Product.

Subscribability has the same available options, minus the **Public** option.

8. Once you have configured **Visibility** and **Subscribability**, clicking the **Next** button will bring you to a final **Summary** page, as shown in the following screenshot. From this screen, you can click the **Edit** button to edit your subscription or click the **Done** button to complete your configuration:

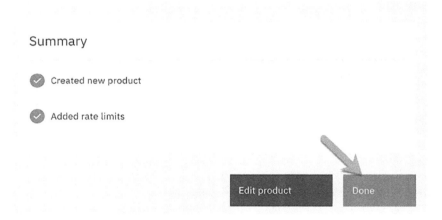

Figure 10.8 – Product creation summary screen

With that, you have created a new Product and Plan that can have new or existing APIs added to it. In the next section, we will delve into more detail on Plan creation parameters and **Rate Limiting** in particular.

Creating Rate Limits

In the previous section, we showed you how to create a Product and Plan. Although we mentioned the Rate Limit parameters, we did not discuss what they are and how we can configure them for different scenarios. In this section, we will discuss what Rate Limits are, how to configure them, and all of the benefits provided by them.

Rate Limiting allows you to define the maximum number of calls permitted within a specified time limit. This limit can be applied across all APIs within a given Product, only to specific APIs within a Product, or even specific operations within specific APIs. There are two different types of Rate Limiting that you can configure. First, there are consumer Rate Limits, which are imposed at the consumer level. These are the Rate Limits that are defined within a Plan. Since each consumer must subscribe to a specific Plan, the Rate Limits that are defined are imposed on each consumer individually.

The second type of Rate Limiting is provider Rate Limits. These are Rate Limits that are defined at the API level to protect the backend provider from being overloaded. These Rate Limits are not consumer-specific and apply to all consumers accessing the API collectively.

Regardless of where these Rate Limits are defined, some general concepts apply across the board.

When you're defining a Rate Limit, there are two areas of Rate Limiting that you are defining:

- **Rate Limit**: Defines a set number of calls to an API over a given period for consumers where the limit applies. This period can be over minutes, hours, days, or weeks. For example, you may want to allow 100 calls within an hour for a specific API. This limit can be defined as a **hard limit** or a **soft limit**. If a hard limit is specified, all the calls exceeding the defined threshold for the given period will be rejected. If a soft limit is specified, any calls over the threshold for the given period will generate a log message and the calls will be allowed.

- **Burst limit**: A threshold that's defined over a small period to protect against sudden bursts of traffic. This limit is meant to protect your infrastructure from large, unexpected bursts of traffic. This limit is always a hard limit where all the requests exceeding this threshold are rejected.

Although the Rate Limiting concept is fairly general and straightforward, you can apply this feature in several different ways within your Products and APIs to provide the granularity that you require. Let's take a look at the different places and ways you can implement this.

Defining Rate Limits in a Plan (consumer Rate Limits)

The first type of Rate Limiting we will look at is consumer Rate Limiting. As consumers subscribe to a Product to gain access to the APIs within it, they will also subscribe to a Plan. It is within this Plan that the Rate Limits are defined. You can think of this as a way to provide different levels of access to your APIs to different consumers. Perhaps you want to provide three different levels of access to your APIs. You could have a Bronze Plan where you provide limited access to your APIs, allowing very few calls in a given period. This could be for introducing your consumers to the API for a trial. Then, you might have your Silver Plan, where you raise the limit of calls allowed. This could be for your average customer, where you still need to impose some limits. Finally, you could define your Gold Plan, which would be dedicated to your VIP customers and where access to your APIs is unlimited.

As you can see, defining your Rate Limits at the Plan level is very consumer-centric. Careful thought and consideration should be given to how you would like to create these Plans and limits for different levels of consumers.

In the previous section, we discussed how to create a new Plan while creating a new Product. We briefly touched on Rate Limits when we configured the Plan as it is part of the Plan's configuration. Let's take a deeper dive into this specific configuration within your Plan.

Defining Rate Limits for all the APIs within a Product

To begin, you must navigate to an existing Plan within the API Manager Web GUI. To do this, you must navigate from the API Manager home screen to the Product that contains the Plan by going to **Develop APIs and Products | Products tab | <Product name>**. Once you have navigated to the Product that contains the Plan you would like to edit, from the left navigation menu, click **Plans**. From this screen, you will see all of the Plans currently contained within this Product. The following screenshot shows our **Member** Product, where you can see that it contains the **Default** Plan, as well as a **Bronze** Plan:

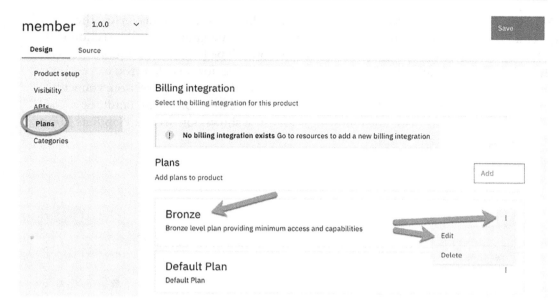

Figure 10.9 – Existing Plans within a Product

To edit your existing Plan, simply click the ellipses to the right of the Plan and select **Edit**, as shown in the preceding screenshot.

> **Note**
>
> There are additional Rate Limits that the preceding screenshot doesn't show. You can find GraphQL Rate Limits, Assembly Rate Limits, Assembly count limits, and Plan APIs by scrolling down. Except for the GraphQL Rate Limit, you will learn about these later in this chapter. For GraphQL Rate Limits, see *Chapter 9, Building a GraphQL API*.

You should now be on a screen where you can edit the existing Plan. Just below the general Plan information, you will see two sections for Rate Limits. The first section will be for the Plan Rate Limits, while the second section will be for Plan burst limits. This is where you can configure your Rate Limits and burst limits for this Plan. You can add as many Rate Limits as you would like here. At the top of the Rate Limit section, you will notice a radio button to indicate whether you would like to enforce Rate Limits or allow unlimited calls for this Plan. The following screenshot shows our **Bronze** Plan Rate Limits, where we allow 100 calls within 1 hour:

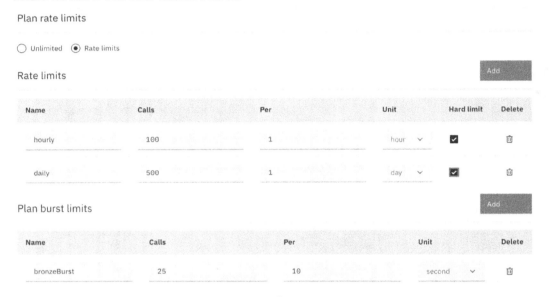

Figure 10.10 – Plan Rate Limits

We also added a second Rate Limit that will only allow 500 calls within 1 day. So, even though we are allowing 100 calls in 1 hour, once the consumer reaches 500 calls within 1 day, they will be rejected. Notice that we have the **Hard limit** option checked for both limits. It is because of this that any transactions exceeding the defined threshold will be rejected. Otherwise, they would only generate a log message.

Below the Plan's **Rate Limit**, we have configured a burst limit that will allow **25** calls within **10** seconds. This limit will always reject client calls that exceed the configured threshold and will be enforced, regardless of whether the client has exceeded the Rate Limit defined for it.

Once you have completed your Rate Limit configuration, you must click the **Save** button at the top right of the screen to save the Product configuration.

Defining Rate Limits for select APIs within a Product

We have just demonstrated how you can configure Rate Limits within a Plan that will apply to every API within the Plan. You can also apply this Plan to specific APIs. This option is at the bottom of the Plan configuration, under the **Plan APIs** section. By selecting the **Customize the Plan API list** radio button, you can choose which APIs within the Product that the current Plan will apply. The following screenshot shows our **Bronze** Plan, which contains **Member API** within the API list once the **Customize the Plan API list** radio button is selected:

Figure 10.11 – Applying a Plan to select APIs

Clicking the ellipses to the far right of an API will allow you to remove it from the Plan list or even select the operations within the API you would like to include in the Plan.

Now that you have seen how to include or exclude Rate Limits in a Plan, let's take a look at how we can override Plan Rate Limits for specific operations.

Defining Rate Limits for specific API operations in a Product

As we progress through our discussion on defining Rate Limits, we will continue to get more and more granular with our Rate Limit configuration. The most granular method within a Plan is to apply Rate Limits to specific operations defined within an API. From within the Plan configuration, you can select an API and then an operation from within that API to override all the other Rate Limit configurations within the Plan. Unlike applying the Rate Limits to only certain APIs within your Plan, this configuration will only override any other Rate Limit configuration you might have. This means that any Rate Limit configuration within the Plan will apply to all other APIs and operations. This override configuration will only apply to the specified API operations.

To configure a Rate Limit override for a specific operation, you must scroll to the bottom of the Plan configuration page, where you will find a check box labeled **Override Plan Rate Limits for individual operations**. Checking this box and clicking the **Add** button, as shown in the following screenshot, will bring you to a screen where you can select a specific API operation and configure the Rate Limit to apply to it:

Figure 10.12 – Selecting Override plan rate limits for individual operation

Now, you can select an API, and then an operation within that API, to apply a new Rate Limit to override any existing Rate Limits. The following screenshot shows our Plan overriding the **addPatient** operation within the **member-api1.0.2** API with a new Rate Limit of 25 calls per 1-hour time period. You will configure this new Rate Limit in the same manner that you configured the Rate Limit for the Plan:

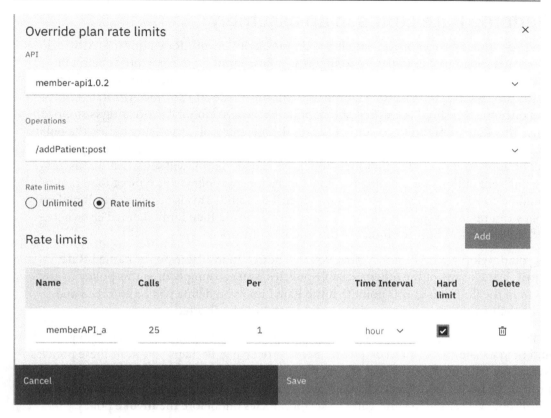

Figure 10.13 – Configuring the Rate Limit for a specific API operation

Once your new Rate Limit override is complete, click the **Save** button to save it.

With that, you have seen the power and flexibility of configuring Rate Limits within your Plans. You can configure many different Rate Limits per Plan at the Plan level and/or the operation level. You can apply your Plan to certain APIs so that the Rate Limits only apply to them. Given this flexibility, you should be able to accommodate almost any situation to apply Rate Limits at a broad or granular level.

Defining Rate Limits in an assembly

As if you didn't have enough flexibility and granularity in your Rate Limiting, APIC allows you to get even more granular by configuring a Rate Limit on the assembly of an API itself. When you configure a Rate Limit on the assembly, you are marking a specific point in the process to apply this Rate Limit. This will supersede all Plan and API Rate Limits and once the threshold is reached at that point in the assembly, all processing is stopped. Since this Rate Limit policy is configured on the assembly, all processing before the policy will be executed. However, any processing that's configured after the Rate Limit policy will be aborted if the Rate Limit threshold is exceeded. As you will see when we discuss the different options for defining the Rate Limits for this policy, this type of Rate Limiting could be considered a consumer or a provider Rate Limit. This is because your Rate Limit policy can refer to a specific Plan or a consumer-agnostic Rate Limit defined elsewhere. Let's take a look at how to configure this and how it works.

As you develop your API policy flow, you will notice an available policy named **Rate Limit**. Just like any of the other available policies, you can simply drag this policy to any point in the flow. It is at this point that the Rate Limit you define will be enforced and if the defined threshold is exceeded, its execution will be aborted.

Let's look at an example where you have some processing logging in your assembly and then a GatewayScript to do some custom processing. Perhaps you want these policies to execute with no Rate Limit at all, but you need to rate-limit calls to the backend to protect the backend server from being overwhelmed. In this case, you can add your **Rate Limit** policy after the **log** and **gatewayscript** policies but before the **invoke** policy, as shown here:

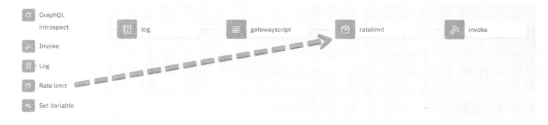

Figure 10.14 – Adding a Rate Limit policy

As you can see, we added this Rate Limit policy just before the invoke policy so that if our defined Rate Limit is exceeded, our policies before the Rate Limit policy will execute, but the transaction will be aborted before the invoke policy.

Now that you have decided where you would like to enforce the Rate Limit, you will need to configure the Rate Limit's threshold and its parameters. Unlike the previous Rate Limit configurations we have discussed, when you are configuring a Rate Limit policy, you will simply refer to a configuration that specifies these parameters. This is specified in the Rate Limit policy configuration, under the **Source** dropdown. Let's take a look at the different options for referring to a defined Rate Limit configuration within your policy:

- **Plan default**: The rate and burst limits that are defined in the Plan default that the calling client or application is subscribed to.

- **Plan by name**: Apply the rate and burst limits as defined in a previously configured Plan.

- **Gateway by Name**: Refers to an object defined on the API gateway of the **apigw** type called **apiconnect**.

- **Catalog by Name**: Refers to the rate and burst limits defined on the API Gateway within the **api-collection** object. This particular object on the API Gateway represents the APIC Catalog.

As you can see, there are several different locations where you can have rate and burst limits defined that you can refer to from your Rate Limit policy. Each requires configuration and can be defined in very different manners. Let's take a deeper dive into how we can configure these limits and then refer to them from our Rate Limit policy. The first option that you can select, **Plan default**, does not require any further discussion as it simply refers to the default Plan Rate Limit for the consumer, which you already know how to define.

Configuring the Plan by name option

As we have already discussed configuring Plans and Rate Limits within this chapter, you should already be familiar with how that works and how to configure them. If you need your Rate Limit policy to refer to the count and burst limits defined within specific Plans, this is the option you would choose as you can reference these limits within your Plan from your Rate Limit policy. However, these limits will not be the same limits you defined in your Plan previously as there are specific rate and burst limits you must use in the assembly that's defined within your Plan. These are appropriately labeled **Assembly count limits** and **Assembly burst limits** and can be found at the bottom of your Plan's configuration, as shown in the following screenshot:

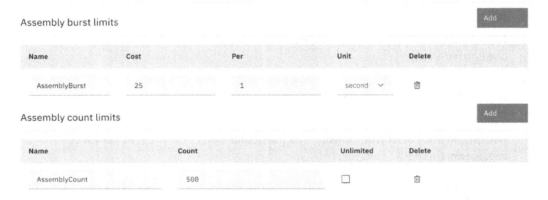

Image 10.15 – Assembly burst limits and Assembly count limits

Here, you can see that we have added a burst limit named **AssemblyBurst** and a count limit named **AssemblyCount**. A count limit is the total number of requests allowed to the API. Once you have defined and saved these limits, you can reference them by name within your Rate Limit policy. The following screenshot shows a Rate Limit policy where we have added a count limit and burst limit referring to the assembly limits we have defined by name:

Figure 10.16 – Rate Limit policy referring to assembly limits

With that, you've learned how to refer to different limits within different Plans by referencing them in your Rate Limit policy. Now, let's learn how to define different types of limits on gateways and reference them from within a Rate Limit policy.

Configuring the Catalog by Name option (provider Rate Limits)

The **Catalog by Name** configuration option within your Rate Limit policy allows you to configure the rate, burst, and count limits that are available to the entire Catalog. Because of this, these Rate Limits are consumer-agnostic and apply to all consumers. This is why we would classify these as provider Rate Limits. The configuration for this will be made on the DataPower Gateway itself and not within APIC. There is a specific object on the DataPower Gateway that represents your APIC Catalog. This object type is called **api-collection** and you will have one to represent each Catalog you have configured. If you have access to your DataPower Web management console and are familiar with navigating around, you can log into your DataPower Gateway and view this object for each Catalog. As shown in the following screenshot, each Catalog object will contain information about the Catalog, including any rate-limiting parameters:

Figure 10.17 – API Collection

If you do not have access to your DataPower Gateway or simply do not have the background to comfortably navigate that space, don't worry. You won't need to access the DataPower Gateway itself to add your Rate Limit parameters to this object. You will accomplish that by adding a gateway extension to your gateway, which is just the CLI commands to configure this object.

To begin, you must prepare a text file with a `.cfg` extension that will contain the CLI commands to alter the API Collection object for your Catalog. The following is an example of the commands that will create a Rate Limit for a Catalog:

```
top; config;
switch apiconnect;
api-collection middleware_sandbox_collection
  assembly-rate-limit  50per1hour 50 1 hour on off on on off
    off "" "my.vars.max-amount - my.vars.used-amount"
exit
```

In this example, we can see that the first two lines are standard CLI commands, telling your DataPower Gateway where to start and which domain it should be applied to.

The third line specifies the API Collection object name that you will be altering. This is the unique object for your Catalog, whose format is `<provider organization>_<Catalog name>_collection`. This is the line that follows the `api-collection` name that will define your Rate Limit. This should be evident as it begins with `assembly-rate-limit`. The first argument in this line will be the name of your Rate Limit. We called ours `50per1hour` to indicate what the Rate Limit is doing.

Following the Rate Limit name are your three critical parameters. In order, they are for defining your number of allowed requests, intervals, and units of time. As shown in our example, the Rate Limit name indicates we are allowing 50 requests every 1 hour.

For the most part, you can leave the remaining parameters as shown. However, we have included the last parameter to demonstrate its weight capability. This parameter can be a JSONata expression that calculates the weight for each particular transaction. This is useful when you are configuring Rate Limit policies to consume and replenish the balance of the current limit.

You can configure a count limit in the same way using slightly different CLI commands. A count limit is simply the maximum number of allowed requests to the API. The configuration commands for this are as follows:

```
top; config;
switch apiconnect;
api-collection middleware_sandbox_collection
  assembly-count-limit  max100 100 1 on on off off off off ""
    "1" on
exit
```

As you can see, you are setting `assembly-count-limit` to the same `api-collection` object that you set the Rate Limit on previously. The first two parameters in this command are the ones you will be interested in. These are the count limit name and the count limit itself.

For more detail on the CLI commands and parameters, you can visit the IBM documentation center at `https://www.ibm.com/docs/en/datapower-gateways/10.0.1?topic=acc-assembly-rate-limit`.

You can now save your two sets of commands in separate .cfg files and add them both to one .zip file. The next step will be to add these to your gateway as gateway extensions.

You can add your new gateway extension to your gateway from the Cloud UI by navigating to **Configure topology** from the home screen.

You should now see the list of available gateway services. Now, click the ellipses to the far right of the gateway service and then **Configure gateway extensions**, as shown in the following screenshot:

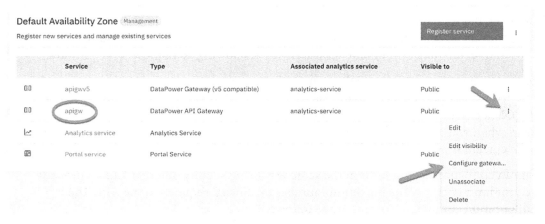

Figure 10.18 – Configure gateway extensions

You should now be viewing the **Configure gateway extension** screen, where you will see a currently configured gateway extension if one has already been configured, or an empty list if one was not previously configured. Note that you can only have one gateway extension configured per gateway service, so if one has already been configured, you must edit it and add to it by modifying your .zip file so that it contains all the possible configurations for your extension. If this is your first time configuring a gateway extension, your list will be empty and you can simply click the **Add** button at the top right, as shown in the following screenshot:

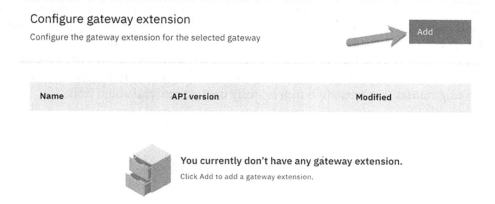

Figure 10.19 – Uploading the gateway extension

From here, you can simply upload or drag the `.zip` file containing all of your `.cfg` files for your rate and count limits, as shown in the following screenshot. Alternatively, you can choose the **Upload** option and navigate to the `.zip` file on your local computer:

Figure 10.20 – Uploading the .zip file

Once your `.zip` file has been uploaded, clicking the **Save** button will complete your gateway extension. If you did not have a gateway extension configured previously, you should now see one in the list. If you did have one configured previously and you updated it to contain your rate and count limit configuration, you should see that the gateway service was just updated in the **Modified** field.

Now that you have defined your Catalog limits within your gateway, you can refer to them within your Rate Limit policy. To do this, select the **Catalog by Name** option within your Rate Limit policy configuration. You will then see some buttons that you can use to add a Rate Limit and also a count limit. Once you've clicked these, you can reference the rate and count limits that you defined in your gateway extension. The following screenshot shows a configuration that refers to our previously defined rate and count limits:

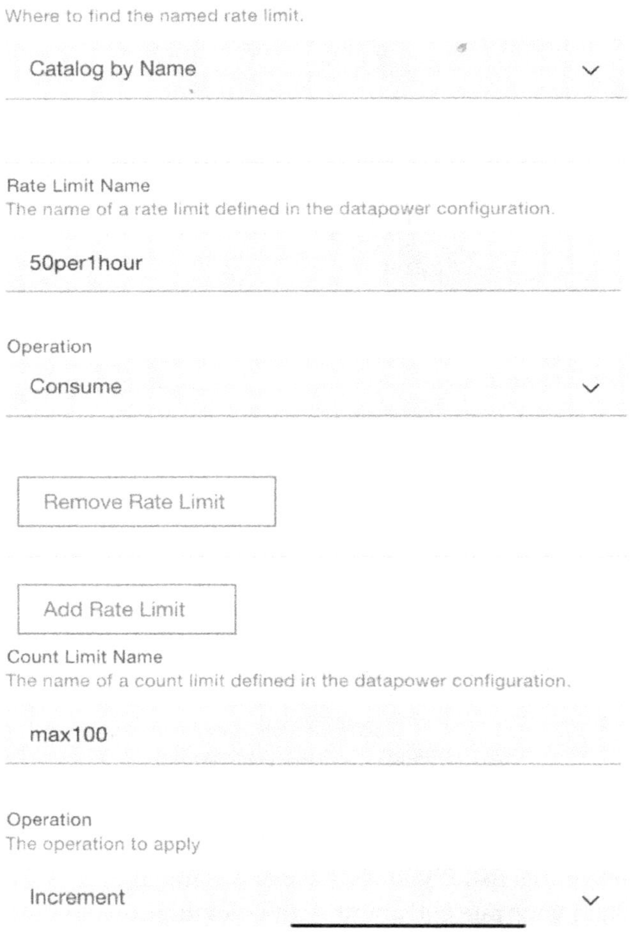

Where to find the named rate limit.

Catalog by Name ⌄

Rate Limit Name
The name of a rate limit defined in the datapower configuration.

50per1hour

Operation

Consume ⌄

Remove Rate Limit

Add Rate Limit

Count Limit Name
The name of a count limit defined in the datapower configuration.

max100

Operation
The operation to apply

Increment ⌄

Figure 10.21 – Configuring the Catalog rate and count limits in the policy

As you can see, there is one additional field that's exposed for your rate and count limits called **Operation**.

For your Rate Limit, this value can be **Consume** or **Replenish**. This indicates whether or not this Rate Limit will decrement or increment the available count in the Rate Limit. This comes into play when you have two Rate Limit policies on the same assembly, where your weight calculation will be used to replenish the count on a **Replenish** policy after a previous **Consume** policy. The default value for this field is **Consume**.

For your count limit, the operation field has two possible values. They are **Increment** and **Decrement**. As you might have guessed, one will increment the count by the weighted value, while the other will decrement the count by the weighted value. The default value here is **Decrement**.

Once your rate and count limits have been added to your policy, you can save your assembly, and you will be finished with your assembly Rate Limit policy. Your API will now Rate Limit based on the Catalog Rate Limit configured on your gateway service. Configuring the last option, **Gateway by Name**, is very similar to configuring the **Catalog by Name** option. Let's briefly discuss how this is accomplished and how it is different.

Configuring the Gateway by Name option

The **Gateway by Name** option within your Rate Limit policy is very similar to the **Catalog by Name** option, in that the configuration for your rate and count limits are defined on the gateway itself. The only difference here is that these limits are available across the entire gateway, not just a specific Catalog. To accomplish this, the DataPower API Gateway object for your specific gateway will be modified to configure your rate and count limits.

The process for configuring these limits is the same as the process we discussed for the **Catalog by Name** configuration. You will create your .cfg files, add them to a .zip file, and add them to your gateway extension. The syntax will be slightly different for your CLI commands. The following is the syntax for creating a Rate Limit for your **Gateway by Name** policy to refer to:

```
top; config;
switch apiconnect;
apigw apigw
  assembly-rate-limit  50per1hour 50 1 hour on off on on off
off "" "my.vars.max-amount - my.vars.used-amount"
exit
```

As you can see, the commands for configuring a Rate Limit for your gateway are very similar to the commands that are used to create your Catalog limits. The only real difference is the DataPower object that you are altering. In a gateway configuration, you are configuring the `apigw` object, which represents the entire gateway.

For more information on the command syntax for count and burst limits, please visit the IBM documentation site at `https://www.ibm.com/docs/en/datapower-gateways/10.0.1?topic=c-api-Plan-commands`.

With that, we have covered the different methods for referring to rate, burst, and count limits within a Rate Limit policy. As you have seen, the last option for configuring these limits provides the highest level of granularity by implementing these limits at a single point within your assembly. By providing these different options for configuring the actual limits, you can configure them once and share them across a single Plan, Catalog, or gateway. With all of these available configuration options, you should be able to apply a Rate Limit policy for any given scenario that you should come across. Once you have configured your API and are ready to be discovered, you will need to publish it. In the next section, we will discuss how to do this and the various options that are available to do so.

Publishing and configuring Catalogs

As we discussed in *Chapter 3*, *Setting Up and Getting Organized*, a Catalog is simply a logical partition for all of your API development within APIC. You can organize this in any way you see fit, though a common and useful way to use your Catalogs is to partition your different development environments. A Catalog will contain your Products, which will contain your APIs, so your Catalog will be one logical unit that will contain everything related to your APIs. This partition also carries over to the **Developer Portal**, where your APIs and Products are discovered, and is also part of the URL that's used to request an API. So, you can see how Catalogs provide the segregation and isolation required for different environments. You can refer back to *Chapter 3*, *Setting Up and Getting Organized*, to understand the importance and methods for organizing your Catalogs.

When setting your Catalogs, you need to understand what type of environment it will be used in. For example, will it be in a development environment, where it will be frequently changing and requires few approvals or restrictions? Or perhaps it will be used to house Production APIs and Products where rigid approvals are required and changes are less frequent? This is important to establish since it is within the Catalog configuration that you will set these permissions and approval requirements.

When you first create your provider organization, it will contain a default Catalog named **Sandbox**. As its name implies, this is a sandbox Catalog for you to experiment with and get started. This is not a Production-ready Catalog and should only be used for development activities.

To create and configure your Catalogs, from your API Manager home screen, click the **Manage Catalogs** tile. This will list all the currently configured Catalogs in your provider organization. To create a new Catalog, click the **Add** button in the top-right corner and select **Create Catalog**. At this point, you can give your Catalog a name and click the **Create** button.

Once your Catalog has been created, click on it, and then click the **Catalog settings** tab to reveal some basic settings for your new Catalog, as shown in the following screenshot:

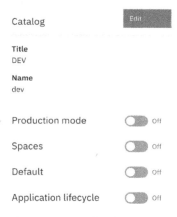

Figure 10.22 – Catalog settings

As you can see, there are a few toggle switches that you can configure:

- **Production mode**: Setting this switch to **On** will indicate that this is a Production Catalog. When you're publishing Products to this Catalog, it will check for any conflicts and reject the publish action if any are found.

- **Spaces**: Enables spaces for this Catalog. You learned more about spaces in *Chapter 3, Setting Up and Getting Organized*.

- **Default**: When **On**, this option makes this Catalog the default Catalog so that calls to APIs within this Catalog can be called with a shorter URL by omitting the Catalog's name. Only one Catalog can be the default Catalog.

- **Application lifecycle**: Switching this to the **On** position enables life cycle approvals. This means that you can configure what life cycle actions require approvals.
  This is the one setting that will change in each environment. For example, your development environment will likely not require approvals as you need to empower developers to rapidly develop their APIs through the life cycle in the development environment. In your Production environment, however, you will likely require approvals for every life cycle action.

Once the **Application lifecycle** toggle is set to the **On** position, you can click the **Lifecycle approvals** link from the left navigation pane and then the **Edit** button to specify which actions will require approvals. The following screenshot shows this screen, where we have set this Catalog to require approvals for all actions:

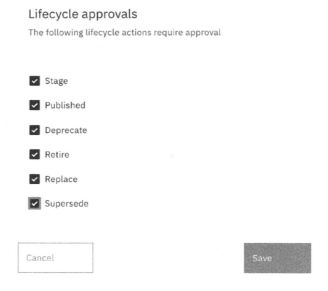

Figure 10.23 – Lifecycle approvals

One final configuration you must add before you can publish your Catalog is that you must add a gateway service. This can be accomplished by clicking the **Gateway services** link from the left navigation menu when viewing your Catalog settings. From here, you can add the appropriate gateway service.

Now, you know how your Catalog configuration should change based on the environment that it is to be used for. Now, let's take a look at how we can add our APIs and Products to a specific Catalog and publish them.

Publishing to your Catalogs

As you develop your APIs and save them, you will eventually need to publish them so that they can be discovered and invoked. As you may recall, we mentioned how your Catalog is a logical segregation within your APIC environment and that each Catalog will also be segregated within the **Developer Portal**. Because of this, you need to decide which Catalog you would like to publish your API and Product to so that it can be discovered, subscribed to, and invoked from the correct **Developer Portal**. Again, this very well may be how you segregate your different **software development life cycle (SDLC)** environments.

Once you are happy with your API development and are ready to publish it to the appropriate Catalog, from the API development screen, click the ellipses in the top-right corner and select the **Publish** option, as shown in the following screenshot:

Figure 10.24 – Publish

On the next screen, you will be asked to specify a Product to add this API to. Remember that all APIs must be part of a Product, which should be a logical grouping of your APIs. From here, you can create a new Product or select an existing Product from the list. The following screenshot shows us publishing our Member API and adding it to an existing Product called **Member 1.0.0**:

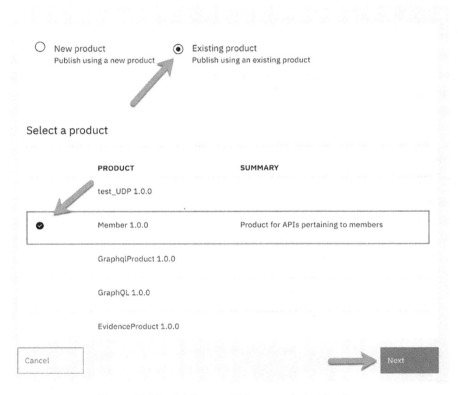

Figure 10.25 – Adding an API to an existing Product

As you can see, versions are shown as part of the Product names, so you will want to be sure to add the current version of your API to the correct Product version. For more information on versioning, please refer to *Chapter 11, API Management and Governance*.

Clicking the **Next** button will bring up a popup indicating that the Product has been updated. This will render the next screen, where you will indicate which Catalog you would like to publish your API to. The following screenshot shows us publishing our API to the **DEV** Product. As you may have guessed, this is our development Product, so we have set it to not require approvals:

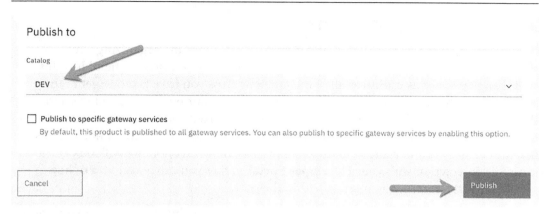

Figure 10.26 – Choosing a Catalog to publish to

When you publish your API, the Product is published to all the gateways defined within the Catalog by default. As you can see, you have the option to publish to specific gateway services.

> **Note**
> Gateway policies for the DataPower Gateway (v5c) are different from the APIC Gateway (apigw). If you have both gateway types defined in your deployment, you should use the **Publish to specific gateway services** option.

Once you've selected the Catalog to publish to, click the **Publish** button. You will see a popup, indicating that the Product has been published. Remember that it is the Product itself that is published.

With that, you have seen how to publish a Product containing your API to a specific Catalog. There are other methods you can use to publish your Product, such as navigating to the Product list and selecting the **Publish** option next to it. One last piece of configuration that you must complete is creating a portal for your Catalog. This is required so that consumers can discover your APIs and Products. This configuration can be found and completed by navigating to the **Catalog Settings** tab and then clicking the **Portal** link from the left navigation menu. For this configuration, you must simply select a portal service that you would like to use from the dropdown menu. A portal URL will automatically be generated for you.

With that, you have looked at the basic configuration for a Catalog and how to publish Products to it to make your APIs and Products discoverable within the **Developer Portal**. In the next section, we will take a look at the consumer's experience and see how they can discover, subscribe to, and use the APIs you have made available to them.

Consumer interaction

You have put in all of this work designing, developing, testing, and organizing your APIs and they are finally published and waiting for the world to discover them! It is this interaction that is just as critical as all of the previous steps you have taken to get to this point. It is how the end consumer sees and interacts with your published Products and APIs that will make all the difference. Are they well organized and easily discoverable? Did you use proper descriptions and keywords for them to search on? All of this matters when the consumers are trying to discover your APIs. In this section, we will look at the consumer's interaction with your **Developer Portal** so that you can get an idea of how they will navigate, search, and interact within the portal.

While developing your APIs within APIC, you are working within a provider organization that you are a member of. Just like we have provider organizations containing developers who provide the APIs, we also have consumer organizations. A consumer organization is just what you might think it is – an organization containing members who wish to consume your APIs. For a developer to subscribe to a Product or Plan, they must be part of a consumer organization. You can create a consumer organization in one of two ways: you can create it yourself, or you can invite a user to become a consumer organization owner. That user would then accept the invitation and create the consumer organization.

No matter which way you create your consumer organization, you must have the proper privileges to do so. You can refer to *Chapter 11, API Management and Governance*, to learn more about permissions. Assuming you have the proper access to create a consumer organization, you can create the consumer organization within the **Developer Portal** itself or the API Manager. Since we have been working in the API manager, let's take a look at how to create a consumer organization there. You must begin by navigating to the Catalog that you wish to add the consumer organization to. Remember, each Catalog provides a **Developer Portal**, so you will need to configure the consumer organizations at the Catalog level, as follows:

1. From the API Manager home screen, navigate to the **Manage Catalogs** | `<Catalog to add the consumer org>` | **Consumers** tab. From this screen, click the **Add** button at the top right and select **Create organization** or **Invite organization owner**, as shown in the following screenshot:

Figure 10.27 – The Create organization or Invite organization owner menu options

2.   If you select **Invite organization owner**, you will add the email address to send the invitation and the recipient will follow the link to create the new consumer organization.

If you select the **Create organization** option, you will be presented with a basic configuration screen for your new organization, where you will provide the organization's name and the owner of the new organization. This user can be a new or existing user. You can also select the user registry where the owner's user ID can be found. The following screenshot shows an example of a new consumer organization that's been configured for our **DEV** Catalog, which is where we are adding a new user:

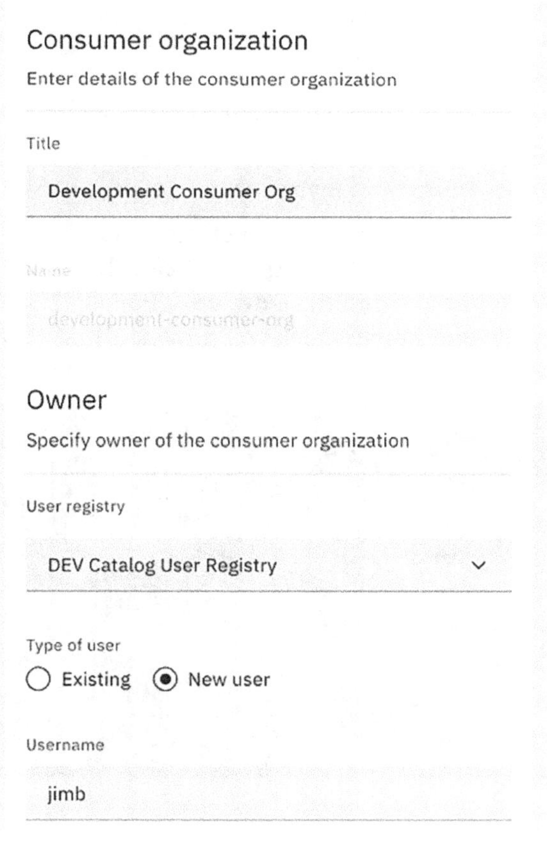

Consumer organization

Enter details of the consumer organization

Title

Development Consumer Org

Name

development-consumer-org

Owner

Specify owner of the consumer organization

User registry

DEV Catalog User Registry   ∨

Type of user

◯ Existing   ⦿ New user

Username

jimb

Figure 10.28 – Creating a new consumer organization

When you're creating a new user as the owner of your new consumer organization, you will need to provide additional information, such as the user's email address and password. Once this has been filled out and submitted, you will receive confirmation that the consumer organization has been created. At this point, you will see your new consumer organization in the list, as shown in the following screenshot:

Figure 10.29 – Consumer organization creation confirmation

1. Now that your consumer organization and user have been created, the owner can log into the **Developer Portal** and invite users to join the consumer organization.

2. Once your consumer organization has been created and your users have been invited, you can start to discover APIs. As you already know, you can discover and subscribe to APIs and Plans within the **Developer Portal**. When you first log into the **Developer Portal**, your home screen will provide you with a single snapshot containing some valuable information, as shown in the following screenshot:

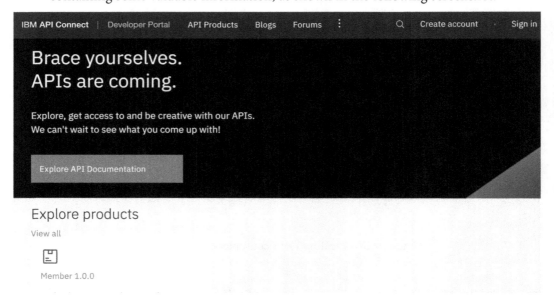

Figure 10.30 – Developer Portal home screen

In the top-right corner, you can see the consumer organization you are logged into. If you happen to be a member of more than one consumer organization, you can switch to them via this dropdown link. From this dropdown link, you can also view information about your current consumer organization or create a new consumer organization, provided you have the proper privileges.

In the center of the screen, you can see a snapshot of some Products that you can explore. Clicking the **View All** link under the **Explore Products** heading will reveal all the Products within the consumer organization. This is a nice, single pane of glass where you can get an overview of the consumer organization.

3. Before you start discovering your Products and APIs, you will need to create an application. You can think of an application as a collection of Products that you subscribe to. To create an application, from the **Developer Portal** home screen, click on the **App** link from the top navigation menu. This will bring you to a list of the current applications that you have created. If this is your first application, this list will be empty, as shown here:

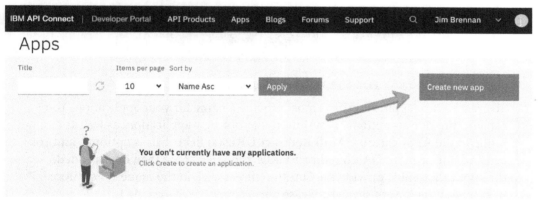

Figure 10.31 – Application list

From this screen, you can click the **Create new app** button at the top right to navigate to a screen where you can provide some basic information for your new application, as shown here:

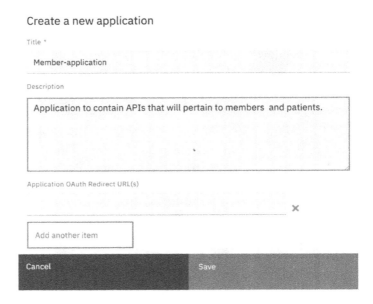

Figure 10.32 – Create a new application

1. To create a new application, you must provide a name for your application. You can also provide an optional description for your new application, as well as filling in the **Application OAuth Redirect URLs(s)** field. This is optional, and you can add one or more OAuth redirect URLs, though if more than one is added, the application must provide the OAuth redirect URL in the request. For more information on OAuth security, please see *Chapter 7, Securing APIs*.

2. Clicking the **Save** button will complete your new application configuration
    and bring you to a confirmation screen, which will let you know that your new
    application was created successfully, as shown in the following screenshot. More
    importantly, this screen will provide you with your client ID and secret for the APIs
    within your application. Be sure to copy these now as they will only be displayed
    once. If you fail to copy these, you will need to reset the client ID and secret to be
    able to obtain it:

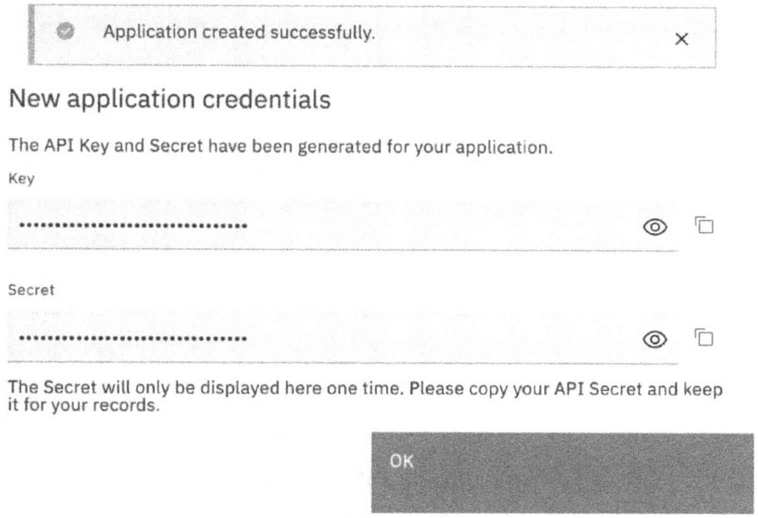

Figure 10.33 – Application creation confirmation

3. Clicking the **OK** button will bring you to the main screen for your new application.
    This screen will show your client ID and any Product and Plan subscriptions. Since
    this is a new application, this list will be empty.

Now that you have created a new application, you can begin to discover and subscribe to new Products and Plans. You can browse the available APIs in a few different ways. If you are viewing your application home screen, you will see a link after the **No subscriptions found** text saying **Why not browse the available APIs?**. Clicking this link will bring you to a list of available Products that you can subscribe to. Alternatively, you can click the **API Products** link via the top navigation menu. A third way to get to the available Products screen is from the **Developer Portal** home screen, as we saw in *Figure 10.30*. No matter which method you use to get to the available Product list, you will be presented with a list of Products that you can subscribe to, as shown in the following screenshot:

API Products

Figure 10.34 –API Products

As you can see, our list only contains one Product. As you develop and publish new APIs and Products, this list will grow. As you can see, you can also filter the list and sort it by using the filter and sort criteria fields at the top of the screen. Now, you can subscribe to any of the available Products, as follows:

1. Once you find a Product in the list that you would like to subscribe to, clicking on it will bring you to the following screen:

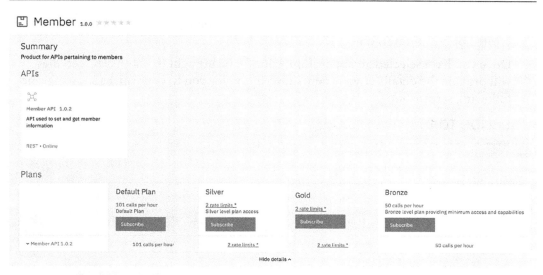

Figure 10.35 – Product summary and subscribe screen

2. From here, you will see all of the available Plans you can subscribe to, for this Product. As you may recall from earlier in this chapter, we configured different Plans for our Product. As a consumer, you can now select the Plan that you would like to subscribe to. Of course, depending on the different approval settings you have set, this subscription will need to be approved.

3. Once you've determined the Plan you would like to subscribe to and clicked the **Subscribe** button, you will be brought to a screen where you will need to select the application you would like to add this subscription to, as shown in the following screenshot:

Subscribe to Member 1.0.0

◉ Select Application   ○ Subscribe   ○ Summary

Select an existing application or create a new application

Member-application

Application to contain APIs that will pertain to members and patients.

Cancel

Figure 10.36 – Select an existing application or create a new subscription

4. From this screen, you can either select an existing application that you have already created or create a new one.

5. Once you have selected an application, you will be brought to another screen that will provide the details of your new subscription for you to confirm, as shown in the following screenshot:

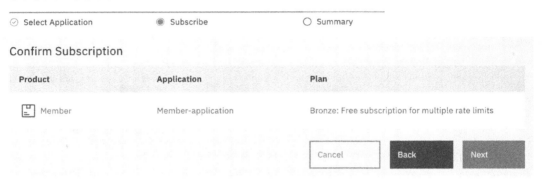

Figure 10.37 – Confirming the new subscription details

6. Clicking the **Next** button will confirm your new subscription and you will be shown a summary, as shown here:

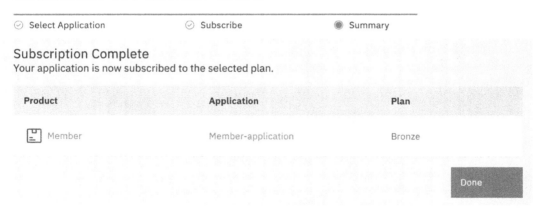

Figure 10.38 – Subscription summary

7. Your subscription is now complete. Clicking the **Done** button will finish this process and return you to the Product screen. You now have a new application containing a new subscription and are ready to begin invoking the APIs within the Product. This same process can be followed to create new applications to contain new subscriptions, though you can continue to add subscriptions to your existing application.

You have now completed the end-to-end process for enabling your consumer experience. This consisted of three basic steps, as follows:

1. Create a consumer organization.
2. Create a new application.
3. Subscribe to Products and APIs.

After following these steps, your consumers are free to discover, subscribe to, and invoke your Products and APIs.

Summary

In this chapter, we have covered a lot of ground by discussing the process of publishing your Products and APIs, as well as the different options that are available for doing so. We took you from creating the Plans, which consist of Rate Limits and the different options for configuring them, to how to configure and publish to a Catalog. Finally, you enabled consumer interaction by creating a consumer organization where applications can be created and Products can be subscribed to.

All of these steps are critical to enabling and enhancing the consumer experience. After all, it is this very experience that will draw consumers to your APIs and produce a robust, consumer-friendly API factory, ensuring a successful digital transformation and modernization effort.

In the next chapter, we will discuss API management and governance, which will expand on this so that you can manage, maintain, and organize your APIC environment to ensure continued success in the future.

11
API Management and Governance

It was inevitable that we would eventually have to get to that oh-so dreaded word **governance**. You might cringe a little thinking about all the red tape and processes that seem to just slow our productivity and time to market. Although governance is important to all software development, it is even more so when we are implementing our API strategy. In the past, developing software had its process to follow from the design phase to code and unit testing, to QA testing, and finally production. Once it hit production, as long as it worked no one seemed to really know or care what process was followed to get there. As long as it worked. We are in a very different world now with our API strategy and digital transformation. Just having software that works is not enough. If your APIs are not easily searchable, discoverable, and consumable, it doesn't really matter that they work. Imagine shopping on Amazon but nothing is categorized and when you search for a product, it doesn't get a search hit on what you were looking for. Then, when you find a product you want, there is no description for it. And finally, you go to make a purchase and the item has been discontinued! That would indeed be a frustrating process and Amazon would likely not be around much longer.

So, what is governance and how is it going to save the day for us here? Webster defines it as *"the act or process of governing or overseeing the control and direction of something."* Well, that doesn't tell us much except that we need someone to oversee our process. But what process? Who is defining this process? Who is implementing and following the process? What happens if the process isn't followed? So many questions here to answer and, hopefully, you'll find all of the answers within this chapter as we cover the what, the who, and the how of our API governance strategy.

This will be accomplished by discussing the following topics:

- Understanding the API and product lifecycle
- Assigning roles
- Managing versions
- Segregating environments
- Applying standards

Technical requirements

To fully implement the governance processes discussed within this chapter, you should have a fully functional API Connect installation to work with. Even if you do not, this chapter will provide you with the understanding of the *who*, *what*, and *how* of the governance process that will be critical in providing a successful framework for your digital transformation.

Understanding the API and product lifecycle

As new APIs are developed, existing APIs are modified, and legacy APIs are retired, they will be brought through a continuous lifecycle that will continue as the API is modified due to bug fixes or enhancements. This process will continue on until the API is eventually retired. Following the same process for all of your APIs is critical to the success of your API strategy. In this section, we will discuss the lifecycle process, who will implement it, and how.

The "what" of the lifecycle

When we talk about API governance, we must first understand exactly what we are governing. For the most part, we are governing the movement of our APIs and products as they move through the different phases of the API lifecycle, which includes your **software development lifecycle (SDLC)** environments. This is such a critical component to understand and govern as this will directly impact how our products are made available to the end consumers as well as how and when they will become unavailable.

There are essentially four stages of our product lifecycle that they will move through—**Stage, Published, Deprecated,** and **Retired**. Although there is a logical succession of these phases, these are not always one-direction movements as they move from Stage to Retired. It is entirely possible that you may have a deprecated or retired product that you want to republish. Or maybe a published product that goes right to retired. Different circumstances will bring about different movements of your products. *Figure 11.1* shows this product lifecycle and all of the possible movements from one stage to another. Also, note at which stages the product is visible and available in the developer portal.

> **Info**
> You can find additional details about the lifecycle in the IBM documentation: `https://www.ibm.com/docs/en/api-connect/10.0.1.x?topic=products-product-lifecycle`.

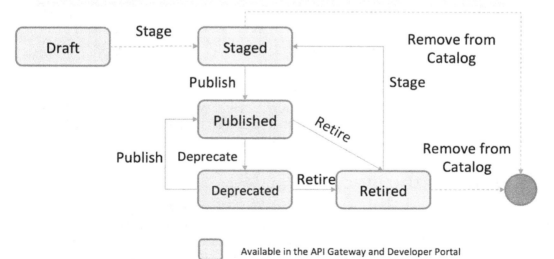

Figure 11.1 – Product lifecycle

In an ideal world, your products will move smoothly in a defined and deliberate manner through the following states:

1. **Draft**: The initial state of your API and product as you begin developing them.

2. **Staged**: A pre-published state where the product is not published to a particular Catalog and hence, not present within the portal. This state is merely a *waiting state* waiting to be published.

3. **Published**: This is the state where your product is ready to be subscribed to and consumed. Once published, your product, and the APIs within, are made available in the portal.

4. **Deprecated**: A state that a product should be in before **retired**, signifying to the current consumers that this will soon be retired and no longer available. In this state, the current consumers will still be able to invoke the APIs within the product but no new subscribers are allowed.

5. **Retired**: The product is no longer visible within the developer portal and the APIs can no longer be invoked.

6. **Deleted**: Technically not a state but the product is removed from the Catalog.

Of course, we rarely work in an ideal world so we must be prepared for the exceptions. This is why *Figure 11.1* shows alternate paths from different states. Let's suppose you have a deprecated or retired version of a product that has been replaced with a newer version. Suddenly you realize that the newer version has some issues and you need to bring back the deprecated or retired version. From either of these two states, you can bring your product back to a published state. From there, you can then take your initial published product straight to a retired state.

You have now seen the lifecycle that your APIs and products will follow as they move from development through retirement. You might imagine when newer versions of your products come about how this will flow, but this doesn't all happen by itself. There is no magic here, so let's take a look at how this all happens.

The "how" of the lifecycle

As you develop new, and update existing, APIs and add them to your products, they must be published to make them visible to consumers to subscribe to and invoke. By now, you should already know how to stage and publish your product, which is what you would do for the initial version. Once you start introducing subsequent versions, your process will change.

Let's take an example of an API named **Member API**, which is contained within the **Member** product in our example health care organization. We have developed, staged, and published version 1.0.0 of this product to our **Production** Catalog. It is now time to create a new version of this API and product.

To create a new version of this API, you will follow the following steps:

1. Navigate to the **Develop APIs and products** screen in **API Manager** and make sure you are on the **APIs** tab. From here, you will locate your current version of your API.

2. Click the ellipsis on the far right of that row, and select **Save as New Version**. *Figure 11.2* shows us creating a new version of our **Member API**.

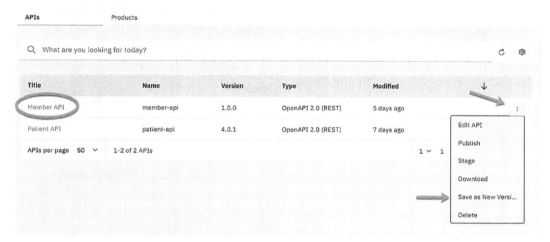

Figure 11.2 – Save as New Version

3. As you are prompted for the new version, you will enter the next version number and click on **Submit**. In our **Member API** example, we created a new 1.0.1 version. You can now see this in *Figure 11.3*:

Title	Name	Version	Type	Modified	↓	
Member API	member-api	1.0.1	OpenAPI 2.0 (REST)	8 minutes ago		⋮
Member API	member-api	1.0.0	OpenAPI 2.0 (REST)	5 days ago		⋮

Figure 11.3 – Version 1.0.1 created

4. Once you have your new version created, you can modify it with the required changes.

5. Once your changes are complete, you will need a new version of your product as well since that is what our consumers will subscribe to. You can follow the same procedure that we followed for creating a new version of your API for the product.

You will notice that when you create a new version of your product from the current version, the current API also comes along with it. You will need to remove this version from the product and add the new version. This is accomplished by navigating to the **API and product development** page, going to the **Products** tab, and clicking the product and version you want to edit. Once you have selected the product and version you want to work with, you will click the API's link on the left of the page. You will see that the new version of your product has the previous version of the API still in it. Clicking the **Edit** button will then bring you to a screen where you can unselect the old version of the API and select the newer version of the API as shown in *Figure 11.4*. Once these changes are made, clicking **Save** will complete the change.

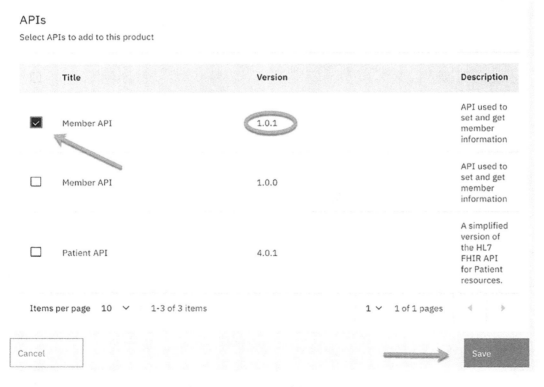

Figure 11.4 – Select a new version of the API within the product

You now have a new version of your API and product that is ready to stage. As you have done before, you can now stage this version of your product to the appropriate Catalog. As with all of your APIs and products, be sure to download and add this to your source control management system.

At this point, you should have a current, published version of your product and a staged version of your newest version. You could just go ahead and publish your staged version and then deprecate your current version in two separate steps but you do have a simpler option available. If you navigate to the Catalog where you have your products, you should see each version of the product and its current state. In our **Member API** product example, we have a published **Member** version **1.0.0**, and a staged **Member** product version **1.0.1** in our **Production** Catalog. At the far right of each product row, you can click the ellipsis of your published product and see all of the available options, including three that can be used to change the state of the product, which are **Deprecate**, **Retire**, and **Supersede**. You will also see a **Replace** option but that technology will not change the current state. *Figure 11.5* shows these options for our published **Member** product version **1.0.0**:

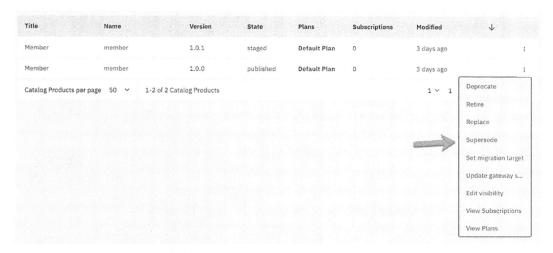

Figure 11.5 – Product options from within the Catalog

The **Deprecate** and **Retire** options should be self-explanatory but the option we are interested in here is the **Supersede** option. This is the one option that we can use to publish our staged version and deprecate our current version because that is exactly what this option does. It will publish the chosen superseding product, giving it the same visibility options as the superseded product. It will then deprecate the superseded product. Once a product is superseded, all current subscribers can continue to consume the APIs within the product, however, no new subscribers will be allowed. Although this is a convenient, one-click option to publish one version and deprecate the current version, it should not be used if your new version's interface has major changes as it could have a negative impact on your existing consumers.

If you go ahead and click the **Supersede** option for your published version of the product, you will then be asked to select which version of the product you want to supersede this version. Again, the version selected will become the published version and the version to supersede will become deprecated. *Figure 11.6* shows us selecting version **1.0.1** of our **Member** product to supersede our published version **1.0.0**:

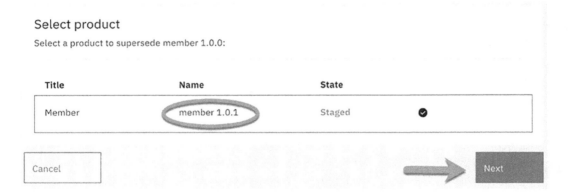

Figure 11.6 – Superseding Member product 1.0.0

Clicking the **Next** button will then bring you to a screen showing the superseded and the superseding products. From this screen, you will also select which plans you will want to migrate to the superseding product. In *Figure 11.7*, you can see that we selected the **default-plan** that was used in our superseded version to migrate:

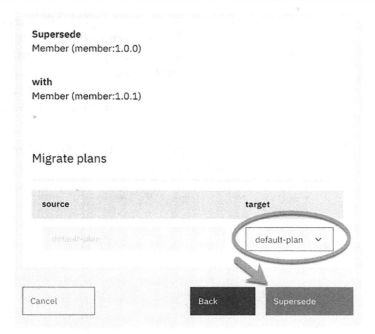

Supersede
Member (member:1.0.0)

with
Member (member:1.0.1)

Migrate plans

source	target
default-plan	default-plan ⌄

Cancel   Back   Supersede

Figure 11.7 – Select plans to migrate to superseding product

Clicking on the **Supersede** button from this screen will complete your supersede operation. You will be brought back to the Catalog page showing all of the current products where you started. You will need to click the refresh icon to see the products reflecting their new states, as shown in *Figure 11.8*, where our **Member 1.0.0** product is in a deprecated state and our **Member 1.0.1** product is now published:

Q What are you looking for today?   **Refresh** ➡ ↻ ⚙

Title	Name	Version	State	Plans	Subscriptions	Modified	↓	
Member	member	1.0.1	published	Default Plan	0	3 minutes ago		⋮
Member	member	1.0.0	deprecated	Default Plan	0	3 minutes ago		⋮

Catalog Products per page 50 ⌄   1-2 of 2 Catalog Products   1 ⌄ 1 of 1 pages ◁ ▷

Figure 11.8 – Member products after superseding

You can continue to use this option each time a new version of your API and product comes out, but your list of deprecated products will just keep growing. Eventually, you will need to remove some of the older versions. You can do this by using the **Retire** option you saw earlier in the product options from within your Catalog. As a good housekeeping rule, you should only keep one API in each state. Before retiring a product version, you should make sure that all consumers subscribed to that version of the product are notified as this version will no longer be accessible.

To retire a version of your product, you would simply select the **Retire** option next to the product in the **Catalog** screen as shown in *Figure 11.9*.

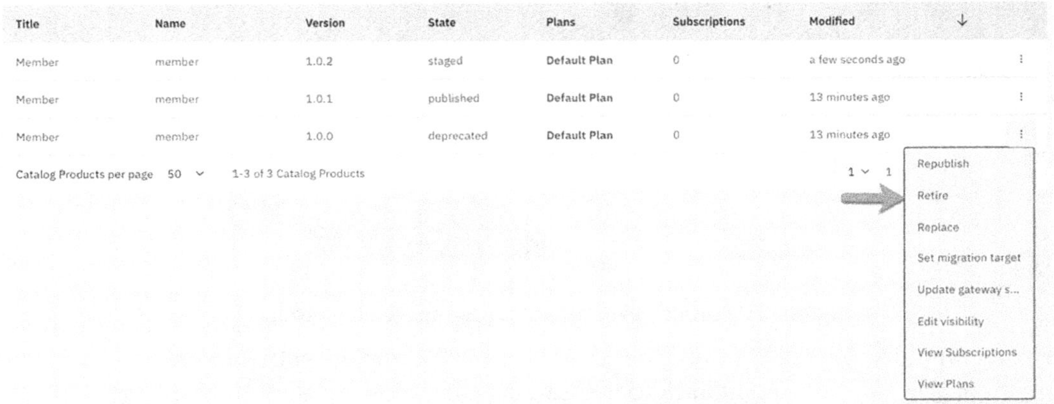

Title	Name	Version	State	Plans	Subscriptions	Modified	↓
Member	member	1.0.2	staged	Default Plan	0	a few seconds ago	⋮
Member	member	1.0.1	published	Default Plan	0	13 minutes ago	⋮
Member	member	1.0.0	deprecated	Default Plan	0	13 minutes ago	⋮

Catalog Products per page   50 ⌄   1-3 of 3 Catalog Products

1 ⌄ 1

Republish
Retire
Replace
Set migration target
Update gateway s...
Edit visibility
View Subscriptions
View Plans

Figure 11.9 – Retire a product version

You will then be shown a confirmation screen where you will confirm that you want to retire this product. Once the product is retired, all subscriptions to it become inactive and it is removed from the developer portal.

In *Figure 11.9*, note the options available for the product version that is currently in a deprecated state. From this state, you can also choose to republish this product if you wish.

As you can see, the products in a retired state will still show in your Catalog within API Manager so you must again stay diligent and delete them when you are sure that they are no longer needed. This is the final stage in the API and product lifecycle. Once you delete a product, it is removed from only the Catalog. The product and APIs within it still exist, however they are no longer part of a Catalog. If this product was again needed, you would need to re-stage it to the Catalog and start the lifecycle over again.

In this section, we demonstrated the API lifecycle showing the APIs and products moving freely with a simple click of an option to bring them to the next state. In a development environment, this may be the most efficient way to migrate from state to state as there will likely be many changes to a particular API throughout the development process.

As you move through higher environments such as **quality assurance (QA)** and production, you will most likely not allow a developer to move an API and product through the different states without approvals. This is why you have the ability to require approvals at each stage of the lifecycle. In the next section, we will discuss this process.

API lifecycle approvals

As we have already mentioned, as part of your governance process, you should have some approval process when migrating a product from one state to another. The required approvals for the different states will likely be different for each environment. API Connect provides the ability to define which state of a product will require approval by Catalog. To view and edit the required approvals for a specific Catalog, from within **API Manager** you should navigate to **Manage Catalogs** | <Catalog name> | **Catalog settings** and click the **Lifecycle approvals** tab on the left navigation menu. By default, there are no approvals enabled for a new Catalog and you will see a message indicating that when you first view this screen, as shown in *Figure 11.10*:

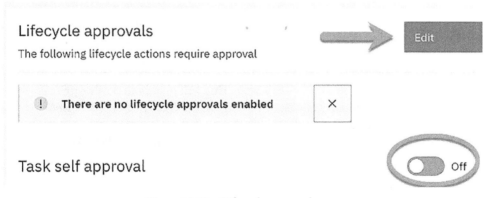

Figure 11.10 – Lifecycle approvals screen

You will also notice a toggle for switching **Task self approval** on and off. With this option set to on, it allows the task originator to approve the task. Again, this may be desirable in lower environments for efficiency, but will most likely be off for your higher environments.

To enable specific approvals for this Catalog, you will click on the **Edit** button, as shown in *Figure 11.10*, which will present a screen with a list of tasks that you can select to require approval for. As mentioned earlier, there are no tasks requiring approval by default so none of the tasks shown will be checked. *Figure 11.11* shows this screen, where we selected all tasks to require approval:

Figure 11.11 – Select required lifecycle approvals

You may need to only select a few of these as well. This will be dictated by your approval process defined for this particular Catalog. Clicking on the **Save** button will save these selections and return you to the lifecycle approval screen showing the tasks that now require approval. You can now see in *Figure 11.12* that all of our lifecycle tasks require approval:

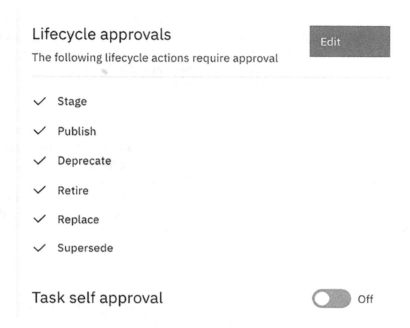

Figure 11.12 – Required lifecycle approvals

Now that you have established which tasks require approvals, and who can provide them, whenever a task that requires approval is submitted, the task is queued up in the Catalog until it is approved.

Let's take a look at a new version of our Member product V1.0.3 that we have created and staged to our **Production** Catalog. Since approvals are enabled for this state, the product will be placed in a pending state. It will also be placed in the **Production** Catalog's tasks. You can view all tasks awaiting approval by navigating to **Manage Catalogs | <Catalog name>** and clicking the **Tasks** tab. From here, a user with the appropriate approval permissions can click the ellipsis to the right of the approval task and click **Approve**. *Figure 11.13* shows our product **member:1.0.3** being approved:

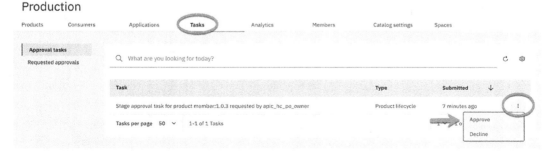

Figure 11.13 – Approving a stage request

This same process will be followed for each task that requires approval. As you can see in *Figure 11.13*, you can also choose to **Decline** the approval request. The stages of the API and product lifecycle that you will require approval for will depend on your overall governance strategy and how much control and how many checkpoints you will require. This will most likely also vary by environment since lower environments will usually require fewer controls.

We have now covered the API and product lifecycle from end to end and demonstrated how it is executed in API Connect. With this knowledge, you should now be capable of moving your APIs and products through each state from stage to deletion with the ability to re-stage or publish from almost any other state. We also showed how you can require approvals for each of these stages of the lifecycle by Catalog. But who will provide the approvals and how do you delegate the permissions to do so? Now that you know the *what* and the *how*, let's take a look at the *who*.

Roles

With all of the continuous work involved with developing APIs, migrating them to a different environment, and approvals, it is essential that you define who will be performing each task in the lifecycle. Assigning roles for individuals and groups will provide a fluent process with appropriate checks and balances throughout. In this section, we will discuss the different roles that can be established and assigned.

The "who" of the lifecycle

We have discussed the entire API and product lifecycle in the previous section, including what it is and how to move your APIs and products through these stages and states. Before we talk about who will perform these steps along the way, we must discuss who can do what within your API Connect environment. That is, what permissions are given to who. This is a major part of our governance process as it enforces the separation of duties throughout your lifecycle and management process.

As with any governance process, it is only as good as how well it is enforced. You can have the best processes defined; however, if they are not enforced at each step along the way, it is likely to fall apart. You wouldn't want to have the same person or group performing every task within our process. This could lead to disorganization and a total breakdown of your standardized process. To avoid this, we must assign roles for all things from the Cloud Manager to the API manager, to the developer portal. Again, this is at the heart of your governance process, so you should give careful consideration to which roles and responsibilities will be given to specific people and groups. Throughout this section, we will be discussing roles and permissions separately, where roles are the high-level classification for a given user and permissions are the capabilities the role will provide to the user. Let's take a look at the different areas' roles and the permissions they provide through our API Connect components.

Cloud Manager roles

When we talk about the API Connect topology and components, the cloud manager is at the core of everything. This is where your underlying runtime/services, provider organization, user registries, and many other critical components are defined and managed. Because of this, the number of groups and people that can make these changes should be limited. Of course, you can – and should – also segregate the roles and responsibilities within the cloud manager by providing different roles to different users. Let's take a look at the different roles available within the cloud manager.

By default, there are six different roles within the cloud manager that a user can be assigned. Within each of these roles, six different permissions are granted. Of course, you have the ability to create your own custom roles and change the defaults, however, the default roles will typically provide all of the different roles with the appropriate permissions that you will need. It will at least be a good start as you separate your user and group responsibilities and permissions.

To view, edit, and add roles, you can log into the cloud manager and navigate from the home screen to **Configure Cloud** and click the **Roles** link on the left sidebar menu. *Figure 11.14* shows the default roles with the **Administrator** role showing the permissions granted to this particular role. Each role can be expanded to show the permissions assigned to that particular role.

Figure 11.14 – Cloud Manager default roles

To add a new role, you can simply click the **Add** button, provide a name for your new role, and assign the permissions. Again, it is likely that the default roles will suffice, however, if you need to create a new role, it is this simple. You can also edit the current roles by clicking the ellipsis to the right of the role and selecting **Edit**. From here, you can edit the permissions assigned to the role. You can also delete a role by clicking the ellipsis and clicking **Delete**.

Now that you can see all of the default roles and permissions, what do they actually mean? Knowing the different roles available is helpful but it is the permissions assigned to each role that really dictates what a user can do. As we mentioned earlier, there are six different permissions available. These are as follows:

- Cloud Settings
- Member
- Org

- Provider-Org

- Settings

- Topology

If these available permissions look a little familiar to you, they should. Each permission listed correlates to a specific link or option within the cloud manager UI. A complete listing of permissions and descriptions can be found on the IBM documentation site:

```
https://www.ibm.com/support/knowledgecenter/SSMNED_v10/com.
ibm.apic.overview.doc/overview_apimgmt_users.html
```

Understanding the default roles provided and the granularity of permissions available is key when determining who will perform each task within your cloud manager. You want to be sure to provide the proper role(s) to each user to enable them to do the job assigned, but also be cautious not to provide permissions that will allow them to perform actions that they should not do. If you think about the criticality of all of the settings and configuration in the cloud manager, you will begin to realize the impact on the entire API Connect infrastructure, hence the reason for strict access controls to be in place and enforced.

Next, we come to API Manager roles and their permissions.

API Manager roles

As critical as the Cloud Manager roles are, the tasks performed within it are, for the most part, *set it and forget* types of tasks. Tasks like configuring your gateway services, user registries, and provider organizations are done at the initial configuration and setup and are rarely modified. It is in the API manager that your day-to-day operations and work happens. This is where the API lifecycle is managed. So, when we talk about governance, we need to focus largely on governing the API lifecycle and the groups that will implement, enforce, and approve each step within. This again will come down to assigning proper permissions to each user so that they can perform the assigned task, while restricting them from having permissions for any others.

Within API Manager, there are seven different roles defined by default. Again, it is likely that you will find one of these roles suitable for each user, however, you can create additional roles if needed. Just like you saw with the cloud manager roles, each role within API Manager has an assigned set of permissions. Because there can be so much happening within API Manager, there is an extensive list of permissions available that can be assigned to roles. This list can also be found in Table 4 on the IBM documentation website where you can see a complete description and the capabilities of each, as well as the roles that have them assigned:

```
https://www.ibm.com/support/knowledgecenter/SSMNED_v10/com.
ibm.apic.overview.doc/overview_apimgmt_users.html
```

The seven different roles provided correlate to a group that will likely perform different tasks within API Manager in order to provide a separation of duties and controls throughout. To see the available roles and the corresponding permissions, from within the API Manager home screen, click the **Manage settings** tile and then click the **Roles** link in the left navigation menu. *Figure 11.15* shows the default roles available:

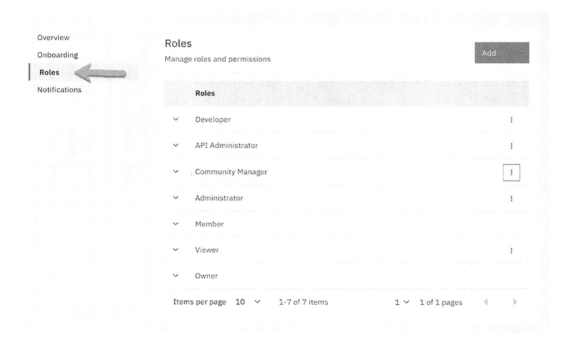

Figure 11.15 – API Manager default roles

Clicking on each role will display the set of permissions assigned to it. As with the cloud manager, you can also edit or delete each role that is not immutable by clicking the ellipsis to the right of it. You can also add new roles by clicking the **Add** button on this screen.

As you can see, each role name is fairly descriptive, depicting the type of role and responsibility any user assigned will have. Let's take a look at each role and its capabilities:

- **Developer**: This will be the role assigned to the developers that are creating the APIs. This role provides permissions to view everything within API Manager as well as to manage permissions for all things related to API and product development such as product creation and approvals, subscriptions and approvals, and so on. It is important to note that if a user is defined at the provider org level, that user will have these permissions for all Catalogs within the provider org. If the user is only a member of a Catalog or space, the permissions will only apply to that particular Catalog or space.

- **API Administrator**: As its name implies, this role should be assigned to the user(s) who will manage the API and product lifecycle. This role provides full permissions to all products as well as approvals; however, it only has view access to non API/product settings such as **Topology**, **Settings**, **Org**, and so on. Typically, a user assigned this role would be responsible for publishing, approving, and managing the entire API and product lifecycle.

- **Community Manager**: Users assigned this role are generally responsible for managing the relationship between the provider organization and the consumers of the APIs. This role will provide view privileges to everything within API Manager as well as permissions to manage tasks that relate to the consumer/developers such as the consumer org, subscriptions, subscriptions, and so on.

- **Administrator**: This role has the same set of permissions as the **Owner** (explained ahead). This is another role that can be assigned to someone who is required to have full permissions within the organization.

- **Member**: A member of the provider org will have view permissions for the organization only.

- **Viewer**: This is a role assignment for a user that should only have view permissions for all things within API Manager.

- **Owner**: As its name implies, this role is assigned to the person who was assigned as the organization owner when the organization was created. This role will have full permissions to access everything within the organization.

The seven roles available within API Manager, as well as any custom roles you might create, are available to assign to any user within the given provider organization. As you add members to your provider organization, you will assign one or more of these available roles. The roles assigned at the provider organization level will be applicable to that user in all Catalogs within the provider organization. You can, however, add additional roles for a given user at the Catalog level, but you cannot remove roles for a user that were assigned at the provider organization for a specific Catalog. It is also possible to add a member to a specific Catalog and not at the provider organization level, which will then provide the opportunity to assign roles only for that given Catalog.

As you can see, the default roles provided cover most roles that you will need within your organization. These will likely change for each environment as you will likely not want the same permissions for your development environment as you would for your production environment. As you define the different responsibilities and workload segregation within your organization, you will likely apply many of these default roles to your users, and if you do not find one that fits your exact needs, you can always edit these roles, or create new ones. At this point, you should have a good idea of what user roles are available in API Manager as well as how they can be assigned to provide the proper restrictions and permissions to each user and group that will play a part in your API lifecycle.

Now that we have discussed how to manage user roles and permissions for your cloud manager and API Manager, there is only one piece left to discuss. That is our developer portal. In the next section, we will cover in the same level of detail the available roles and permissions, as well as the user responsibilities for each within your developer portal.

Developer portal roles

As we move into discussing the different roles and permissions applicable to the developer portal, we should take a moment to step back and think about what the developer portal is, what types of tasks are performed, and who is involved in performing these tasks. As you know, the developer portal is where you socialize all of your APIs and products to be discovered. This is where consumers/developers will go to discover and subscribe to your products. With that understanding, it is safe to assume that the major actor in the developer portal will be the developer. Hence the name *developer portal*! By developer, we mean the developer of the applications that will be subscribing to and consuming your APIs and products. Developers will also require the ability to perform work within their developer organization such as building applications. Of course, you will have the person or people that will administer the portal itself. Other than that, there will not be the need for many more additional roles within the developer portal.

Let's take a look at the small set of default roles and permissions available.

So, when we talk about roles and permission, we are talking primarily about how they pertain to the members within the consumer organizations.

There are a total of four default roles available within the developer portal and only eight different permissions available. Again, for a complete list and explanation of these permissions, please refer to Table 5 in the IBM documentation site mentioned earlier in this chapter. The four roles available are as follows:

- **Owner:** This is the role assigned to the user assigned as the owner of the consumer organization. This user is specified when the consumer organization is created within API Manager for the specific Catalog. As the owner, this user will administer the consumer organization, therefore will have full permissions to it.

- **Administrator:** This role can be granted to a user when inviting them to the consumer organization from within the developer portal. This user will be responsible for administering the consumer organization, therefore will have the same permissions as the owner role.

- **Developer:** This role is intended for the developer who will be responsible for apps within the consumer organization. This role will provide view and manage permissions for application development as well as view permissions for everything else within the consumer organization.

- **Viewer:** As its name implies, this user will only have view permissions to everything within the consumer organization.

Of the four roles available, one is automatically assigned when creating the consumer organization. That is the owner role. Once the consumer organization is created, the owner can invite additional members to it. This is accomplished within the developer portal by clicking the consumer organization name in the top-right corner and then selecting **My organization,** as shown in *Figure 11.16*:

Figure 11.16 – Navigating to My organization

This will bring you to the organization settings, where you can click the **Invite** button to invite new members. You will then be presented with a form where you will enter the email address of the member to invite and select the role that you wish to assign to this new member. As you can see in *Figure 11.17*, there are three roles available to select from:

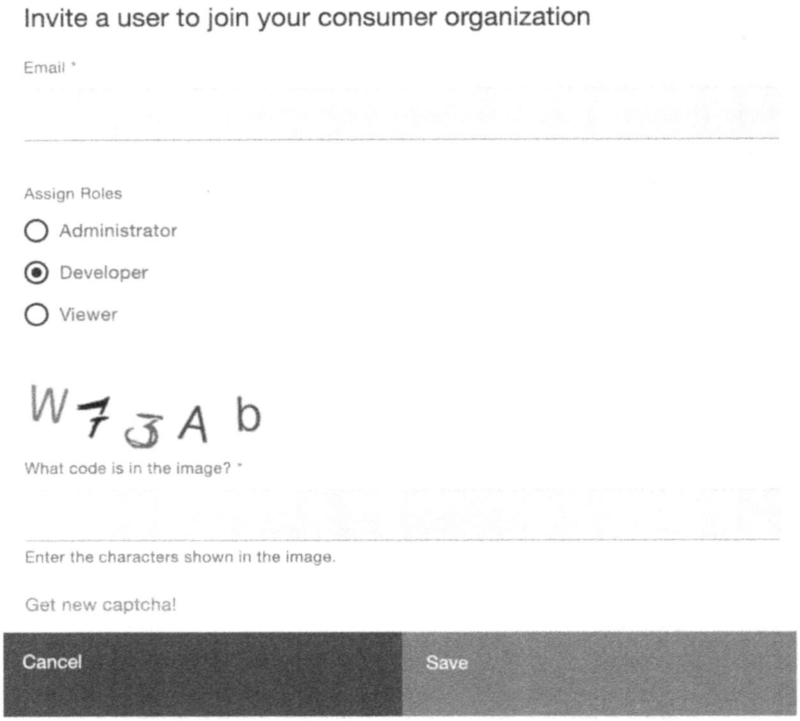

Invite a user to join your consumer organization

Email *

_____

Assign Roles

○ Administrator

◉ Developer

○ Viewer

W7ᴣAb

What code is in the image? *

_____

Enter the characters shown in the image.

Get new captcha!

| Cancel | Save |

Figure 11.17 – Inviting a new consumer org member

Once you complete the form, including the captcha, click on **Save**, and an invitation will be sent to the new member. Once the invitation is accepted, the new member will be included in the consumer organization with the role and corresponding permissions selected.

You have now seen how to assign permissions to users within your cloud manager, API Manager, and developer portal by using pre-defined and custom roles. This will establish the "who" in your governance process. Let's now drill down a little deeper into how you can establish and maintain a meaningful versioning scheme that can also be enforced by using the roles defined for each user and group.

Managing versions

API development is typically not a *set it and forget it* type of process. As bugs are discovered and new features are introduced to your APIs, you will need to roll out a new version of them. Since you are likely dealing with many consumers of your APIs, both internally and externally, you would not expect them to just adopt this new version as soon as you roll it out. Of course, each consumer will have their own quality assurance process and timeline so you will need to leave prior versions accessible for each consumer until they adopt the most current one. Also, some versions may have a direct impact on consumers, such as an interface change, which would require them to make a change to adopt this new version, while other versions may only be an implementation change where the consumers are not directly impacted.

Understanding that changes to your APIs are likely and versioning them is critical, we must realize that a consumer does not subscribe to an API itself, rather to a product that is published. Because of this, we must not only version our APIs as we make changes, but we also need to version our products appropriately. So, when we think about it, when an API changes, so does the product that it belongs to. A product version can change when one of the APIs contained within it changes, or perhaps a new API is added, or an old one is removed. With all of this in mind, it is critical to pay extra attention to which APIs will be a part of which product.

Sticking with the theme of planning ahead, versioning is yet another area that will greatly benefit from a little pre-planning before you even begin coding your APIs. As mentioned earlier, you should think about how your APIs will be included in products so that you are versioning them at the product level. You will want to package all APIs in a single product as an offering of related APIs, however, keep in mind that if one of the APIs changes, the product version will also change. Because of this, you might think about keeping the number of APIs within a product to a minimum. You wouldn't want to ask many consumers to constantly adopt a new version of a product when the APIs that they use are rarely changing.

Version numbering scheme

Another thing you should establish before developing your APIs is the versioning methodology or scheme. I'm sure you have seen versions of software that follow different patterns. Some have a single dot in the version (n.n), some have two (n.n.n), and some may even have three dots (n.n.n.n).

This is not just some random versioning format that has been implemented, but a well-thought-out scheme where each dot says something about that version. They could represent a major release, a minor release, or a bug fix.

For example, a simple one-dot scheme might simply be `<major>.<minor>`, where a two-dot scheme could be `<major>.<minor>.<fix>`, and finally a three-dot scheme could be `<release>.<major>.<minor>.<fix>`.

Also, keep in mind that it will be extremely useful for the consumers to be able to identify whether your new version contains an interface change or just an implementation change. Or, whether it is backward compatible with the prior version. This would give them a good idea of whether they need to make a change on their end. So perhaps only a major version change would contain an interface change and all others would be implementation only. Which scheme is right for your organization is a decision you should make based on factors such as how often you are pushing out major and minor updates, whether your consumers are all internal or external, and maybe just the level of complexity you wish to manage to provide a more granular versioning scheme. How you choose to version your APIs is completely up to you but keep in mind that the more dots you have in your versions, the more granular they will be, hence the more versions to manage at one time. You might notice that the API Connect product itself uses the two-dot versioning scheme.

So how many versions should you keep around? That question has an *it depends* answer. It depends on how many levels are in your version numbering scheme, how many consumers you have, and how quickly they can migrate to the new versions. A general rule of thumb is to keep three of the most granular versions for each of the higher-level versions. So let's say you have a simple one-dot versioning scheme that represents `<major>.<minor>`. In this case, you will keep three major versions, and then three minor versions for each major version. This would give you a total of nine versions in production at a given time. If you extrapolate that out to the two- and three-dot approach, you can see how you would have a lot of versions to manage at a given time.

Now that you have decided on your versioning scheme, you have to think about how to implement it in your API and product lifecycle. Again, we must think of the implementation with the consumer in mind. We wouldn't want to just retire a version of a product without a proper warning and opportunity for the consumers to test and adopt the new version. If you recall our API and product lifecycle we discussed in this chapter, you know that there are basically two states that our products can be in where a consumer can consume the APIs within. These are the published and deprecated states. So, when we say that a good rule of thumb is to keep a maximum of three versions for each level in your versioning scheme, we mean in a consumable state. If you are following this three-version rule of thumb, you might have one version in a published state, one in a superseded state, and one in a deprecated state.

Versioning scheme for a healthcare organization

Let's take the fictional healthcare organization we created in earlier chapters and suppose we created and published an API named **Member API**. We will be following the default two-dot versioning scheme as <major>.<minor>.<fix>. This will mean that our first version of this API is created as 1.0.0. Now, the time comes where a bug is discovered and we need to implement a fix.

You already know how to create a new version of your API and product so once you do that, you can straight away publish your product if you want to have two versions in the published state. As you will recall from the *Understanding the API and product lifecycle* section, if you are planning on deprecating the current version, you can also use the **Supersede** option to publish your new version and deprecate the existing one.

As time goes by, so will bug fixes and updates to our APIs as we follow our versioning scheme and communicate these changes to our consumers. Following our API and product lifecycle, they would go from published to deprecated and finally retired. Taking a look at *Figure 11.18*, you can see our **Member API** has gone through a couple of revisions and updates and now has three different versions. We have 1.0.2 as our most recent version in a published state, version 1.0.1 is our *n-1 version*, also in a published state, and version 1.0.0 is in a deprecated state.

As mentioned in the previous section, we will need to eventually retire some of the older versions to make room for the newer ones. Recall that we had advised a maximum of three versions under each of the higher versions in our versioning scheme. As we hit our limit for a particular version branch, we wouldn't want to just retire that version without any warnings to our consumers. We also wouldn't want to have consumers adopting a particular version of an API just before we retire it. This is why we need to deprecate versions as they come closer to being retired. When we deprecate a version, we notify current consumers that it will soon be retired and do not allow new consumers to use it. But when do we deprecate a version? If we take our limit of three versions per higher version, as a particular version becomes the third and final version, this is when we would want to deprecate it. So, we would have the most current version, the n-1 version, and lastly, we would have our deprecated version.

Figure 11.18 shows our **Member API** with three different versions. Since we have hit our maximum number of versions for our 1.0 branch, we have version **1.0.2** being the current version, version **1.0.1** is our n-1 version, and version **1.0.0** is our deprecated version.

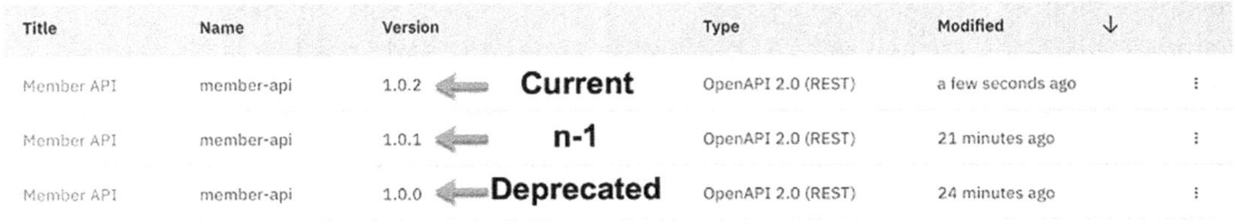

Title	Name	Version		Type	Modified	↓
Member API	member-api	1.0.2 ⬅	**Current**	OpenAPI 2.0 (REST)	a few seconds ago	⋮
Member API	member-api	1.0.1 ⬅	**n-1**	OpenAPI 2.0 (REST)	21 minutes ago	⋮
Member API	member-api	1.0.0 ⬅	**Deprecated**	OpenAPI 2.0 (REST)	24 minutes ago	⋮

Figure 11.18 – Member API with three versions

Of course, when our next version comes along, we would retire our version **1.0.0** and all other versions would move up one level, making room for our new and current version.

As you establish your versioning scheme and procedure, you will most likely not want to enforce it within your development environment as you will be making many changes to your APIs as you develop, publish, and test. If we were to apply versioning to every change, imagine the version you would ultimately go to production with. The same applies to your other non-production environments. As you go through your SDLC process for a change to your APIs, the outcome should be a single new, tested version in your production environment. The fact is, your versioning is really for your consumers, who will only see the APIs and Products in your production environment where you should be diligent about reflecting every change as a new version. To account for this churn in your non-production environments, API Connect provides an option at the Catalog level named **Production mode**. This setting can be found by navigating from the home screen in API Manager to **Manage Catalogs | <Catalog> | Catalog settings**, as shown in *Figure 11.19*:

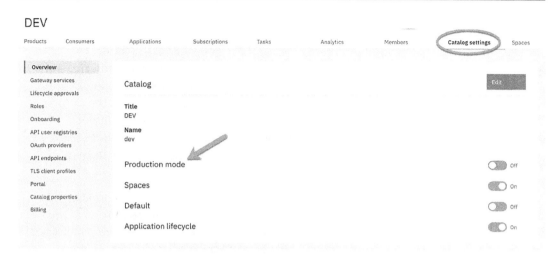

Figure 11.19 – Production mode within a Catalog

As you can see, this is an option with an **On/Off** toggle switch. When this option is set to **Off**, your APIs can be published even if there are conflicts. For example, you can repeatedly publish the same API with the same version over and over. You can see how this option would be needed for your non-production environments. Conversely, when this option is in the **On** position, you must resolve all conflicts before the API and product can be published. This would then prohibit you from publishing APIs and products with the same name and version as an already published API or product. You can see how this stringent enforcement would be desirable for a production environment. Of course, there might be situations where you would like to overwrite an existing version. For example, you may have APIs that are strictly internal with only one consumer so this can be managed with the same version.

We have discussed several different versioning schemes in this section and demonstrated one in particular. As each organization is different in its own way, you may choose to adopt one of these options or possibly create your own. For example, you may only have a handful of consumers that are all internal so your versioning scheme may not need to be very granular. In this case, you may choose the simple one-dot versioning or maybe even just overwrite the same version with your changes. Or, perhaps you are following a specification, as we saw with FHIR in *Chapter 6*, *Supporting FHIR REST Services*, where the version of your API represents the actual FHIR version you are following. Or, maybe you will use a hybrid of them. The point is, your versioning scheme should be consistent, representative of the types of changes, and easily communicated and understood by your consumers. As with all governance, you must come up with a plan that works for your organization and follow it.

Segregating environments

In the beginning stages of your digital transformation journey with API Connect, you will be having discussions about vendor products, licensing, physical hardware, and so on. All of this will be dependent on how many different environments you have within your **SDLC**. Each environment will have its own set of requirements, such as the number of servers required, high availability, and security. As you know, each environment will serve a different purpose within the SDLC, but should not have a physical dependency or interaction with other environments. That is, each environment should be its own self-contained, independent entity.

When we talk about environments being independent and isolated, this could mean different things for different environments. For example, you may only require a logical separation between your development and QA environments where they share resources, whereas your production environment will likely be physically isolated from all other environments. You might also have a performance test environment that will need to be more production-like so that you can get more realistic performance metrics when running your tests. Or, maybe you might have an entirely segregated PCI environment where it is on separate hardware segregated on an isolated network. The point is, you will need to segregate your environments from each other in one way or another. Let's take a look at the different ways we can segregate each environment from the others.

Logical segregation

Let's face it, having all of your environments physically segregated so that no one environment can interfere with another sounds ideal. I'm sure you have had your share of experiences where a developer is trying something new in the development environment that brings down the entire infrastructure or its runtime halts all QA testing. Or maybe there is a performance test being conducted that is meant to push the limits of the infrastructure and that brings down all other environments. As much as it would be nice to not have to worry about this, the reality is, standing up separate hardware, acquiring new licensing, and managing separate infrastructures for each environment can get very costly. This is why you will likely combine some of your lower environments into one single API Connect installation and provide logical segregation so that each environment can act and perform independently.

As we covered the logical components of your API Connect environment in *Chapter 3, Setting Up and Getting Organized*, you saw how we create logical segregation at several different levels. As a refresher, take a look at *Figure 11.20*, which shows how the API Connect environment is organized within each provider organization.

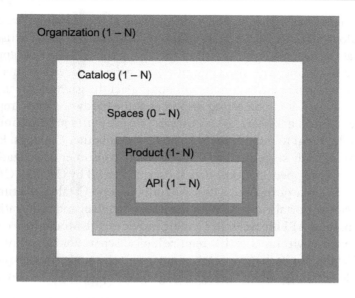

Figure 11.20 – API Connect organization structure

Each layer shown in *Figure 11.20* depicts another level of isolation and segregation. Of course, you are not likely to name your products and APIs by the environment, so we are talking about provider organizations, and Catalogs when discussing opportunities for environment segregation.

At the topmost level of our hierarchy shown in *Figure 11.20*, is the provider **Organization**. We know that you can configure multiple provider organizations within one cloud manager, so this is your highest level of segregation that you can have within one API Connect installation or cloud manager. At this level, you assign a single provider organization owner who has total control over the provider organization setup, who invites members, creates Catalogs, and so on. At this high level in the hierarchy, you can choose to create a provider organization for each environment, or you can configure this at the top level of your organization. You may also configure separate provider organizations for subsidiaries or branches of your company. The important thing to remember when making this decision is that at the highest level of isolation within the cloud manager, members of one provider organization cannot view or manage anything within another provider organization, including Catalogs, spaces, products, and APIs. This level of isolation should help guide you in making your decision as to how you want to organize and configure your provider organizations.

Moving on to the next layer in our organization, we look at the **Catalog**. You can have one or more Catalogs within a provider organization. You can also provide user access and permissions at the Catalog level, as well as specifying the API lifecycle approval process as we have seen. Since you also configure your gateway service at this level, any runaway API running on one environment will not impact the gateway or runtime on another. Unless, of course, it is exhausting resources at the hardware level, impacting everything on the physical hardware. So as you can see, the entire product and API lifecycle from development to publishing is isolated from all other Catalogs. Even the developer portal is logically segregated by Catalog where consumer organizations can be created, apps developed, and products subscribed to, all by Catalog. Of course, there is only one developer portal, but a user will only see one Catalog at a time as if it's a separate developer portal entirely. With the level of control and segregation of the access, permissions, and API lifecycle the Catalog provides, this would be a convenient level to configure your environments that require logical separation. Creating a separate environment or Catalog for each of your lower environments under one provider organization would certainly satisfy the logical segregation you might be looking for.

Throughout this and the previous chapter's discussions on topics such as API development, the developer portal, the API and product lifecycle, users, roles and permissions, and gateway services, we noted how each can be set up and controlled down to the Catalog or space level. Components that are not overridden at this level, such as user registries, will be inherited from the provider organization.

At the lowest level of this structure, we have **spaces**. Although spaces are another level of isolation within the Catalog, they inherit most of their configuration from the Catalog itself. Most importantly, the developer portal. Because of this, spaces would not be a good candidate for isolating your different environments. You can refer to *Chapter 3*, *Setting Up and Getting Organized*, to learn more about spaces.

What's New in V10.0.1.5?

There are new capabilities introduced in v10.0.1.5 that provide more flexibility with Catalogs and spaces. Visit What's new in the latest release (Version 10.0.1.5): `https://www.ibm.com/docs/en/api-connect/10.0.1.x?topic=overview-whats-new-in-latest-release-version-10015`.

As you are digesting the different options presented here to logically segregate your environments, let's think about one more option, which is a hybrid approach. Until now, we have been talking about one layer of segregation, which is the environment within our SDLC. Mostly, this logical separation is for our lower environments, which would not include production because we will likely have some physical segregation for that. So, we are dividing our environments into production and non-production for obvious reasons. What if we take that one step further and divide our non-production environments as well? As we mentioned, your development environment will likely have fewer controls, more changes, and more issues. Because of this, you might think about separating your development and testing environments from each other. This is where a hybrid approach might come in handy. Perhaps you might provide this development and testing environment at the provider organization level, and then the different environments within them at the Catalog level. *Figure 11.21* shows how that might look:

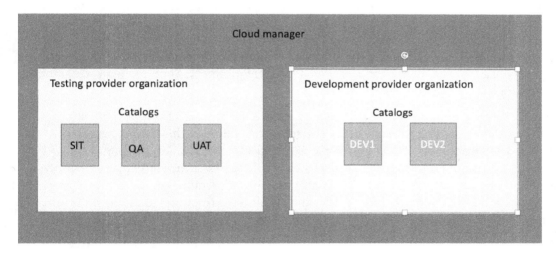

Figure 11.21 – Provider org and Catalog environment segregation

We have presented different levels of logical segregation for each environment within our SDLC, which will satisfy all the appropriate logical segregation for the entire API lifecycle. So, which one should you use? There is not a one-size-fits-all answer here. It is entirely up to you. You must decide which level of segregation fits into your organization. API Connect provides different options and much flexibility for you to fit your needs. With this comes decision-making on how to make this segregation work for your organization.

Physical segregation

Of course, logical segregation is not always the answer for all environments. Remember the cowboy developer experimenting with some new idea that is consuming all of the machine's resources? Or maybe just a small piece of code going into an infinite loop! Either way, any new and untested piece of code has the ability to wreak havoc on the hardware and infrastructure it resides in. And your code is not fully tested until it goes through your entire testing process, including performance testing. The point is, any of your lower environments have the potential to inflict unintended harm on the environment impacting all other environments on the shared infrastructure. This is why we wouldn't want to cohabitate our production environments with our lower environments. Besides, your production environment is likely to be more robust, providing high availability, failover, and security from a physical network perspective.

Physical segregation might also be required for other environments where you will need a production-like configuration such as for performance testing where you want to see how the environment will behave under a production load. Or, perhaps you have an entirely isolated production environment that is more secure, such as for PCI processing. Or, maybe a separate internal and external facing environment. There are many situations where logical segregation will not suffice.

Physical segregation of your environments provides the benefits of complete isolation where it shares nothing with any other environment. Although this may be necessary for some environments, it does come at a cost. Basically, this is a new installation of API Connect. This not only requires new licensing but think about the entire planning process when setting up a new API Connect installation. You will require new hardware, a new network and DNS configuration, new certificates, and the resources to perform the initial install and configuration. Although this sounds like a lot of work, it is necessary to achieve complete segregation for your most critical environments.

As we conclude this section on environment segregation, you can see that you have several options, each with its own set of pros and cons. Understanding the cost and benefit of each along with your hardware requirements will help guide you through planning and implementing your own environment segregation strategy. Keep in mind that you can certainly mix and match these as you see fit. For example, you could use Catalogs to segregate your pre-production environments within one provider organization and have a completely new installation of APIC for your production environment. This would give you the cost-effectiveness and simplicity of managing multiple environments within one API installation for non-production while providing the reliability and physical segregation required for a production environment.

Let's take a look at a high-level table of pros and cons for each method of segregating your environments.

Environment segregation method	Pros	Cons
Catalog	• Cost-effective • A single configuration under one provider org • A single view of all Catalogs/ environments	• Users have access to view environments. Possible security risk. • Difficult to manage all in one place. • Risk of destabilizing all environments if one presents issues.
Provider Organization	• Cost-effective • Separate access and roles for each provider org • Can manage each environment separately	• Risk of destabilizing all environments if one presents issues
New APIC install	• Provides isolation for each environment • Can scale each environment as needed • User access managed independently • Issues in one environment will not impact others	• Cost of new APIC license and hardware for each environment • Time and money spent configuring each installation • Managing each installation separately

Once you have your segregation strategy established, you must establish a set of standards to configure and manage each environment. In the next section, we will discuss the types of standards you should establish, as well as the importance of them.

Applying standards to your API environment

Up to now, we have discussed the governance process for organizing and managing your API lifecycle, including roles and permissions, approvals, and environment setup. These are all critical to the governance process, but they are mostly enforced by the initial configuration of your API Connect environment. That is, once you decide on how to implement these, and configure your installation appropriately, the restrictions put in place will enforce the rules. As you know, your API factory is a living and ever-changing cycle. Over time, you will be creating new APIs and changing existing ones. As this happens, there are certain aspects of the configuration that are not as easily controlled. These are the standards that should be established, monitored, and enforced to make your API organization a success.

As we have mentioned before, API creation is very different from our old application development where we could just develop some code or application and as long as it served its purpose, everyone was happy. If you think about some of your old running code in your production environment today, does anyone really know or care what you named your modules or functions within? Comments and descriptions were only helpful to someone debugging your code. In the API world, this all changed. Of course, your API still needs to work as designed, but there is just as much focus on the discoverability of your APIs. Are they easily searched? Are they self-descriptive? Are they grouped logically? Unlike your traditional application code, your APIs and products are all on display for the world (or at least your consumers) to see! Your thought process needs to be more consumer-centric. Because of this, you need to keep your API house clean by applying and enforcing some strict standards. Let's take a look at some standards to think about when developing your APIs.

Discoverability

Keeping the consumer-centric theme in mind, you need to think about how the consumer can discover your APIs. Of course, they will be published to the developer portal within a product, but you cannot expect a consumer to scroll through all products and APIs to find the one they need. Consumers will likely be using the search feature within the developer portal to discover the APIs that might suit their needs. This search functionality is only as good as the metadata associated with your APIs; specifically, the description you provide when creating your APIs. Although this is an optional field when creating your APIs, it should be a hard requirement for you to always provide a complete description, keeping in mind keywords and phrases that might be used in a consumer's search criteria.

In addition to consumers discovering your APIs and products via the search feature, they also have the ability to filter APIs and products by category. Adding categories to each of your products and APIs will make it much easier for your consumers to narrow their search results by category to see any and all products and APIs that may be of use to them in a specific category. Again, the easier it is to discover your products and APIs, the more successful you will be.

As you begin developing your APIs and adding them to your plans, it is important to take the time to consider this logical grouping of APIs in your products. With the consumer in mind, think about how this grouping would be useful and make sense to them. Consider the logical grouping of your APIs within a product somewhat of a logical category in and of itself. As your consumers search for and discover your products, they would likely find more than one API within a product useful if they are logically grouped together. Think about the different types of use cases each consumer might have, how different APIs fit into that, and also how your versioning strategy could coincide with the product as a whole.

Naming standards

Defining, implementing, and enforcing naming standards within application development has always been a challenge. We define a set of loose standards, which are more like guidelines, but then rarely enforce them. Again, since you are advertising your products and APIs, it is a good idea to at least have some consistency, as well as conveying meaningful information within the name.

It is likely that most, if not all of your APIs will be RESTful APIs. Since your API name will become part of the URI used to invoke it, and REST has a strict URI naming convention, there is no need to come up with your own naming convention for your API names.

A major part of the REST architectural style is the emphasis on HTTP methods. The method used to invoke a REST service will dictate the action of the verb that is to be taken. Think about the GET, PUT, POST, and DELETE methods and how they describe the verb or action of the request. Since the verb is already built into the HTTP method, there is no need to specify it within the API. So if you had an API that provided the ability to get some data, you would not use the verb `get` within the API name because that would become part of the request URI itself. Instead of naming an API `getPatientInfo`, you might just call it `patientInfo` or just `patient`. You see, the REST guidelines for URIs focus more on the noun than on the verb. There is much more to REST naming conventions that can be found on the internet for your reading pleasure. One very informative article can be found at `https://restfulapi.net/resource-naming/`.

The point here is not to go into the complete REST naming convention but to understand how your API name makes its way into the URI itself, so when naming your APIs, you should consider the URI guidelines.

Aside from having an already established naming convention to help guide your API names, how you name your APIs and products should be more about providing short, yet descriptive names for them. And just as important, be consistent with the names themselves. Of course, you should be consistent with the case convention used within your names, but more importantly be consistent with naming your resource or noun your API deals with. For example, if you have a hotel organization and some APIs refer to people staying in the hotel as guests and some APIs refer to them as customers, it will become difficult for potential consumers to see all within a single search as they may be searching for one or the other.

In addition to the REST naming conventions that specifically deal with the HTTP request/response conventions, you also need to consider the OpenAPI Specification, which defines the standard for interface to these RESTful APIs. The OpenAPI Specification is meant to provide a language-agnostic interface to your APIs. Having this specification is critical to providing all consumers, computer and human, with a way to easily identify the capabilities of your APIs. This specification covers many aspects of your API interface, including documentation, metadata, data types, and so on. This specification should guide you to develop and implement many standards within your API development and implementation process.

The latest OpenAPI Specification can be found at `https://swagger.io/specification`.

As you work your way through establishing your naming conventions for your organization's products and APIs, you will hopefully take some help from other architectural standards such as REST, as well as deriving some from your particular industry. Whatever way you come up with your naming standards, the most important part is being consistent with them and finding a way to enforce them through a strict approval process throughout your API lifecycle.

In this section, we have introduced the concept of applying standards throughout your API environment, development, and lifecycle process. As we have only presented a few important standards to consider, there will likely be many more that your organization will want to implement. This may be driven by existing standards throughout your organization or perhaps new standards that you see as important and beneficial. No matter how many or how few standards you implement, it is critical to have a set of standards that makes for a well-structured, stable environment and process. This list may evolve as your API environment and process matures. The most important part of standards is that they are followed and enforced.

Summary

As we conclude this chapter on the dreaded *governance* word, you should now have a better understanding of the what, how, and who of the governance process. You now know what you have to govern, how to accomplish it, and who should govern the process. Possibly one of the most important takeaways that you hopefully got from this chapter is the *why*. We all have heard that governance is important and I'm sure most would agree with it.

As we have pointed out throughout this chapter, the need for strict standards and governance within your API environment is more critical than ever. Hopefully, with these guidelines, examples, and explanations, the word *governance* is not as scary and dreaded as it once was. If defined and configured properly, it will become part of the everyday life of your API lifecycle and you will see the benefits of having a clean API house where your consumers can discover, subscribe, and consume your APIs easily.

12
User-Defined Policies

API Connect (APIC) is designed to provide a user-friendly, drag and drop experience for creating APIs within the Assembly screen of the **graphical user interface (GUI)**. This provides a very efficient means to create, expose, and secure your APIs without having to know all of the underlying details. In most cases, the work is done for you under the hood. APIC provides a plethora of policies that can be used within the processing of your API requests and responses, ranging from transformations to custom code (XSLT or GatewayScript), to validation policies. Although this drag and drop functionality provides a simple interface to get your APIs created quickly, there may be times when you might feel a little restricted and want to customize your own policies. Perhaps you have a set of built-in policies that you use throughout many of your APIs and would like to package up as one reusable policy. Or maybe you are very familiar with the DataPower product itself and how to develop services. Knowing that your API runtime is actually a DataPower gateway, you might be tempted to use the built-in DataPower capabilities within your APIs. All of this customization is possible in APIC by using **user-defined policies (UDPs)**. These customizable policies provide you with the ability to create your own APIC policies that will show up on the palette for use when developing your APIs.

Throughout this chapter, you will learn the different types of UDPs that are available, their different use cases, and how to configure and install them.

In this chapter, we will cover the following topics:

- Introduction to user-defined policies
- Creating a new user-defined policy
- Implementing your new policy in the Designer
- Deploying your policy

By the end of this chapter, you should have all of the knowledge required to create your own UDPs and make them available to be used within other APIs in your organization.

Technical requirements

The examples presented in this chapter will require an installation of API Connect or LTE to configure your own custom policies. For configuring your global UDPs you will also require access to a DataPower Gateway. An intermediate to advanced level of experience of working with the DataPower Gateway and service development will be required to follow along and build your own UDPs, which will involve using the built-in DataPower functionality.

The configuration files created within this chapter can be found in the GitHub repository here: `https://github.com/PacktPublishing/Digital-Transformation-and-Modernization-with-IBM-API-Connect/tree/main/Chapter12`.

Understanding user-defined policy types

Before we dive into the configuration of user-defined policies, it is important to understand the different types available, as they are quite different in their use cases, configuration, and implementation. There are two different types of UDPs that you can create, which will provide a level of customization that should satisfy most, if not all, of your requirements when the built-in policies just won't do the trick:

- **Catalog scoped user-defined policies**: This UDP is defined at the Catalog level and is only available to APIs within that Catalog. Only built-in APIC policies can be used within this type of policy.

- **Global scoped user-defined policies**: Once created, this type of UDP is available for use within all Catalogs and Provider Organizations. This type of UDP can also incorporate DataPower functionality.

Throughout this chapter, we will discuss these two types of UDPs along with each of their particular use case, benefits, and, of course, how to configure them.

Creating catalog scoped user-defined policies

As you continue down your API development path, you might find that you and other developers within your organization continuously use the same set and sequence of policies within all, or many, of your APIs. Perhaps this is a standard laid down in your development organization that all APIs must implement. To ensure consistency and increase efficiency, reusability is always a good choice. Creating a Catalog scoped UDP will allow you to wrap all of these built-in APIC policies in one reusable policy that will show up on the assembly palette when developing APIs within the defined catalog. Remember, a Catalog scoped UDP can only use built-in APIC policies and is only available to APIs with the defined Catalog.

Before defining a UDP, you would likely want to design and test the policy flow that you want to make reusable. To demonstrate this, let's take a look at an example UDP for our healthcare organization that we might want to build. For this example, we have identified a common logging framework that we would want to use for every API we create within our Catalog. For our logging, we want to first redact some sensitive data that we cannot log. From there, we can send the logging data to be logged. Finally, we have some custom logging that we implement via a custom GatewayScript. This custom code will utilize an input parameter named env to make some logging decisions, such as the logging destination. This parameter will be passed into the UDP for each implementation.

To begin creating your new UDP, you will first need to create and test an actual implementation of the sequence of policies you wish to incorporate. To do this, you will create a sample API, configure the policies, and finally test to be sure you have achieved the expected results. For our implementation, we have created the policy flow shown in *Figure 12.1*:

Figure 12.1 – Policies for custom logging flow

In our example policy flow shown in *Figure 12.1*, we have the following policies that will execute in order:

1. log-gather – A log policy set to **gather-only** that will gather all logs
2. redact – A redact policy that is configured to redact sensitive data
3. log-send – A log policy set to **send-only** that will log the redacted log data
4. GatewayScript-CommonLogging – A custom GatewayScript that expects a parameter to be passed in and logs to a remote log server

You have now established a working model for your custom flow. Your next step in the process will be to build your configuration .yaml file that you will use to configure your custom policy.

Building your UDP .yaml file

Now that you have your assembly representing the common flow that you want to make reusable, you can start to build your .yaml file that will define your UDP. The .yaml file that you will build will consist of six sections, which we will discuss in detail now.

Specification version

The first line of your .yaml file will be the specification version section. This is a one-line section where you add the specification version as follows:

```
policy: 1.0.0
```

Information section

The next section will be your information section. In this section, you will provide some basic information about your UDP, such as name, title, version, and description of your UDP, as well as the contact information for the developer or support resources. The following is the format for this section, with the information provided for our example UDP:

```
info:
  title: Redact and Log
  name: redactlog
  version: 2.0.0
  description: Redact sensitive CC and custom  data and log
  contact:
    name: Jim Brennan
    url: https://github.com/apic
    email: jim.brennan@apichealth.com
```

Attach section

The next section in this file will specify which types of APIs this UDP can be used for, such as `rest` or `soap`. At least one type is required here, but both can also be included. The format for this section is as follows and shows that both `rest` and `soap` types can be used for this UDP:

```
attach:
  - rest
  - soap
```

The gateways section

The `gateways` section will specify to which API gateway this UDP can be applied. This value will specify the type of API gateway and not the name of your gateway service itself. The possible options for this section are `micro-gateway`, `datapower-gateway`, and `datapower-api-gateway`. These values will specify the type of gateway that you have implemented in your catalog.

The format of this section as well as the values for our example UDP are as follows:

```
gateways:
  - datapower-api-gateway
```

The properties section

The `properties` section will now begin to define more of the implementation of your UDP. Specifically, this section will define any and all parameters that can, or must, be provided when using this particular UDP. If you recall in our example, we had a GatewayScript that requires a parameter named `env` to be passed into it. It is in this section that you define such parameters. Here, you will define each parameter, providing its name, properties, and whether or not it is required.

Let's now take a look at the format of this section by showing how this would be represented in our example UDP, which requires one parameter of the `string` type:

```
properties:
  $schema: 'http://json-schema.org/draft-04/schema#'
  type: object
  properties:
    env:
      label: Environment
```

```
      description: Runtime Environment
      type: string
  required:
    - env
```

The values for the $schema and type fields are static and will remain the same for all of your UDPs.

The following fields will be specific to your particular UDP definition and defined parameters. You can define one or more parameters within the properties section.

- Properties: <name of parameter> Specify the name of the parameter you are defining. This will be the name referenced at runtime when accessing the parameter.

- label: The display name you wish to give your parameter. This value will show in the assembly when implementing and configuring the UDP.

- description: A brief description of the parameter.

- required: A list of all the defined parameters that are required when implementing this UDP.

Assembly section

Until now, we have defined mostly metadata regarding our UDP, which included the name of the UDP, where it can be used, and what is required when using it. It is here in the assembly section that we define the implementation. More specifically, we define the policies and their configuration here. If you recall at the beginning of this chapter, before we began building the .yaml file, we discussed building the actual implementation of your soon-to-be UDP by creating and testing an example of the assembly in a test API. That is essentially our working model that will be used to create a reusable artifact. Our example in *Figure 12.1* showed our logging flow that we would like to make into a reusable policy. It is the result of this working configuration that will provide us with the assembly section for our UDP.

If you go back to this working example in the Gateway tab where you created the flow, you can click on the </> icon, which will show you the implementation .yaml file for this API. If you scroll down to the assembly section of this .yaml file, you will find all you need for the assembly section of your UDP .yaml file. You can simply copy this section from here and paste it right at the end of your UDP .yaml file.

Figure 12.2 shows the source for our example that we configured and showed in *Figure 12.1*. Note that the entire configuration for this flow is contained within this assembly section as well as the actual GatewayScript itself.

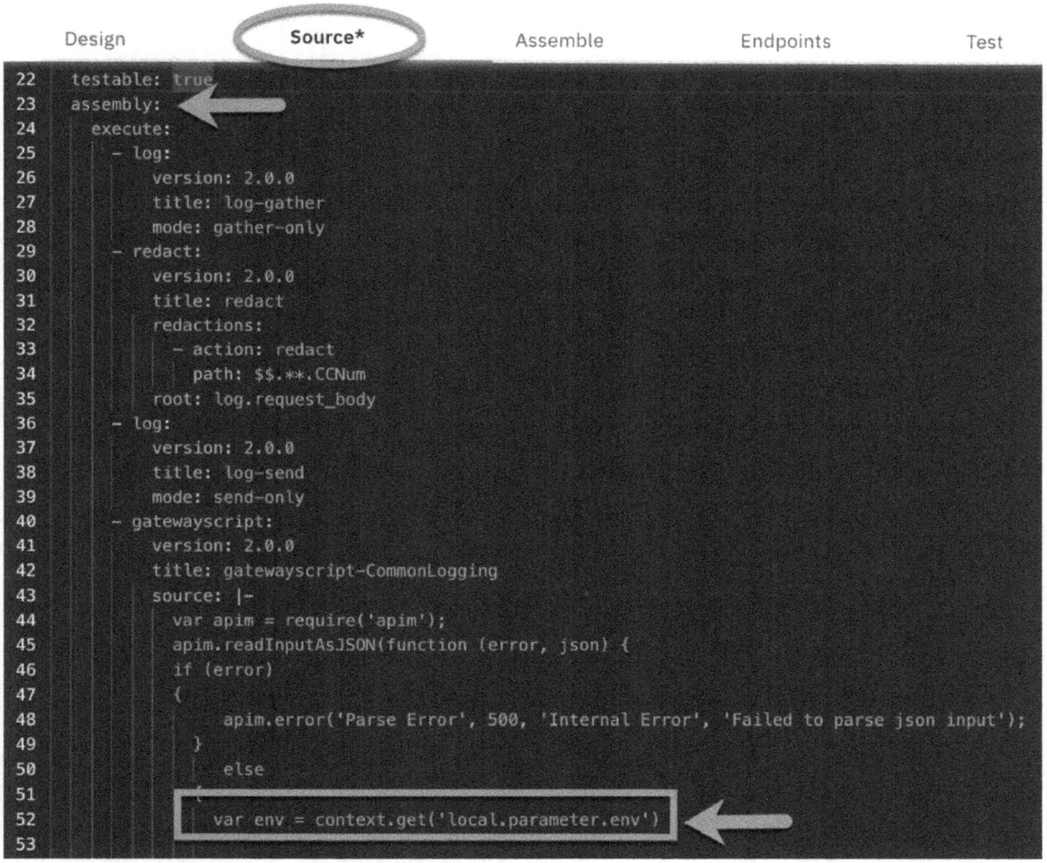

```
Design              Source*              Assemble          Endpoints          Test
22   testable: true
23   assembly:
24     execute:
25       - log:
26           version: 2.0.0
27           title: log-gather
28           mode: gather-only
29       - redact:
30           version: 2.0.0
31           title: redact
32           redactions:
33             - action: redact
34               path: $$.**.CCNum
35           root: log.request_body
36       - log:
37           version: 2.0.0
38           title: log-send
39           mode: send-only
40       - gatewayscript:
41           version: 2.0.0
42           title: gatewayscript-CommonLogging
43           source: |-
44             var apim = require('apim');
45             apim.readInputAsJSON(function (error, json) {
46             if (error)
47             {
48                 apim.error('Parse Error', 500, 'Internal Error', 'Failed to parse json input');
49             }
50               else
51             {
52               var env = context.get('local.parameter.env')
53
```

Figure 12.2 – Assembly section of the working policy flow

You will also notice in this GatewayScript the reference to the `env` parameter that is required to be passed in, as shown in the preceding screenshot.

Completed .yaml

And there you have it! A completed `.yaml` file that fully encapsulates all of the components required to wrap an entire assembly flow into one reusable policy. The following code block shows the completed `.yaml` file with the contents of the assembly section omitted for brevity:

```yaml
policy: 1.0.0
info:
  title: Redact and Log
  name: redactlog
  version: 2.0.0
  description: Redact sensitive CC and custom  data and log
  contact:
    name: Jim Brennan
    url: https://github.com/apic
    email: jim.brennan@apichealth.com
attach:
  - rest
  - soap
gateways:
  - datapower-api-gateway
properties:
  $schema: 'http://json-schema.org/draft-04/schema#'
  type: object
  properties:
    env:
      label: Environment
      description: Runtime Environment
      type: string
  required:
    - env
assembly:
```

Now that we have completed the `.yaml` file that defines our UDP, it is almost time to implement it. The final step for preparing your configuration file is to simply generate a `.zip` file containing your `.yaml` file. For our example, we created a `logging_UDP.yaml` file and zipped it into a `logging_UDP.yaml.zip` file. Now we are ready to implement our custom UDP!

Installing your UDP

As we stated at the beginning of this chapter, a catalog UDP has two limitations. First, it must only use policies and functionality available within APIC. Second, it must be implemented at a catalog level. Given that it must be implemented at the catalog level, you must perform the following steps for every catalog where you wish to have this UDP visible:

1. From within the **API Manager** home screen, navigate to the **Manage catalogs |** `<catalog name>` **| Catalog settings** tab. From this screen, click the **Gateway services** link on the left navigation menu to bring you to the configured gateway services for this catalog.

2. For the gateway service as regards for which you wish to implement this UDP, click the ellipses to the right of it and click **View policies**, as shown in the following screenshot:

Figure 12.3 – Selecting the gateway service to implement the UDP

3. You should now be presented with every built-in and custom user-defined policy available to this catalog and gateway service. In the top right of the main screen, click the **Upload** button to upload your new UDP.

4. You should now be looking at an **Import policy** screen where you can simply drag your `.zip` file, which contains your `.yaml` file, into the top box on the screen, as shown in the following screenshot:

Figure 12.4 – Uploading the UDP configuration file

Alternatively, you can click the top box to navigate to your local filesystem and upload the file. Once the file is uploaded, click the **Upload** button to complete the installation.

5. You will now be brought back to the **Policies** screen where you should see your new UDP at the top of the list. The following screenshot shows our new UDP, named **redactlog**, successfully installed:

Figure 12.5 – UDP successfully installed

You have now successfully installed your custom UDP in your APIC catalog and it is available for use. Let's now take a look at how we can implement this in your APIs within the catalog.

Implementing your UDP

Now that you have your UDP successfully installed, let's take a look at how we can use and test this new custom policy in our API flow. If you create a new or go to an existing API within the catalog where you installed your UDP, you can navigate to the **Gateway** tab as if you were creating your policy flow for your API assembly.

Something with which you are familiar with is the built-in policies available that you can drag onto the palette to create your assembly. If you scroll to the bottom of these available policies on the left, you should see the **User Defined** category with your newly created UDP listed there. You can now drag this custom policy onto your palette in your assembly, as you would any built-in policy. *Figure 12.6* shows our newly created UDP, which requires a parameter to be passed in. As you can see, once this is dragged onto the palette, you are prompted for the parameter value on the right side of the page. The value entered here will be passed to the policy that can be accessed at runtime by the policy configuration.

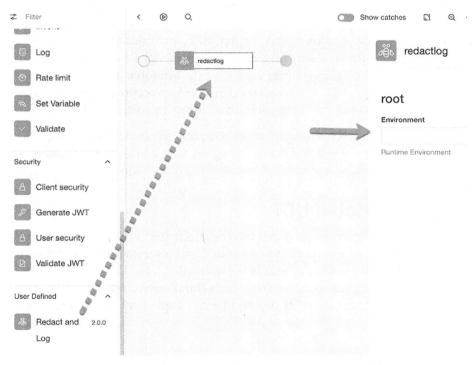

Figure 12.6 – Implementing the UDP in your API assembly

And there you have it! You have taken a sequence of several policies and wrapped them into one reusable policy available to all APIs within the catalog. You have now seen how you can create custom, reusable policies as well as how to configure them to accept parameters that can be utilized by the underlying configuration. This feature can prove very beneficial by allowing custom configurations to become reusable assets, thereby reducing development and testing time, as well as providing consistency in your API quality.

Now that you know how to create a catalog UDP, the next step is to take a look at the second type of UDP, the global UDP, which can be shared across any Provider Organization and utilize not only APIC features but also the built-in DataPower functionality.

Global scoped user-defined policy

As you should be aware by now, APIC uses a DataPower Gateway as its runtime for your APIs. Many API developers using APIC often have previous DataPower development experience and naturally ask the question *"Why can't I use any of the robust features and functionality that the DataPower product offers?"* This is a legitimate question, and many developers will take it upon themselves to create separate services on DataPower to implement this functionality, but this is not the way you should accomplish this. Much like you saw in the previous section of this chapter, APIC provides a way to package all of your DataPower-specific functionality into one shareable policy that can be shared and reused by all of your APIs by simply dragging it onto the assembly flow like any other policy. Unlike the Catalog UDP we discussed in the previous section, any global UDP created will be available to any API being developed in any Provider Organization.

In this section, we will walk you through how this is accomplished by providing a real working example. Keep in mind that a strong working knowledge of DataPower configuration is strongly recommended when venturing into this area.

Planning your global UDP

As a savvy DataPower developer, you might now be chomping at the bit ready to create your DataPower policies, rules, and actions on your DataPower device to be shared and used across your APIs. This is not necessarily the case with global UDPs. You are not creating DataPower configs the way you would for a DataPower service; however, you will create specific DataPower assembly objects that can implement certain DataPower functionality.

To make it clear as to what you can and cannot do within a global UDP, let's take a look at the specific objects that are created on your DataPower gateway service when creating a global UDP.

These steps are not required for creating your UDP; however, they will provide you with a better understanding of what is created for you, an understanding of what is available to you, and finally, the **command-line interface (CLI)** commands required for your UDP. If you have access to a DataPower Web GUI, you can simply log in and create an **Assembly Function** object. As you can see in *Figure 12.7*, you will provide a name for your **Assembly Function** as well as the ability to complete the **Title**, **Description**, **Scope**, and **Parameters** fields to be passed to it. All of these parameters will be discussed further when we configure our UDP configuration file.

From here, you will want to create a new **Assembly** object by clicking the + symbol as shown in *Figure 12.7*:

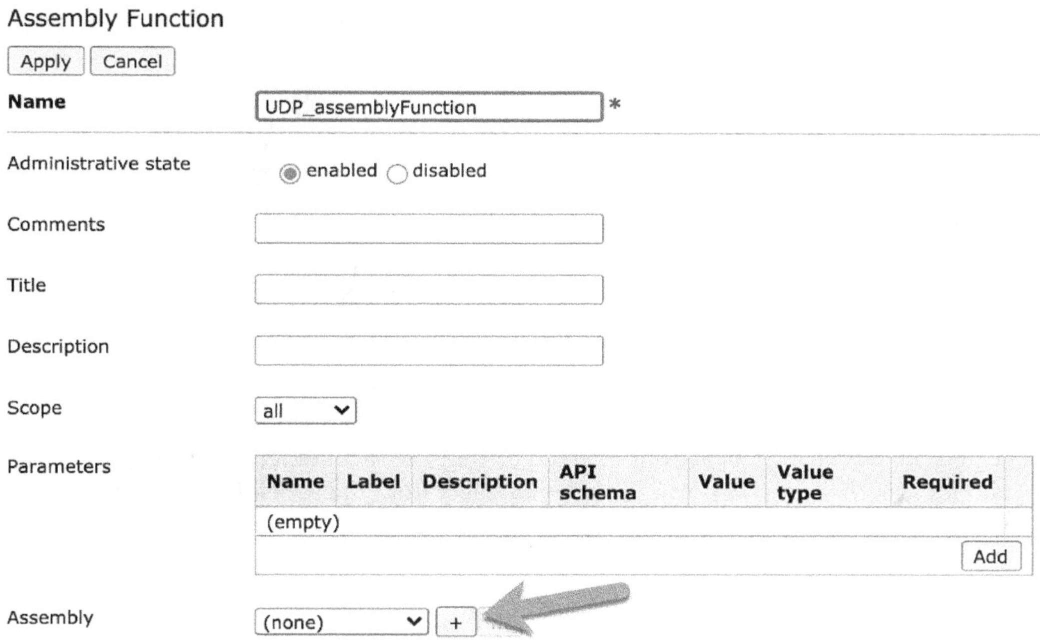

Assembly Function

| Apply | Cancel |

Name   UDP_assemblyFunction   *

Administrative state   ⦿ enabled ◯ disabled

Comments

Title

Description

Scope   all ⌄

Parameters

Name	Label	Description	API schema	Value	Value type	Required
(empty)						
						Add

Assembly   (none) ⌄ +

Figure 12.7 – Creating an assembly function in DataPower

Then, your assembly object definition will simply require a name and the creation of a rule, as shown here in *Figure 12.8*:

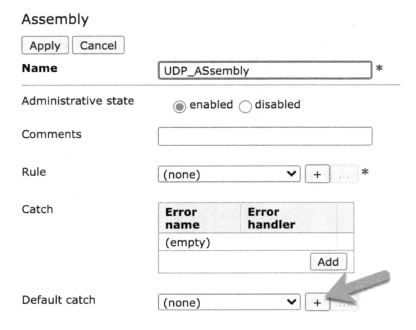

Figure 12.8 – Assembly object creation

From your **Assembly** configuration, clicking the + symbol will allow you to create a new rule for your Assembly. Within your rule configuration, you will specify which actions you would like to configure within your Assembly.

As you can see in *Figure 12.9*, by clicking the + symbol next to the **API actions** field, you will be presented with a list of actions to select from. Note that not all actions presented are valid for this configuration. You should only select from the **Assembly** actions.

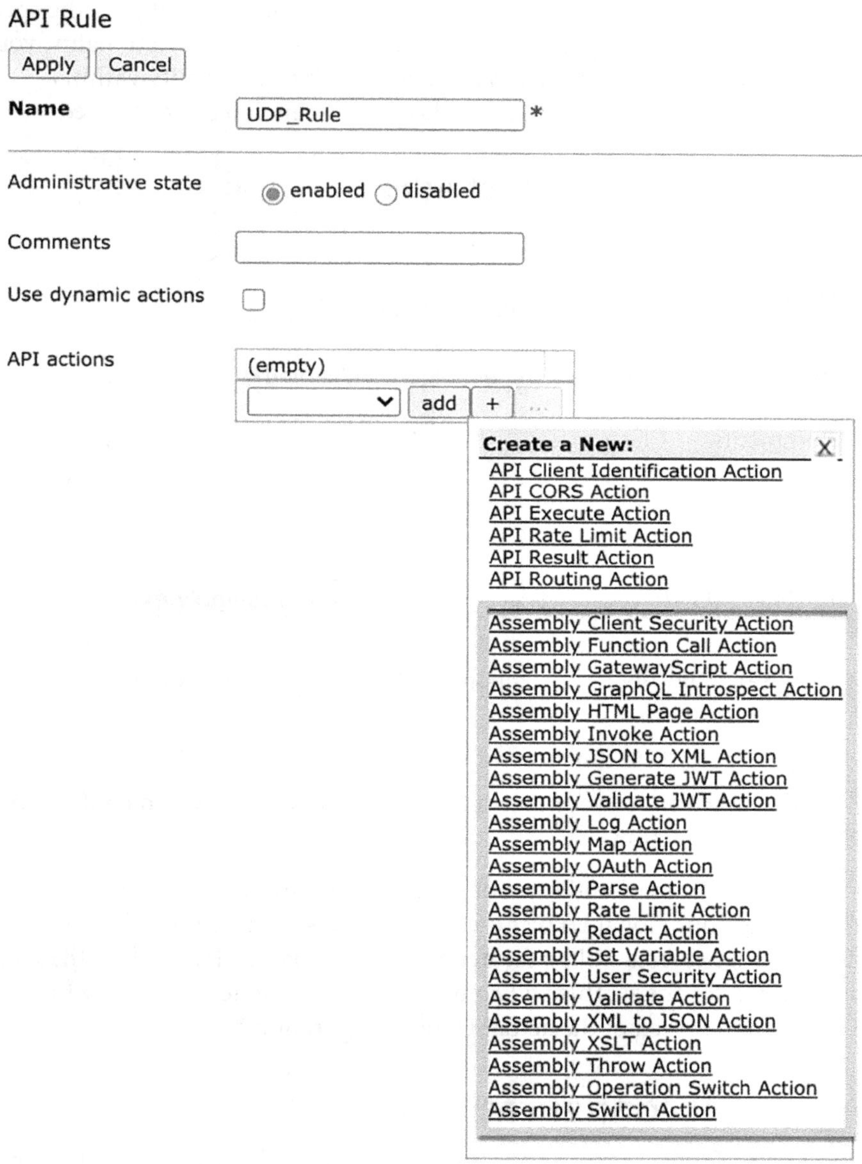

API Rule

Apply | Cancel

Name   UDP_Rule   *

Administrative state   ● enabled ○ disabled

Comments   [   ]

Use dynamic actions   ☐

API actions   (empty)

                    [∨] [add] [+] ...

Create a New: X

API Client Identification Action
API CORS Action
API Execute Action
API Rate Limit Action
API Result Action
API Routing Action

Assembly Client Security Action
Assembly Function Call Action
Assembly GatewayScript Action
Assembly GraphQL Introspect Action
Assembly HTML Page Action
Assembly Invoke Action
Assembly JSON to XML Action
Assembly Generate JWT Action
Assembly Validate JWT Action
Assembly Log Action
Assembly Map Action
Assembly OAuth Action
Assembly Parse Action
Assembly Rate Limit Action
Assembly Redact Action
Assembly Set Variable Action
Assembly User Security Action
Assembly Validate Action
Assembly XML to JSON Action
Assembly XSLT Action
Assembly Throw Action
Assembly Operation Switch Action
Assembly Switch Action

Figure 12.9 – Selecting actions for the rule

Once you select the action to implement within your rule, you will configure the specific parameters for that action. To implement most DataPower-specific functionality, you would select the GatewayScript action and implement this functionality within a GatewayScipt file. The following screenshot shows how this can be configured:

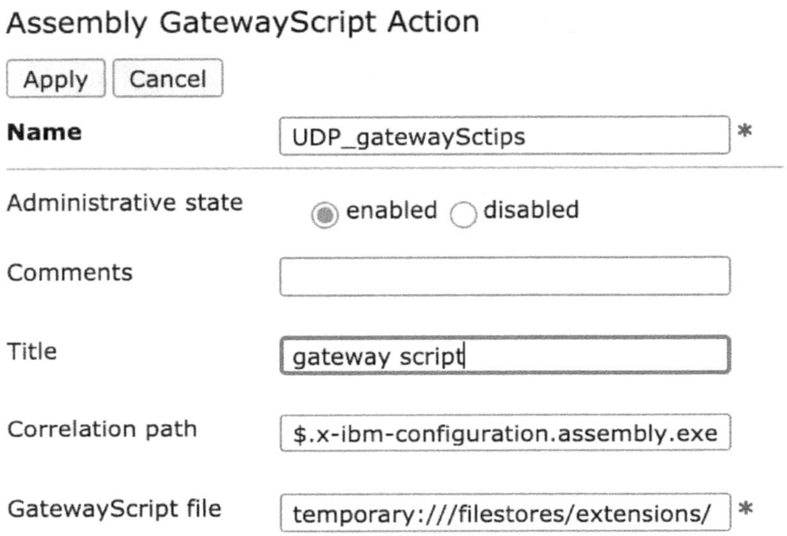

Assembly GatewayScript Action

Apply   Cancel

Name   UDP_gatewaySctips   *

Administrative state   ⦿ enabled ◯ disabled

Comments   [   ]

Title   [gateway script   ]

Correlation path   [$.x-ibm-configuration.assembly.exe]

GatewayScript file   [temporary:///filestores/extensions/] *

Figure 12.10 – Configuring Assembly GatewayScript Action

You can add multiple actions to your rule in the same way you did when configuring the Assembly GatewayScript action.

This process of creating these DataPower objects is not necessarily what you would do when configuring your global UDP; however, it is a good way to get a better understanding of what objects will be created for your global UDP as well as the CLI commands generated for them. This will come in handy in the next section, where we discuss the format and syntax of the actual UDP configuration file.

Creating a UDP configuration file

Much like the catalog UDP, we will need a configuration file that will define the complete UDP. For a global UDP, we will need to create this using DataPower CLI commands in a specific format that will ultimately be uploaded to your APIC environment. If you are familiar with DataPower CLI commands, this will look very familiar to you. If you are not, don't worry; there is a simple method for generating them (hint, you just did it in the previous section when creating DataPower objects!).

Since the configuration file for a global UDP is more DataPower specific, you will not create a `.yaml` file; you will be creating a `.cfg` file that DataPower will understand. The sections that you will define in this file might start to look familiar to you as they will be the CLI commands that would be used to generate the DataPower objects we discussed previously. In fact, if you are familiar with the DataPower filesystem and config files, you can actually copy these CLI commands from the `.cfg` file on DataPower after you create the objects within the DataPower Web GUI.

For now, let's take a look at the contents of a `.cfg` file for a global UDP. Let's take an example where we would want to set a variable and then generate a digital signature for the incoming request message for our API.

To begin our configuration file, we will define the actions required to accomplish the set variable and the digital signature. Let's take a look at the configuration of the set variable action first. If you refer back to *Figure 12.9*, where we showed all the available actions for a global UDP, you will see **Assembly Set Variable Action** listed. In our example scenario, we want to set a variable named `env`, with the value being specified when implementing the UDP in your assembly. This will be our first action in our configuration file and is defined as follows:

```
assembly-setvar udp-dsig-setvar_1.0.0_set-variable_0
   reset
   title "set-variable"
   correlation-path "$.x-ibm-configuration.assembly.
      execute[0]"
   variable
      action set
      name "env"
      type string
      value "$(local.parameter.environment)"
   exit
exit
```

In this configuration CLI, most of it is static values, with a few values you have to provide, such as the name of the action, `title`, `correlation-path`, the name of the variable to set, and the value of the variable. In this example, you can see that the value will be passed in as a parameter from the API configuration.

Our next action to define will be the action to generate a digital signature for our request message. Since APIC does not provide this ability, we will have to utilize the DataPower functionality to accomplish the digital signature generation. If you refer back to *Figure 12.9*, you will notice that there is no action available to generate a digital signature. As we mentioned earlier, you will need to rely on implementing most DataPower functionality within your GatewayScript itself. In our example, we would utilize the crypto function, `createSign()`, to generate the digital signature. Since GatewayScript and the available functions for DataPower functionality are vast, we will not be covering that in the book. A working knowledge of this topic is strongly recommended before implementing this functionality.

For now, let's assume we have created a GatewayScript file named `digitalSig.js` that contains the functionality to digitally sign the incoming message. Within our `.cfg` file, we must now define this action much as we did for our `set-variable` action. The following CLI is used to accomplish this:

```
assembly-gatewayscript udp-dsig-setvar_1.0.0_gatewayscript_1
  reset
  title "gatewayscript"
  correlation-path "$.x-ibm-configuration.assembly
    .execute[1]"
  gatewayscript-location temporary:///filestores
    /extensions/gateway-extension/digitalSig.js
exit
```

As you can see in the CLI shown for the GatewayScript action, you need to provide the `title`, `correlation-path`, and `location` fields of your GatewayScript file. When you deploy your global UDP, you will provide the GatewayScript file within the package that will be deployed to the `temporary:///filestores/extenstions/gateway-extensions` directory. This is the directory you will provide in your configuration file.

Now that you have provided the configuration for your actions, you need to define the additional objects within the assembly. Again, these will be the same objects that we demonstrated within the DataPower Web GUI previously. Next, we will define the API rule, which will list the actions previously defined in the order in which they should execute. The CLI for this configuration is as follows:

```
api-rule udp-dsig-setvar_1.0.0_main
  reset
  action udp-dsig-setvar_1.0.0_set-variable_0
```

```
   action udp-dsig-setvar_1.0.0_gatewayscript_1
exit
```

The next section in the configuration defines the assembly that will simply contain a reference to the API rule defined previously as follows:

```
assembly udp-dsig-setvar_1.0.0
   reset
   rule udp-dsig-setvar_1.0.0_main
exit
```

The final section of our configuration is the assembly function. This is where you define the actual UDP and the information that will be displayed on the API assembly. This is where you will provide a title and summary to be displayed as well as define any input parameters. In our example, we require one input parameter, so we will need to define that here. The following commands define this assembly function for our example:

```
assembly-function "udp-dsig-setvar"
   reset
   summary "udp-dsig-setvar_1.0.0"
   title "udp-dsig-setvar_"
   parameter
      name "environment"
      description "Runtime Environment"
      value-type string
   exit
   assembly udp-dsig-setvar_1.0.0
exit
```

As you can see, we are defining a UDP named udp-dsig-setvar, which requires a parameter named environment to be passed, which is of the string type.

There you have it. That is the entire .cfg file required for our custom global UDP. We can save this file to our local filesystem with a .cfg extension to be used when packaging up our global UDP to be published to our API gateway.

The next step in making our global UDP available for use in your APIs is to package up all of the artifacts and publish the UDP. In the next section, we will demonstrate exactly how we can accomplish this.

Packaging and publishing your global UDP

Now that you have all of the components for your global UDP, you will need to package them up into a .zip file to be published to your API gateway. To do this, simply create a .zip file that contains the .cfg file as well as all the requisite policy implementation files. In our example, we would include the .cfg file and the GatewayScript file.

Once you have all of your required artifacts in one .zip file, you are ready to deploy this to your API gateway. Since this is a global UDP and not specific to any Provider Organization or Catalog, we will publish this to our API gateway via the Cloud Manager UI. Once signed into the Cloud UI, click the **Configure Topology** tile on the home screen. This will bring you to your availability zone, where you will see all services registered. We are interested in gateway services, so click the ellipses to the right of your gateway service and click **Configure gateway extensions**, as shown in *Figure 12.11*:

Figure 12.11 – Configure gateway extension

From here, on the next screen, click the **Add** button, which will bring you to a screen where you can either drag your .zip file you created to the screen or upload it, as shown in the following screenshot:

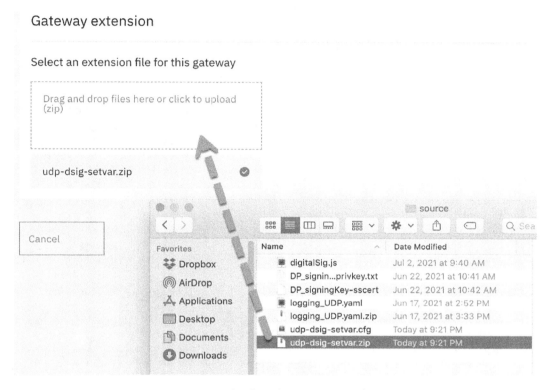

Figure 12.12 – Uploading the .zip UDP configuration

Clicking the **Save** button completes the configuration and brings you back to the **Configure Gateway Extensions** screen, where you should now see a gateway extension configured.

> **Note**
>
> There can only be one .zip gateway extension file uploaded to a given gateway service. If more UDPs are required, the gateway extension must be deleted and all UDPs should be configured in one .zip file.

Your new global UDP is now configured in APIC; however, it is not immediately uploaded to the gateway service and installed. To have this take effect immediately, you can log in to your DataPower device in the configured APIC domain and navigate to **API Connect Gateway Service**. The easiest way to find this is to type API Connect Gateway Service in the search field and click it when it appears. From this screen, you can disable and then enable the API Connect Gateway Service to force a refresh and bring in your newly created UDP. Once the gateway service is enabled, you should see the new UDP in the dropdown next to the **User-defined policies** field. You can now select this and click the **add** button, as shown in *Figure 12.13*:

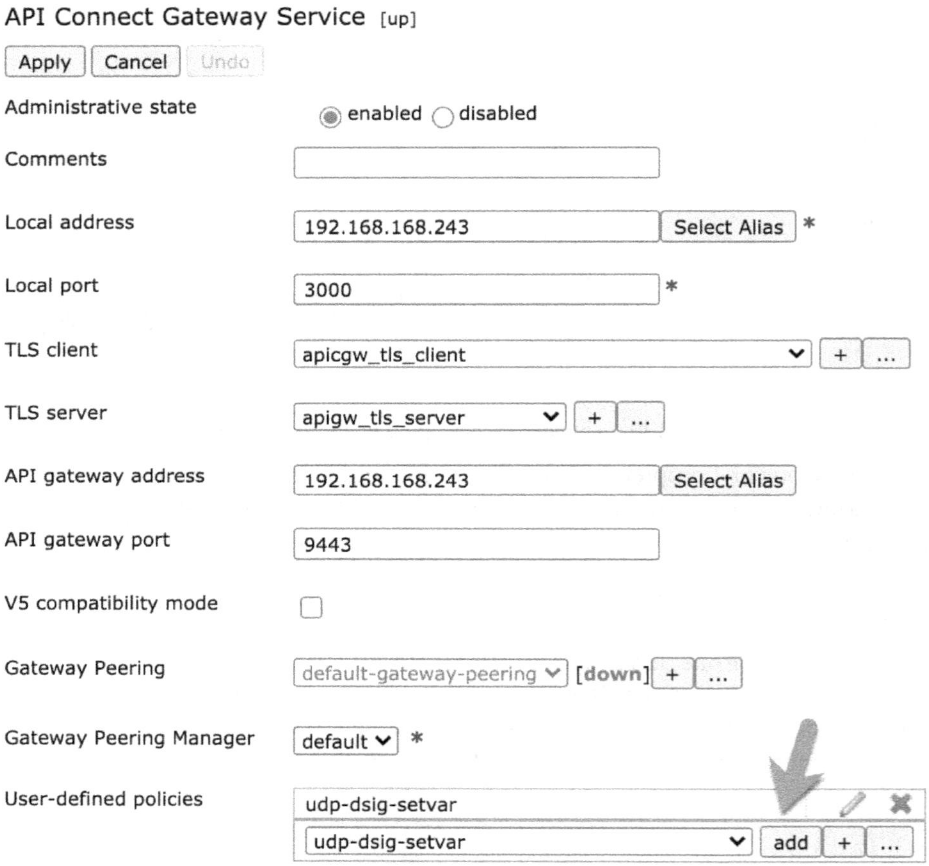

Figure 12.13 – Adding a UDP to a gateway service in DataPower

Once you have added the UDP to your gateway service, you can apply the changes and save the configuration.

Your new global UDP is now configured, installed, and ready for use. If you configure an API in any Provider Organization or Catalog, the new UDP should appear in the **Policies** palette within the **Gateway** tab. *Figure 12.14* shows our new UDP available and being used in the assembly. Note the required parameter that we configured in our configuration file.

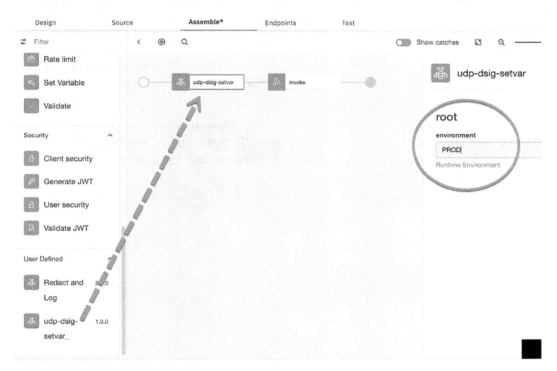

Figure 12.14 – Global UDP available in the Policies palette

You have now seen how you can configure a global UDP and hopefully understand its benefits. Much like a catalog UDP, it can package up one or more actions or policies in a single policy, which will be shareable. The additional benefits of the global UDP are that it can utilize DataPower-specific functionality and is shared across all Provider Organizations and catalogs.

Summary

Having the ability to create, package, and share common functionality across your organization provides many benefits. Not only does it save time in development, but it provides a high level of confidence that it's been thoroughly tested as well as providing consistency. APIC provides the ability through the use of user-defined policies to combine as much functionality as desired into one shareable policy. With the flexibility of Catalog UDPs and global UDPs, you can choose what type of functionality you wish to incorporate and the scope of users that you wish to share the UDP with.

Throughout this chapter, you learned the benefits and differences of each type of UDP as well as how to configure and install them. Through the use and experimentation of these policies, you will certainly learn the endless possibilities, flexibility, and benefits of them.

At this point, you should have a good understanding of how to build some basic and advanced APIs using the features provided in the APIC product. Our next chapter will discuss how you can use some built-in features to perform unit tests that can also be incorporated into your DevOps pipeline.

Section 3: DevOps Pipelines and What's Next

This section introduces a new tool for testing and building unit tests. You will also learn about building DevOps pipelines for API Connect. We conclude the book by discussing how to upgrade API Connect and discuss future solutions for digital transformation.

This section comprises the following chapters:

- *Chapter 13, Using Test and Monitor for Unit Tests*
- *Chapter 14, Building Pipelines for API Connect*
- *Chapter 15, API Analytics and the Developer Portal*
- *Chapter 16, What's Next in Digital Transformation Post-COVID*

13
Using Test and Monitor for Unit Tests

In today's digital world, we are all about speed and efficiency. With **continuous integration (CI)** and **continuous deployment (CD)**, we can be developing our APIs and publishing them at a rapid pace, but we must also not lose sight of quality. Often, we are so concerned with time to market that we may loosen up on our testing efforts. This is why we must also have an efficient yet effective way to test as we progress through our development and publishing efforts. Many tools on the market let us send test requests, build Test Suites, and create assertions, but this same type of tool is already available within the API Connect product. This is the test and monitor tool, which must be installed as a separate add-on. This tool is included in, and integrates into, your API Connect API Manager, allowing you to test your APIs as you develop and publish them.

In this chapter, we will introduce you to the basic functionality of the test and monitor tool so that you will know how to utilize the capabilities within this pre-packaged toolset in API Connect.

In this chapter, we will cover the following topics:

- Configuring unit tests
- Working with environments
- Monitoring tests

Technical requirements

The examples presented in this chapter will require that you install API Connect, along with the test and monitor tool. You should also have at least one API developed and published that you can test within your test cases.

Configuring unit tests

As you develop and change your APIs, you should be unit testing them incrementally with specific unit test cases for the change you have just made or testing the API in general. The API Connect test and monitor tool allows you to quickly generate these unit tests by simply sending a request to the API URL and getting a response. This tool will provide some assertions that you can use as-is, alter, delete, or add to. Let's take a look at how to configure a simple unit test case for an API that you are working on.

Before you start to create your first unit test, you must ensure that you have the test and monitor tool add-on installed. To verify this, log into API Manager. The bottom right tile on the home screen should be labeled **Test APIs**, as shown in the following screenshot:

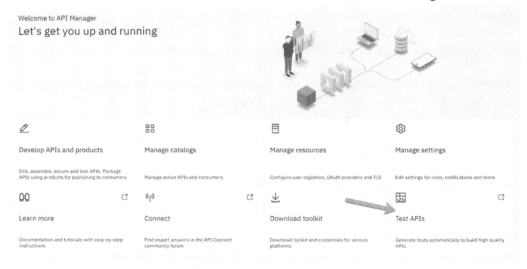

Figure 13.1 – Verifying that the test and monitor tool is installed

Once you have verified that you have the test and monitor tool installed and available, clicking this tile will open the tool in a new tab, where you will be looking at the tool's home screen, as shown here:

Figure 13.2 – The test and monitor tool's home screen

As you will notice when you open the tool for the first time, there will be a current summary of all of your Test Suites, test cases, and other data from past test case executions. Since this is your first time accessing this tool, all of these values will have a zero count.

Before we start configuring these Test Suites and test cases, let's take a look at the HTTP Client. To access this, you must click the **HTTP Client** link on the home screen. Once you click this link, you will be brought to the HTTP Client screen. This layout and functionality may look familiar to you as it is similar to other testing tools present on the market today. From this screen, you can submit an HTTP request to any URL using any of the available HTTP methods. The following screenshot shows this default screen as it is presented to you. Here, we have numbered each of the components of interest when you submit a simple request to your API:

Figure 13.3 – HTTP Client

Let's take a look at each of the fields, options, and buttons that will be of interest to you when you're generating your HTTP request:

1. **HTTP method**: Using the dropdown list in this field, you can select the HTTP method to use for your test request message. The available options are **GET, POST, PUT, PATCH,** and **DELETE.**

2. **Request url**: This provides the request URL for your API to be tested. Do not add query parameters to this field.

3. **Params**: Using this button, you can add query parameters to your request in key/ value pairs. As you add these params, they will automatically populate in your request URL.

4. **Headers**: Provide any HTTP headers to be sent in the request in key/value pairs. Using the + button, you can add additional HTTP headers to the list.

5. **Body**: If applicable, this provides two options for adding a request body to your request. You can provide a raw request body that gives you a generic text area to add your request, or you can choose a URL that's been encoded where you can provide key/value pairs for your request body.

6. **Send**: Use this button to send your request message once all the request fields have been entered.

7. **Response**: The response to your request will be shown here once sent.

As you can see, there are more options available on this screen that we will cover later in this chapter. The fields we've described already will be the ones you need to generate your first HTTP test message for.

Let's take one of our fictitious healthcare APIs, called **Member API,** and send a test request to it. This API has a path for getPatient that accepts an HTTP GET method and requires a query parameter named memberID. This API is secured by a client ID and client secret, so we will need to add these as our request headers. The following screenshot shows our test request containing all of the required information, as well as the response that's received from the API. Since this is an HTTP GET request, there will be no body to send:

Figure 13.4 – Test request for Member API

Taking a closer look at the preceding screenshot, you can see that we have chosen the HTTP GET request and provided the request URL. We added the **memberId** query parameter to our **Params** list, which automatically added it to the request URL as expected. Finally, we added **X-IBM-Client-id** and **X-IBM-Client-Secret** and their corresponding values to the **Headers** list and clicked on the **Send** button.

After submitting this request, the response body message is displayed in the **Response** area. Let's take a closer look at what is provided here. After all, the response is what you are interested in when testing. What should be clear right away is the response body that is returned from the API. This can be seen as raw text or parsed for better readability. Of course, you would be interested in the response body, but there is more information that you would likely be interested in when testing your APIs. If you look at the top right of the **Response** area, you will notice some critical pieces of information. There, you can see the HTTP response code, **Latency** time, **Fetch** time, and **Total Time** taken in milliseconds.

In addition to the response body, HTTP response code, and latency details, you can view all of the response headers by clicking the tab to the left of **Body** in the **Response** area.

As we mentioned earlier, you can also submit a request using other HTTP methods. For example, you could submit an HTTP POST that would require a message body. The following screenshot shows an example of this by submitting an HTTP **POST** request to our **Member API** using the addPatient path, which would expect a POST request and message body. As you can see, you have the same type of response data available as you did for the GET request:

Figure 13.5 – Member API POST request

As you have seen, the API Connect test and monitoring tool lets you send a test request to your APIs and display some useful information about the response. Although this could be useful for some one-off tests, it isn't much more than you could do with some command-line tools you might already have, such as cURL. Let's face it – having to manually enter your test case and visually inspect the response is time-consuming and prone to errors. To be useful and show some value in our rapid development and delivery model, you would need some more repeatable test cases that can automatically check the response data via assertions that you could build on as you develop. Let's dive a little deeper into the test and monitor tool to see how it can accomplish these tasks as well.

Before you start creating test cases, you will need to create a Test Suite to save the test cases in. This project could be a grouping of test cases for one API, one product, or any other way you wish to organize your test cases.

To create a Test Suite, you will need to navigate to the home screen by clicking the **API Management Test APIs** link in the top-left corner of your screen. Once you are back on the home screen, you can click the + or **Create Test Suite** button to create a new Test Suite, as shown here:

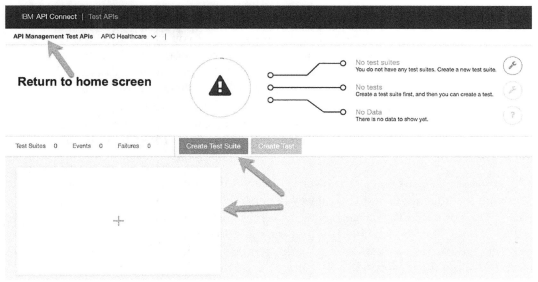

Figure 13.6 – Create Test Suite

Providing a name for your new Test Suite and an optional description, and then clicking the check symbol in the top-right corner, completes the process of creating your new Test Suite. The following screenshot shows us creating a new Test Suite for our Member API:

Add new Test Suite

Test Suite name   **Member API Test Suite**

Description   **Test suite to test the member API**

Notes

Figure 13.7 – Add new Test Suite

Once you have created your new Test Suite, you will see all of the details for it, such as test cases, events, and failures. Of course, these will all be set to zero when you first create the Test Suite until you add some test cases to it. Now, let's learn how you can take your test case, add some assertions, and save it to your newly created Test Suite.

You can create a new test case in several different ways within the test and monitor tool. You can click the **Create Test** button on the home screen or you can create a new test case from within your Test Suite configuration. There is also one more way to create your test case, which will build upon what you have learned so far in this chapter. You can take the tests you create within the HTTP Client and simply add them to a Test Suite. Let's explore this option as we already have some experience here. Besides, this is a good way to fine-tune your test cases and add assertions before adding them to your suite. To do this, navigate back to the **HTTP Client** screen by clicking the **HTTP Client** link at the top right of the screen.

In *Figure 13.4* and *Figure 13.5*, we showed how easy it is to submit a request to your API. This would include the request URL, query parameters, request body, and HTTP request headers. Once you submit your test requests to your APIs, you will also get the response data, which should prove whether your test is a success or a failure. Upon visually inspecting these fields, you can determine this test result's status. As we mentioned earlier, this is not the ideal practice for running and interpreting your test cases. The test and monitor tool, like any good testing tool, lets you create assertions for each test case that can be configured to inspect any piece of the response data available. This lets you determine if the test was a success or a failure. Luckily, you don't have to start this process from scratch. This tool can take your current request and response data and automatically create most of the assertions that you might need to determine the test results. After submitting a test request, with one click of a button, you can *automagically* generate a test case that will contain several assertions based on your current response data. If you haven't guessed already, the button at the top of the **HTTP Client** screen labeled **Generate Test** is this magic button, as shown here:

Figure 13.8 – The Generate Test button

Once you've clicked this button, you will be asked to give the test case a name and select which project to save it to. The following screenshot shows that we have named our test case getMember_success and saved it to our newly created **Member API** project:

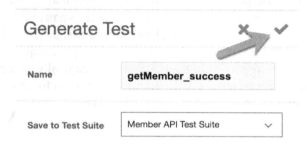

Figure 13.9 – Providing a test case name and Test Suite

If you're ready for the magic, all that is left is to click the checkmark to create your new test case, which automagically creates assertions! Once you've done this, you will be notified that the magic has been completed as you will see a list showing everything that the tool has done and if it was successful. The following screenshot shows that we were successful in creating our new test case and that it created some global parameters, an input set, and assertions:

Figure 13.10 – Completed test case confirmation

Clicking the **Close** button will bring you to your new test case. Depending on the size of your response, you will see that several assertions have been created corresponding to each data element that was in the response. You will also see an assertion for the HTTP response code, response Content-Type header, and any additional response data that the tool was able to derive an assertion for. In addition to the automatically created assertions, you will notice that the test case now has global parameters assigned for the request fields that may change from environment to environment, such as your request URL, client ID, and client secret. The following screenshot shows our automagically generated test case with the global parameters replacing the URL, input set variables for the client secret, client key, parameters, and all of the assertions for the successful response we received from our test request:

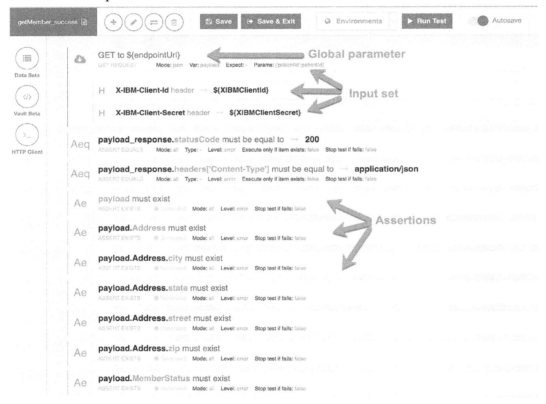

Figure 13.11 – Test case generated for the getMember test case

Clicking the **Save & Exit** button and then confirming that you want to save and exit will save your test case to your Test Suite and return you to a different screen. This will show some interesting information and options for your test case, as shown in the following screenshot:

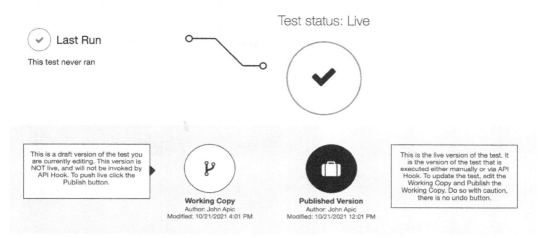

Figure 13.12 – Test case status screen

By taking a look at the test case status screen, you can quickly gather some good information about the status of your new test case. The first thing you might notice in the largest portion of the screen at the top is that your new test case is live and has never run. Below that, you will see that you have two versions of your test case. You will have a **Working Copy** and a **Published Version**. **Working Copy** is the copy that you are editing. Once published, it becomes the **Published Version** type of this test case, which is executed either manually or by an API hook, as you will see in *Chapter 14, Building Pipelines for API Connect*. To publish your working copy, you must click the **Publish** button on this screen. Using the corresponding buttons on this screen, you can also perform other actions, such as editing or deleting the test case, building the test case from an API specification, or clearing your working copy.

So far, you have seen how to generate a new test case from the **HTTP Client** screen using a sample request and response and how to let the tool generate the test case based on the response. There are several other ways to generate test cases within the test and monitor tool, many of which are wizard-driven and intuitive. To explore these additional methods, you can click the **API Management Test APIs** link from the top navigation bar; it should be visible from any screen within the tool. This will bring you to the home screen, which will show all of your current projects, along with information about how many test cases are within them, how many tests were run, success versus failures, and so on:

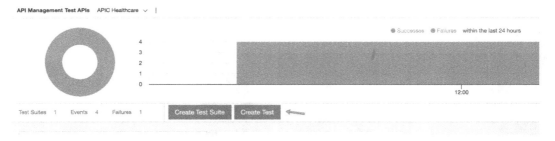

Figure 13.13 – Test APIs screen

Clicking on the **Create Test** button from this screen will start the process of creating a new test case using one of the available methods, as follows:

1. You will be prompted to select an existing Test Suite to save your test case in. Select your Test Suite and click the checkmark button at the top right.

2. Enter your test case's name, description, and any tags you might want to add to it for searchability and click the checkmark button.

3. The next screen should look familiar to you as it shows all of the information about your test case. Since you haven't configured anything for it yet, it will show as incomplete. Click the **Edit** button to start configuring your test case.

4. The next screen you will see is where things get interesting. You will see three different options for creating your new test case, as shown in the following screenshot:

Figure 13.14 – Options for creating a new test case

In addition to the three available options for creating a new test case, you can click the right and left arrows to go through a tutorial on the different screens for creating your new API.

5. The three options you have to create your test case are as follows:

    - **Quick start from an API call**: Generate a test automatically by submitting a request to an existing API. This is the same method you saw earlier using the HTTP Client to generate your test case. With this option, you can select previous requests you have made within the tool.

    - **Quick start from an example unit**: This option will guide you through an example unit test. From this example test, you can alter the request URL, headers, assertions, and so on to customize your test case.

    - **I got this, let's start from scratch**: This option provides a blank palette and provides the most options and flexibility for creating your test case. Selecting this option and then the **Add Request/Assertions** button will reveal an extensive pallet of options for building your test case, as shown in the following screenshot. As you can see, these options look similar to the options in the assembly screen, when you're creating your actual API and providing various logical decisions and connectivity components:

Figure 13.15 – Manual test creation Request/Assertions palette

As you can see, the test and monitoring tool provides you with several ways to generate your test cases. You can choose to have it create it for you or take total control and build it from scratch using any of the options provided to generate a robust test case. No matter which way you choose to initially create your test case, you can customize it to suit your needs once it has been generated.

Once you have generated your test case, you will want to run it to test it out and then continue to test the API it runs against. To view and run your test case, you can click the **API Management Test APIs** link at the top left of the main navigation bar to reveal all of your configured projects. Clicking the **Tests** button within your Test Suite will bring you to a list of all of the tests within that project. From this screen, you can run your test case by clicking the play button to the right of it, as shown in the following screenshot. Once you have executed the test case, you will see the run **results** to the right of it:

Figure 13.16 – Running the test case

As you can see, we ran our new test case, which succeeded.

Although we executed the test case from the test case screen, you didn't get much information about the test results themselves. To reveal all of the details of your test case run, you can click the **See report document** link in the results section. Clicking this will show the full report of the test case's run, including how many assertions passed, how many failed, latency, HTTP response code, and more. From this report, you can also view the actual request details, including request URL and HTTP headers.

In this section, you saw how easy it is to configure robust test cases to test your APIs. These repeatable tests will certainly provide the scope and breadth of your testing for when you're making any changes to your APIs. Of course, as you move your APIs through your **software development life cycle (SDLC)** environments, the test case parameters within your test cases will change. In the next section, you will learn how to configure your test cases so that they can be utilized in every environment, without the need to make changes to the test cases themselves.

Working with environments

As you may recall from when you created your new test case, the test and monitor tool generated some global parameters and an input set. These come in handy when you need to submit the same test case against a new version or environment for an API where the URL, client ID, and client secret would change. You can pass in different values for these global parameters when executing a manual test run by navigating to the edit test case screen for your test case and clicking the **Data sets** button from the left navigation menu. The following screenshot shows our **getMember_success** test case's **Data Sets**, along with its **GLOBAL PARAMETERS** and the dataset. From here, you can modify or delete a parameter, or even add additional parameters:

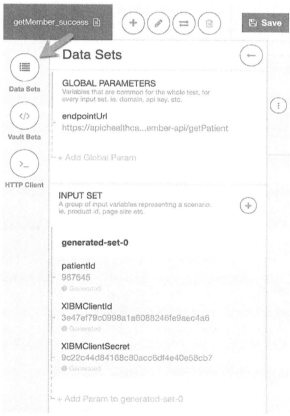

Figure 13.17 – Viewing and modifying parameters

Even with the ability to modify the test case parameters manually, it would still be cumbersome to have to change them each time you want to submit your test case to a new environment. To address this concern, the test and monitor tool lets you create environments. Just as it sounds, an environment is a separate configuration where you can configure new values for each environment and save them appropriately. This way, the same test case can be submitted by simply selecting the environment that you wish to run it for.

To create and configure your environments, from the test case editor, click the **Environments** dropdown at the top of the page and click on **Open Presets Editor**, as shown in the following screenshot:

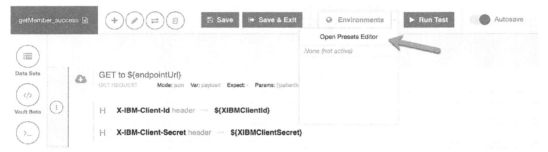

Figure 13.18 – Open Presets Editor

You will then be presented with a configuration screen where you can add your global parameters and values for the environment you are configuring for. Let's take a look back at the first test case that we configured; that is, **getMember_success**. As you may recall, when we had this test case automatically generated for us, it generated a global variable named endpointUrl for the request URL, as well as the XIBMClientId and XIBMClientSecret variables for the client ID and client secret headers. Since these are the variables that will be changed for each environment, these are the key/value pairs we want in our environment configuration. The following screenshot shows three key/ value pairs for our development environment to help us resolve our endpointURL, XIBMCLientid, and XIBMClient secret parameters to the values for this environment:

Figure 13.19 – Development environment configuration

To activate and save this configuration, click the **Preset Active** slider and then the **Save** button. You will be prompted to provide a name for this environment configuration, where you can enter a meaningful name to represent the environment that this configuration corresponds to. The following screenshot shows us naming our new environment configuration DEV. Clicking the checkmark icon in the top-right corner will save the current configuration:

Figure 13.20 – Saving the DEV environment configuration

Now that you have configured, activated, and saved your environment configuration, you can submit a new run for your existing test case by selecting this environment configuration from the **Environments** dropdown. The following screenshot shows the new **DEV** environment selected to be used for the next run of this test. Executing the test will replace the environment variables with the values provided in the **DEV** environment configuration before execution:

Figure 13.21 – Executing the test case using the DEV environment configuration

As you can see, you can configure a new environment configuration for each of your environments, which lets you execute the same test case against multiple environments. This eliminates the need to maintain multiple versions of your test cases to accommodate all of your different environments.

With that, you have learned how to configure and execute your test cases for your APIs across multiple environments. Whether these tests run in an automated fashion, manually, or as part of your DevOps pipeline, if no one looks at the results, then what good is it? You know what they say – *"If a tree falls in the woods and no one is around to hear it, did it make a sound?"* Now, let's learn how to view historical results as all of your test cases run.

Monitoring the test cases

So far, we have spent all of our time discussing creating and executing our test cases. You might be thinking, *where is the monitoring piece of this test and monitor tool?*, and that would be a fair question, considering the name of the tool itself. To answer this question, let's take a look at the available features that allow us to monitor the execution of our test cases.

The test and monitor tool provides several different views you can use to gather historical data on the execution of your test cases. To get a high-level view of how your test cases are doing, you can click the **API Management Test APIs** link at the top of any page within the tool. This will bring you to a single screen where you can glean quite a bit of information on the overall performance of your test cases within your provider organization. The following screenshot shows this high-level screen for our **APIC Healthcare** provider organization:

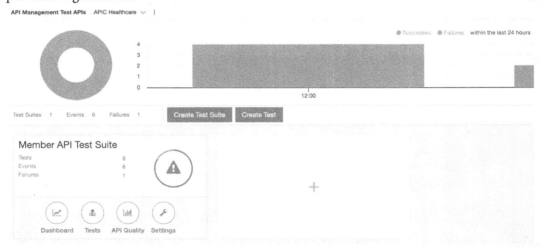

Figure 13.22 – The APIC Healthcare test case's status

Looking at our provider organization test API screen, the first thing you might notice is the large area stretching across the top of the entire screen. Here, you can see an overall success/failure chart for all of your test cases within the last 24 hours. By quickly looking at this, we can immediately tell that some of our test cases require immediate attention. Below this chart, you can see the number of **Test suites**, **Events**, and **Failures**.

Moving further down this page, you can see all of the Test Suites that you have created, along with the number of test cases created within, the total number of events for them, and the total number of failures. In our example, you can see that we only have one project configured consisting of three test cases. To drill down and get a closer look at the performance of your test cases within the project, you can click the **API Quality** button within the Test Suite. This will present a screen showing a breakdown of each test case's successes and failures for a given month, as shown in the following screenshot:

Figure 13.23 – API Quality

Here, you can start to get a better picture of where our failures are occurring. You can view the previous month's data by selecting a different month at the top of the screen.

From what you have seen on the previous two screens, you were able to identify your test case success versus failure ratio at a high level and drill down view this breakdown by test case within a project. From here, you will likely want to drill down into these test case executions even more, which will help you identify what was causing the failed test cases. For this, you can use the dashboard. You can navigate to the dashboard from the home screen by clicking the **Dashboard** button within the Test Suite, or you can get there from the **API Quality** screen by clicking the **Dashboard** icon, which can be found in the top main navigation menu.

Once you navigate to the dashboard, it should become evident that there is a lot of information contained within one screen with several ways to filter and view it. The following screenshot shows the initial view of the **Dashboard** area when you navigate to it:

Figure 13.24 – Dashboard default view

At first glance, you will notice the familiar chart running across the top of the screen showing the number of successes, failures, and tests for the time selected. The default time that's selected is the last 2 days. Above this chart, you will see all of the options available to filter this data down. You can filter by date range, test case, tags configured in a test case, the location the test was initiated from, and events (all events or failures).

Moving along to the main body of this page, you can see each test case that's been submitted that matches the filtered criteria. Each test case execution will show the date and time submitted, the number of warnings and failures, the submitted location, and the status of the test run. Clicking on an individual test case run will provide the options to delete the entry, provide a public link to it, or view the execution report document. Within this document, you can see specific details about the execution, including the request URL, request header, and the request itself. You will also see response latency, any failures, and the overall test results. When troubleshooting test case failures, this report could prove to be your best asset in diagnosing the problem.

The sole purpose of generating test cases is to check the health of your APIs. As you generate your test cases and time goes by, these test cases should be run not only after publishing a new API but regularly to ensure the health of all of your APIs. If you don't monitor these test results properly, they won't provide much value. As you have seen, the test and monitor tool lets you track historical test case execution data, which is critical to detecting and analyzing your API failures.

Summary

Keeping your API environment healthy is critical to the success of your digital transformation. Detecting such issues early could help mitigate the damage that's done by such issues. Whether you use your test cases in your unit test environment as you develop or utilize them as continuous health checks, creating robust, repeatable test cases that provide full coverage of all the functionality of your APIs can ensure that your efficient time to market efforts do not come at a cost.

In this chapter, you learned how to utilize the built-in test and monitoring tool to generate robust test cases as you develop your APIs in a way that they can be used in each environment. Having such a tool at your disposal can significantly improve your development and deployment efficiency by keeping quality at the forefront. Although we have covered a lot of material within this chapter, there is more to discover within this testing tool. It would certainly behoove you to dig even deeper into all of the features provided by the test and monitor tool. Its robust feature set and flexibility will provide you with all of the tools you might need to keep your environment healthy.

In the next chapter, you will expand on this even further by learning how this can be integrated into your DevOps pipeline.

14
Building Pipelines for API Connect

You have been learning over the past few chapters how to develop and manage APIs using the API Manager. Your journey toward digital transformation and modernization has been greatly enhanced by using API Connect's capabilities. As you are speeding through development, there is one more critical aspect of developing with agility, and that is DevOps **CI/CD** (**Continuous Integration/Continuous Deployment**).

Perhaps you have already integrated a DevOps process within your organization. Even if you have, this chapter will provide you with the details on how to integrate API Connect into the flow. DevOps is an ever-evolving practice. Whether you have DevOps implemented or not, this chapter will be of great value.

In this chapter, we're going to cover the following main topics:

- Introducing pipelines
- Choosing between CLIs or Platform API interface use
- Testing and monitoring hooks
- Using Git as your SCM
- Constructing the Jenkins pipeline
- Working on an API Connect sample pipeline

There is much to learn and investigate but, by the end of this chapter, you will have learned the following new skills:

- An understanding of how pipelines work and how they enhance agility, testing, and deployment.

- You will have learned how to use the **apic developer toolkit CLI** and the **API Connect platform APIs** in DevOps scenarios to upload APIs and deploy them to multiple environments. You will learn how to log on to environments, run the `publish` CLI command, and execute platform APIs to accomplish the same task.

- Integrating APIs and Products into Git to serve as an SCM environment. You will learn some basic Git commands to add, commit, and branch code.

- Lay out a plan to build your DevOps Pipeline with API Connect

You'll be learning about a lot of new tooling to accelerate your digital transformation, so let's get started to see how it's done.

Technical requirements

With this chapter, you will be referencing a number of DevOps tools as well as some open source projects. Minimally, you will need to have access to the **APIC CLI** toolkit as well as access to GitHub, a Jenkins server implementation, and RedHat Ansible. You learned about the APIC CLI in *Chapter 2, Introducing API Connect.* Your **API Connect API Manager** must have installed the test and monitor feature, either during initial installation or added on afterward.

You will find an API and product file as well as a Jenkinsfile in GitHub using the following URL: `https://github.com/PacktPublishing/Digital-Transformation-and-Modernization-with-IBM-API-Connect/tree/main/Chapter14`

As you have progressed from *Chapter 13, Using Test and Monitor for Unit Tests*, you have some unit tests to incorporate into the pipelines defined in this chapter. These tests will make your learning of pipelines more realistic and valuable to your company.

Let's begin with an introduction to pipelines.

Introducing pipelines

When you think of a pipeline, you may envision a pipe where liquid flows from one end to another. When put into the context of developing applications, the pipeline is divided into segments of the pipe where activities are performed and then passed downstream. Now, if you think of your current **Software Development Lifecycle (SDLC)**, you probably know the set of steps. If these steps are merely handoffs to another person or team to perform, you can envision the velocity of execution. It could be very slow and may take days. If you have to fill out service tickets to initiate the next task, it may be even longer. Your pipeline should be efficient and expedient. To put it into API Connect terminology, how will you develop APIs and ensure they are properly versioned, built, and deployed to a test server? How will you ensure the APIs are tested before deploying to a QA environment and finally released to production? Those are the steps you should envision in your API Connect pipeline.

Figure 14.1 provides a visual depiction of those stages. This will be the pipeline that you will be building in this chapter.

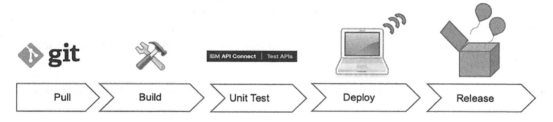

Figure 14.1 – Creating a pipeline

As you can see, there are five stages in your pipeline. Each stage performs some steps to validate the integrity of the pipeline. If a step fails, you can determine whether it's a soft failure or something that requires total rejection of the stage. Let's briefly describe the stages.

- **Pull**: In this stage, there is an assumption that you have developed your API/Product and have checked it into source control. In this example, it is Git. You'll be checking out your code from Git.

- **Build**: In this stage, you are initiating your first deployment to a test environment on API Connect. This could be a separate environment or merely a different catalog within API Connect.

- **Unit Test**: After deployment, you will want to ensure the code deployed minimally passes basic Unit testing. This will minimally ensure your build was properly deployed and that your code is functioning. Of course, you could have created extensive steps to also validate other business functions. That is up to you.

- **Deploy**: After successful unit testing, the deploy step will move your implementation to either a new environment or a different catalog designated for QA testing. Optionally, you can add another testing stage. If the deployment fails, your pipeline puts out an error message and it terminates the processing.

- **Release**: When the appropriate party determines that your API is ready to be implemented in production, the release stage will manage those steps

As you review these stages, you will probably see where you can incorporate API management and governance. Now that you have an idea of what stages you need to implement, now is a good time to learn what API Connect capabilities are available to help build the steps within the stages.

Choosing between CLIs or Platform API Interface use

In *Chapter 2*, *Introducing API Connect*, you were introduced to the API Connect Toolkit CLIs. As you may recall from there, the CLIs provide you with the ability to create APIs and Products as well as publish the APIs to a catalog. There is also another method for performing similar publishing functionality, and that is with the Platform APIs provided with API Connect.

Since you will be using the CLI for your pipeline in this chapter, you'll learn about the CLI calls first.

Using the CLIs

Although you learned about the CLI commands in *Chapter 2*, *Introducing API Connect*, a quick refresher here is appropriate. When you look at the stages for the pipeline, the **Build** stage will be invoking the steps to interface with API Connect. These are the CLI commands that you will be using in our pipeline:

- `apic login`: This is how it is used:

```
apic login --username <username> --password <password>
--server < api-manager host > --realm provider/
default-idp-2
```

- `apic logout`: This is how it is used:

  ```
  apic logout --server <api-manager host>.com
  ```

- `apic publish`: This is how it is used:

  ```
  apic products:publish <yourProduct.yaml> -c <sandbox>
  -o <your org> -s <api-manager host>
  ```

Initially, you may also require the `apic:identity-provider` CLI command to determine appropriate user credentials:

```
apic identity-providers:list --scope admin --server
<api-manager host >
```

There are other CLI commands you may want to add in the future, but for now, these are the only ones that are needed for this chapter.

Let's now move on and learn about Platform APIs so that you have the knowledge to choose the feature you would like to use that is most appropriate for your future DevOps pipeline.

Using the Platform APIs

When you are working with any of the web user interfaces (Cloud Admin or API Manager), they are actually utilizing APIs to perform all the tasks you initiate. Those APIs are the Platform APIs. The Platform API interface is constructed when API Connect is initially installed.

When you are creating and managing APIs within API Connect, there will be situations where you have repetitive tasks performed by various roles. These tasks can be added to your DevOps pipeline if you so desire. In this chapter, you will learn what you need to do to execute one of these tasks. For more information on the available APIs, you can visit the following website and peruse the offerings:

`https://apic-api.apiconnect.ibmcloud.com/v10/.`

The best way to learn about the platform APIs is to learn by example. You'll learn that next.

Platform API execution

For all the interactions with the API Manager interface, you will require the logon credentials. Since we are invoking APIs, you will need to acquire the necessary Client ID and the secret to invoke your app. Here are the steps to do this:

1. You will need to log in to the server. The `apic` login is required to establish a connection to the API Connect server with the proper credentials to make updates to `API Connect`:

    ```
    apic login --username admin --server cloud-admin-
    ui.apicisoa.com --realm admin/default-idp-1
    Enter your API Connect credentials
    Password?
    ```

    Once you are logged in to `cloud-admin` successfully, you need to create a consumer application so you can make API calls to the Platform API interface. You do that by registering the app.

2. You must create an application and also get set a client ID and secret. Registering a new application will require you to create a JSON file that provides the details of the application. That includes setting your own client ID and secret. You need to create an input file to pass with the API call. This input file should be created in JSON format. In this example, you can name the file `app1.json`. Refer to the following code:

    ```
    {
        "name": "app1book",
        "client_id": "app1bookid",
        "client_secret": "app1booksecret",
        "client_type": "toolkit"
    }
    ```

    Once the file is created, you will run the `registration:create` command, passing in the server you are interfacing with and the name of the JSON file you just created.

3. You first need to ensure you are authorized to issue these commands to API Connect. Therefore, you must register with API Connect the credentials that you created in *step 2*. Issue the `apic registration:create` command as follows:

    ```
    apic registrations:create --server [your cloud
    adminserver] app1.json
    ```

A successful response will list out the newly created application name and internal identifier of the application registration. Refer to the following code:

```
app1book [state: enabled]
https://[your cloudadmin
    server/api/cloud/registrations/e40dbb1a-de6a-48c1-
      ba9a-a094b1cc3cb9
```

You don't have to save the internal identifier. It's for internal reference only.

You will use `client_id` and `client_secret` from *step 2* in the next step. You should save the client ID and secret for future use.

4. Now that you have registered your application, the next thing that is required is a bearer token. This token can be used for multiple platform API calls until the token expires (28,800 seconds). The OAuth token is acquired by calling the URL `https://[apimserver]/api/token` and passing in the relevant information to obtain a bearer token. The data within brackets should be replaced with the proper information for your environment. Using the `curl` command, you will make a call to the server requesting a token, as follows:

```
curl -v -k -X POST -d '{"username": "[username]",
"password": "[****]", "realm": "provider/default-idp-
2", "client_id": "[app1bookid]", "client_secret":
"[app1booksecret]", "grant_type": "password"}'
-H 'Content-Type: application/json'
-H 'Accept: application/json'
https://[apimserver]/api/token
```

If successful, you will receive a bearer token that must be included with any Platform API call. A successful message is similar to the following:

```
{
    "access_token":
"eyJhbGciOiJSUzI1NiIsInR5cCI6IkpXVCIsImtpZCI6IjMyNjQO0
UE4RUFFMEQ5Mjk2RTM4MkYwREQ4RUFFGRDVDODc2QTI0QUIifQ.
[...]
-WW_KSdie6Cy6UnEveNWASVDuBg6a6tIXqCTopnPv_5dB-
qk6IFYivAtaZW2rYtDkf6VdMu58cbu6DDBy7UMA7YbsFSdXTwjwvlA
fbY9GDAL4XhqzlLkI5vkm3NdVz0REv_FJxNd5iV1b1TM4nVXO63rEB
Ltc-_hUNBxHfPJwuIHBNo6Qh9d2np3CG1KqSE3Ue5cSoMQIwTXU
AwwPn8oGa2k21hNeCBxE2kbqYDTHEcBwQZUAV_Q",
```

```
        "token_type": "Bearer",
        "expires_in": 28800
    }
```

Now that you have a token, you can proceed to call the platform APIs. The steps you just learned will need to be done regardless of which Platform API you wish to call. Next, you'll learn how to use the publishing Platform API.

Calling the publish Platform API

In the pipeline, we will be cloning a GitHub repository and publishing a product and its APIs to a target environment for user testing. With the publish API, you have two choices of where to publish. You can post to a catalog or space:

- POST   /catalogs/{org}/{catalog}/publish
- POST   /spaces/{org}/{catalog}/{space}/publish

Each POST will require a payload. Since you will be uploading files in a multipart form to API Connect, there is a requirement to properly label the file type you are uploading. Here is how you do it:

- For a product, the part name is product, and the content type is application/yaml or application/json.
- For the OpenAPI(s), the part name is openapi, and the content type is application/yaml or application/json.

You can download the basic-product and basic-test files from the GitHub repository for this chapter to use with this example. This curl command uses the -F argument to produce the MIME content from the files referenced by pathnames following the @:

```
curl -v -k -X POST
  -F "product=@/home/[user]/jenkins/workspace/basic-
    product_1.0.0.yaml;type=application/yaml"
  -F "openapi=@/home/[user]/jenkins/workspace/basic-
    test_1.0.0.yaml;type=application/yaml"
  -H "Authorization: Bearer [GENERATED TOKEN]"
  -H 'Accept: application/json'
  https://[apimserver]/api/catalogs/[organization
    name]/[catalog]/publish
```

If the command was successful, you should receive a `201 HTTP` return code.

Now you have seen two ways in which to publish APIs for your DevOps pipeline. You will be using the toolkit CLIs in the pipeline instead of the Platform APIs, but now you have the additional knowledge of how to use Platform APIs if you find the need to use them instead of the toolkit CLIs.

As you may very well know, constructing a DevOps pipeline may require additional functionality, such as setting up environments, opening up a firewall, and running backups. You can learn how you can automate those tasks with the help of Ansible automation, which is covered next.

Understanding Ansible automation

You might wonder what is Ansible and why Ansible is being discussed in this chapter. Infrastructure as code is becoming a go-to strategy to reduce time to delivery and downtime attributed to human errors. Having infrastructure as code enforces standardization and reduces time spent by resources to handle mundane and repeatable tasks.

Ansible automation is an automation platform that provides enterprise-wide capabilities to automate many repetitive and often manual IT tasks. Being agentless, it is a terrific solution for an organization wishing to automate and free up resources instead of spending countless hours doing tasks manually. While many organizations use Ansible for infrastructure provisioning (which is great), there are so many other areas that can tap into this automation capability, such as security, network configuration, patch management, and, of course, DevOps and DevSecOps.

Digital transformation and modernization may take you down the road of containerization and definitely DevOps. Ansible will a perfect tool to assist with this integration. Because it is agent-less, your company doesn't have to be concerned about applying third-party agents and maintaining them. Ansible works with SSH and a configuration server that triggers the automation tasks. These tasks are built with what are called **playbooks**.

You'll learn how to create a playbook in this chapter, but we will not be incorporating Ansible in the overall pipeline example. However, it will be beneficial to understand where it can fit within your pipeline so that when you are ready to build your own DevOps pipeline, you will know its capabilities. The following is a quick introduction to Ansible that will show you just how easy it is to incorporate Ansible:

1. Install Ansible first. You can follow the instructions from this web page:

   `https://docs.ansible.com/ansible/latest/installation_guide/intro_installation.html`.

2. Assuming you have installed it on a Linux workstation, navigate to `/etc/ansible` and create a file named `hosts`. Here is an example of it:

```
[rhelservers]
192.168.168.220
192.168.168.221
[webservers]
192.168.168.13
192.168.168.14
[jenkinsnodes]
192.168.168.14
```

You can see three groups in the example. Groups are demarcated with brackets and provide an informative name for the servers that fall into that group category. The first group, `rhelservers`, identifies which servers have RedHat Linux installed or targeted to be installed. `webservers` identifies which servers will contain webservers. The last group is `jenkinsnodes`. Jenkins utilizes these nodes to allow the execution of stages and steps so you don't overload the Jenkins primary server. When using the API Connect toolkit CLI to execute commands to manage API Connect, you will need to install the `apic` command on each Jenkins node. Just imagine you have 20 nodes. If new `apic` releases are provided each quarter, that means each quarter you will have to update 20 servers. That's a lot of extra work! This is where an Ansible playbook is a solution.

3. Using an editor, create an Ansible playbook using the following example:

```
---
# Ansible playbook to distribute apic toolkit to
Jenkins Nodes
- hosts: jenkinsnodes
  become_user: root
```

```
tasks:
- name: copy apic toolkit file and set permissions
  copy:
    src: /Users/[your userid]/Downloads/apic
    dest: /usr/local/bin/apic
- name: Changing permission of
  "/usr/local/bin/apic", adding "+x"
    file: dest=/usr/local/bin/apic mode=a+x


- name: verify version
  shell: "apic version"
```

4. Next, you run the playbook by typing the command `ansible-playbook apicdeploy.yml`:

```
ansible-playbook apicdeploy.yml
```

The playbook will run and show successful tasks in green. Red symbolizes errors, and gold represents tasks that made successful updates.

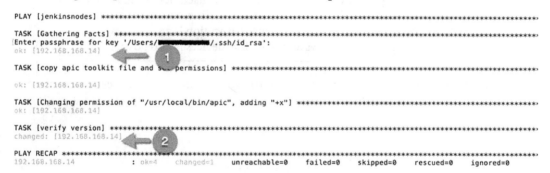

Figure 14.2 – Ansible playbook showing successful execution

As shown in *Figure 14.2*, a green OK represented by the number 1 signifies the task that you defined was executed successfully. A gold change symbol, shown by the number 2, represents that something on the target got updated and sets the status to *changed*.

So, you have now seen a simple use of Ansible. As you might imagine, there are many other situations where you might use Ansible. Here are just a few suggestions on where it can be utilized:

- Installing operating systems or patches
- Applying security patches
- Updating network firewalls
- Building containers
- Performing backups
- Orchestrating API Connect Platform APIs

There are more beyond these. You should consider as much automation as possible to reduce wasted resource time and also reduce errors. Playbooks are easy to build and can be executed throughout your organization.

> **Tip**
>
> The API Connect Toolkit should be installed on the slave machines for Jenkins pipeline execution. If you have a large number of slave machines, using Ansible to distribute new versions of the toolkit and ensuring it is made properly executable and placed in the proper directory for execution is a good idea. Using Ansible to deploy to Windows and Mac slave devices is also possible.

The final capability that you need to learn is how to incorporate your user test cases into your pipeline. This is accomplished by generating hooks into your test cases and we will explore that next.

Testing and monitoring hooks

In *Chapter 13, Using Test and Monitor for Unit Tests*, we learned how to use the API Connect's *Test and Monitor* add-on feature that generated unit tests automatically for your APIs. Having the ability to create test cases is valuable to the overall success of your DevOps process. In the pipeline shown previously in *Figure 14.1*, there is one stage where you execute test cases. You will be utilizing the test cases from Test and Monitor to validate your APIs. Running test cases validates that your unit test ran successfully and allows you to promote to the next stage of deployment.

In order to make those test cases available for inclusion in your DevOps pipeline, you will need to generate hooks for the specified test cases. Here is how it is done:

1. Click on **Test APIs** on your **API Manager** user interface, shown as follows:

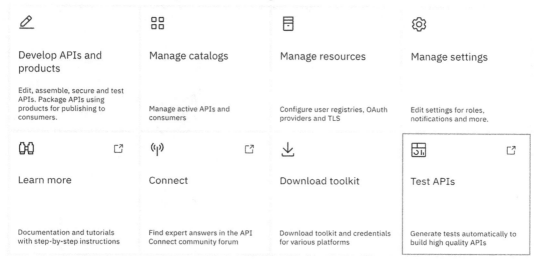

Figure 14.3 – Selecting Test APIs on API Manager

This will launch a new browser tab with the **Test APIs** panel. As you recall, this screen from *Chapter 13, Using Test and Monitor for Unit Tests*, is where you test your APIs. You should have already set up a project from *Chapter 13, Using Test and Monitor for Unit Tests*, called Member API for your test cases. If you haven't set up the Member API project already, you should create one. Otherwise, you can jump to *step 7*.

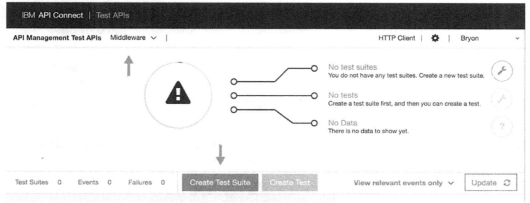

Figure 14.4 – Creating a Test Suite for your unit tests

Click on the **Create Test Suite** button as shown in the preceding screenshot.

2. Fill out the **Test Suite** details with a test suite name for a Member API and then click the checkmark.

    Once completed, you will be presented with an updated panel shown as follows:

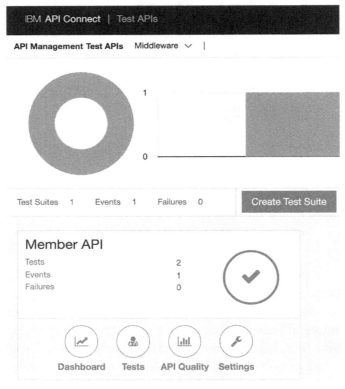

Figure 14.5 – A new Test Suite is created

Figure 14.5 shows the Member API details that may have been created in *Chapter 13, Using Test and Monitor for Unit Tests*. If you are just creating it for the first time, the values for **Tests**, **Events**, and **Failures** will be different. You will need to create the tests using the guidance set out in *Chapter 13, Using Test and Monitor for Unit Tests*. Assuming you have completed those steps and are working with a validated unit test project, you can now begin generating your hooks.

3. In the upper-right corner is a gear icon. This is where you can generate your test hooks. Click on the gear icon to invoke a new page.

Figure 14.6 – Click on the gear to show the API Hooks option

On the left-hand side of the page, you will see the **API Hooks** menu item with two choices – **HOOKS** and **KEYS**. The center pane indicates the provider organization you are generating hooks for and gives you the option to update the time zone.

4. On the left side menu, click on **HOOKS** and the **API Hooks** panel is presented.

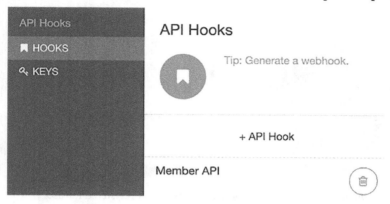

Figure 14.7 – Click on + API Hook to create the hook

When you click on **+ API Hook**, you can add the hook as shown in the following screenshot:

Figure 14.8 – Adding the API Hook

Since **Member API** was the project you created, this will default as the project. Click the checkmark to create the hook. A new URL is presented that you can copy for future reference.

Figure 14.9 – Keep the Hook URL for future reference

You will be using this **Hook URL** later when we start incorporating unit tests within our pipeline. For now, copy the URL and save it.

5. Click the closing **x** and see how it lists your **Member API** hook.

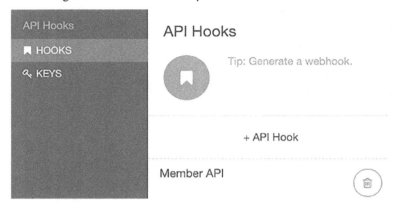

Figure 14.10 – The Member API hook is now registered for use

Now that you have the URL, you can generate an API Key and Secret.

6. Now, click on **KEYS** under the **API Hooks** menu on the left pane, as shown here:

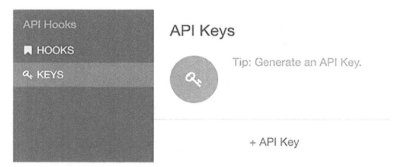

Figure 14.11 – API Keys will provide you with a Key and Secret

You will notice the **+ API Key** button. When you click on that, you will be provided with a new Key and Secret.

7. When you click on **+ API Key**, a new popup will be displayed.

Add API Key   ✕ ✔

The field [name] can't be blank

Name: MemberAPI Hook

Key: 706724d1-de96-43a6-9854-928a8ad17b2f

Secret: 86062657054832da7c24420167f095fb7fd6123db6fb022cada76d26;

Figure 14.12 – Capturing the Key and Secret for your hook

You will need to provide a name (MemberAPI Hook) and copy the **Key** and **Secret** fields so that you can use them when we execute our unit tests within the pipeline. Click on the checkmark to continue.

The original screen reappears and now you will see the key displayed, as follows:

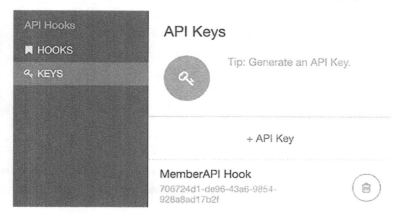

Figure 14.13 – MemberAPI Hook Key created

We have created everything we need to invoke the unit test. We have a URL, the key, and the secret. Now you should run a quick test to verify that the test API exists.

8. Initiate a terminal or command session and execute the following `curl` command using the format shown here:

```
curl -X GET \
  -H X-API-Key:{API-Key} \
  -H X-API-Secret:{API-Secret} \
  <API-Hook-URL>/tests
```

You will set two headers representing X-API-Key and X-API-Secret. Use the URL that you generated earlier and append /tests to the path. An example of the output is shown as follows:

```
curl -k -X GET -H X-API-Key:706724d1-de96-43a6-9854-
928a8ad17b2f
-H X-API-
Secret:86062657054832da7c24420167f095fb7fd6123db6fb022
cada76d262a2e5cb8

https://hub.apicisoa.com/app/api/rest/v1/68d05760-
98be-4ee6-b9a4-d4d188d31ee3867/tests
[{"testId":"5be8d8ab-cd85-4357-914d-cea2fc0096f4",
"testName":"getMember_success",
"testDescription":"",
"lastModified":"2021-03-27T16:29:57Z",
"authorName":"John Apic",
"tags":[]},
{"testId":"2fea53bc-ae07-4755-a0f7-
b42a5ec5f699","testName":"addPatient_success","testDes
cription":"Send successful test request to the Member
API","lastModified":"2021-03-
28T16:45:21Z","authorName":"John Apic","tags":[]}
```

The output that is provided shows you the number of tests available for your project. The last thing you should verify is that the tests actually execute.

9. To run the actual tests, you will use the same URL, Key, and Secret and just change the path from /test to /test/run. Execute the following curl command using API-Key and API-Secret:

```
curl -X POST -k
-H X-API-Key:706724d1-de96-43a6-9854-928a8ad17b2f
-H X-API-Secret:86062657054832da7c24420167f095fb7fd6123db
6fb022
cada76d262a2e5cb8
-H Content-Type:application/json
-d "{options: {allAssertions: false,JUnitFormat:
true},variables: {}}"
https://hub.apicisoa.com/app/api/rest/v1/68d05760-
98be-4ee6-b9a4-d4d188d31ee3867/tests/run
```

Upon execution, you should get test results similar to the following:

```
<testsuite name='Member API' tests='1' disabled='0'
errors='0' failures='0' timestamp='2021-03-29T18:31Z'
time='0'>
  <testcase name='getMember_success'
  classname='5be8d8ab-cd85-4357-914d-cea2fc0096f4'
  time='1' />
```

You can see that there are no errors reported, meaning your test was successful.

Now that you have performed all the necessary steps to add a unit test stage within the pipeline, it's time to explore how to pull your APIs from version control. It is important to have a secure repository for your software revisions. You'll use Git for that purpose.

Using Git as your SCM

There have been many version control systems used in corporate environments. They usually required a centralized server and developers would develop code and check in the code when the code was complete. Examples include Subversion and Team Foundation Server (now called Azure DevOps server). Often, the check-in happened after various deployments have already occurred. Even with more rigid procedures of code check-ins, having a centralized server could lead to a loss of agility if the version control server was down.

In 2005, Linus Torvalds authored Git to support the mass Linux open source community after their version control system became commercialized. Given that the open source community was distributed, a version control system that was fully distributed, simple in design, and allowed parallel branches was needed. And thus, Git became a reality.

Git is very powerful and has many options. You'll be introduced to the features that we will be using in the pipeline next.

Git capabilities

Minimally, you should know the capabilities of Git so you know how to incorporate it within your DevOps pipeline. As a developer, you should understand how to use Git to ensure you are able to create and modify code while maintaining the integrity of the main branch. The features you should be capable of doing are as follows:

- Create and configure a local Git repository: `git init`, `git config`.

- Add files to your local repository: `git add [file]`.

- Associate a remote repository (an example is GitHub): `git config –global user.username [github username]`, `git remote add origin [url for your github repository]`.

- Commit your files with minimal documentation: `git commit -m "description"`.

- Pull and push changes to your remote repository: `git push origin master`.

- Rebase your repository: `git rebase master`.

- Create branches and merge branches: `git branch [new branch name]`, `git checkout [branch name]`, `git merge [branch name]`.

- Clone the repository: `git clone -b [branch] [user credentials] [location]`.

Figure 14.14 shows, at a high level, the activities you will perform when interfacing with a Git repository:

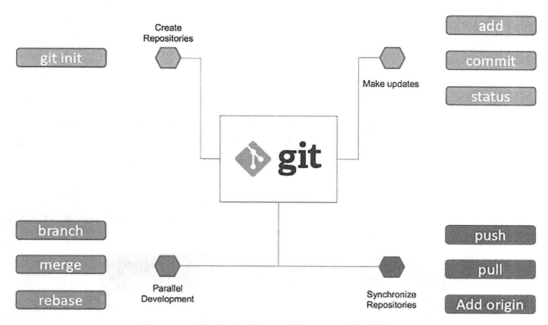

Figure 14.14 – Basic Git capabilities you should learn

As you can see in *Figure 14.14*, **git** is supporting all the capabilities necessary to work within teams. You can create multiple repositories and allow multiple developers to branch code streams, merge changes, and rebase. When you need to synchronize your local repository with GitHub, you have the ability to **push**, **pull**, and add from the origin.

Now that you have reviewed how a developer interacts with Git, you are ready to move on to your pipeline construction using Jenkins.

Constructing the Jenkins pipeline

With all the background information, you are ready to start developing the pipeline stages required for your Jenkins pipeline. While we are utilizing Jenkins to implement our DevOps pipeline, you should be aware that other products can perform similar tasks that you may come across. Since API Connect runs on top of RedHat OpenShift, you should be aware that **Tekton** provides a cloud-native, open source CI/CD platform that makes it easier to deploy across multiple clouds. Tekton runs using Kubernetes so it is a natural fit for companies moving to containers and microservices. To learn more about Tekton, you can visit the following URL: `https://www.openshift.com/learn/topics/pipelines`.

While there are choices in terms of CI/CD, you will be learning how to implement using Jenkins in this chapter. Let's review some of the Jenkins configurations and prepare you for building the Jenkins pipeline.

Jenkins in a nutshell

Jenkins is an open source CI/CD automation tool built on Java. Learning Jenkins will help you build and test your APIs and make it easier to deliver software easier and more often. Jenkins will help you build, test, stage, publish, and deploy all the way up through production.

When you install Jenkins, you will immediately become aware of the many plugins built to run on Jenkins. These plugins will help Jenkins integrate with various tools, such as Git and Ansible. Refer to the following screenshot:

Figure 14.15 – Jenkins Dashboard

When you install Jenkins, you install the master server first. From within the master server, you can perform all CI/CD stages, but that could eventually tax the system. It is the master server that schedules jobs, records build results, and provides status. To alleviate that concern, you can create additional server nodes (called **slaves**) that work on the build stages. When building your pipeline, you can direct execution to specific nodes. The master server will dispatch builds to the additional node and monitor the execution of the builds on those nodes.

With Jenkins, you create various stages of your builds that run either on the master or one of the nodes. Within each stage, you define a number of steps to accomplish.

In *Figure 14.16*, you see a **Build** step. In this step, you can define what you want Jenkins to do.

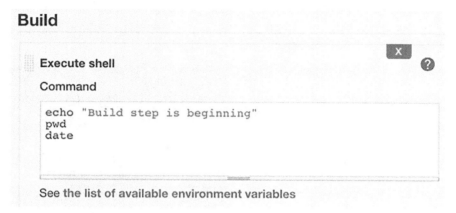

Build

Figure 14.16 – The Build step defines the task to execute

In *Figure 14.16*, you see Jenkins executing a shell script running three shell commands (echo, pwd, and date). The Build step was initially created by clicking on **New Item** shown in *Figure 14.15*. Within a stage, you can create simple commands or complex processing. The combination of the stages and steps creates what is called a **Jenkinsfile**.

To start putting your pipeline together, you will need to decide between the two methods of coding your Jenkinsfile. You have a choice between the scripted or declarative methods to create your Jenkinsfile. Let's understand each of these next.

Exploring the scripted and declarative methods

There are two methods for creating a Jenkinsfile. The scripted method allows you to use the Jenkins user interface to build your pipeline scripts. Scripted development is built using the *Groovy* language. It is very strict in following Groovy syntax, but that is great for complex pipelines. In a scripted pipeline, you declare stage blocks that demarcate the stage that Jenkins will be processing. For instance, you can create a build, test, and deploy stage and execute the steps within each stage. Stage blocks can be combined or kept separately.

The node block tells Jenkins where to execute the steps. You can see in the following scripted pipeline that there are four stages defined and each has a node block that specifies that it should be delegated to run on the node named JenkinsNode:

```
stage('Build') { // for display purposes
    node("JenkinsNode") {
    echo "calling apic";
    sh "apic version"
    }
}
stage('Test') {
    // Run the test
    node("JenkinsNode") {
    echo "Run Unit Test";
    }
}
stage('Deploy') {
    // Deploy
    node("JenkinsNode") {
    echo "Deploy to QA environment";
}
stage('Release') {
    // Release
    echo "Release to Production"
}
```

As you can see, the script is self-documenting.

The declarative method is very similar to the scripted methods but provides more flexibility as it is not as strict as Groovy. In fact, using the declarative method allows you to utilize Git to store your Jenkinsfile and process your pipeline-like code if you so wish.

The declarative method utilizes blocks just like the scripted method, but with the declarative method, you wrap your execution steps within a pipeline block. In the next section, we will learn how to utilize the declarative method.

Perhaps the best way to learn how to integrate API Connect is to actually work through a pipeline in Jenkins. We'll do that next.

Working on an API Connect sample pipeline

For you to build a pipeline in Jenkins, it's a good idea to identify the stages you want to initiate. For this simple sample pipeline, you will be creating a Jenkinsfile that does the following:

- Clones a repository from Git to a workspace on Jenkins server
- Utilizes the API Connect toolkit CLIs to log on to the target servers
- Publishes the product to the Sandbox catalog
- Runs unit tests on the deployed product
- Publishes the product to the Q&A catalog

Although this is a simple example, you should be able to add additional stages and steps to suit your environment.

The Jenkinsfile is provided for your review in the book's GitHub repository here: `https://github.com/PacktPublishing/Digital-Transformation-and-Modernization-with-IBM-API-Connect/blob/main/Chapter14/Jenkinsfile`.

You will be working with Jenkins, and with customizable software packages, there is some initial configuration that needs to be conducted. You will learn that initial configuration next.

Jenkins housekeeping

Assuming you have access to a Jenkins instance, there are a few housekeeping tasks you need to initiate to ensure Jenkins will execute this example. If you do not have access to Jenkins, that is alright. You can read through the section and review the screenshots to have a good understanding of what needs to be done.

First, you should ensure you have loaded the Jenkins plugin for Jenkins pipelines. If you haven't installed the Pipeline plugin, do the following:

On the **Manage Jenkins** page for your installation, navigate to **Manage Plugins**. Find **Pipeline Plugin** from among the plugins listed on the **Available** tab. Select the checkbox for the **Pipeline** plugin.

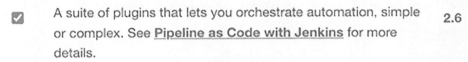

Figure 14.17 – Installing the Pipeline plugin

Select either **Install without restart** or **Download now** and install after restart. The **Pipeline** plugin installation automatically includes all necessary dependencies. Then, simply restart Jenkins.

With the plugin installed, you can now begin the process of building your pipeline.

You also need to set credentials to your remote GitHub repository so your script can make the connection without having the credentials visible. You have to set that up using **Credentials** within Jenkins. You get to that section by clicking on your username in Jenkins and then it should appear on the left. Refer to the following screenshot:

Figure 14.18 – Adding credentials to Jenkins

When you complete this task, you can then reference the credentials in your pipeline using the Jenkins `withCredentials` function, as shown in the following example:

```
withCredentials([[$class: 'UsernamePasswordMultiBinding',
credentialsId: 'github-credentials-repo-book',
passwordVariable: 'password', usernameVariable: 'user']])
```

With that housekeeping task completed, you can now proceed to build the Jenkins pipeline.

Building a Pipeline item

With the plugin loaded, you are now able to create a new item and choose **Pipeline** to begin setting up your pipeline:

1. Click on **New Item** and a new page will display:

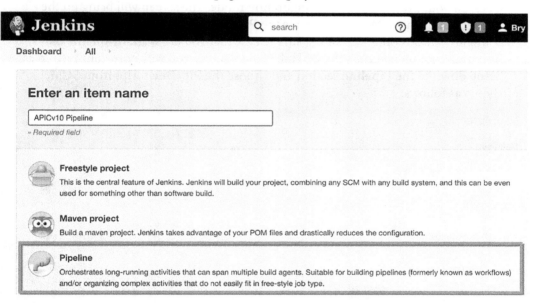

Figure 14.19 – Creating the Jenkins Pipeline

You begin by providing an item name. As shown in *Figure 14.20*, `APICv10 Pipeline` was entered. You will want to select the `Pipeline` project plugin that you installed. If **Pipeline** is not visible, scroll until you find it. Select it and click **OK** at the bottom of the page.

2. Once the page refreshes, you will see the new project in **Dashboard**:

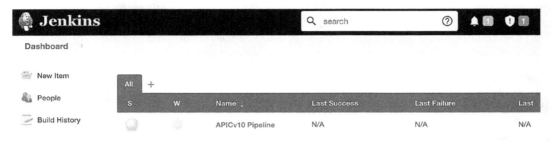

Figure 14.20 – The new pipeline is created

From this point forward, you will be working exclusively within the **Project** pipeline. Now you need to configure the project to utilize **Pipeline** as **Code** by referencing your GitHub repository and providing your credentials.

3. Click on your project to enter configuration mode. The screen will bring up the configuration panel. When you create the **Pipeline**, you have the capabilities to configure various options. For the simple case that you are working on, the only important option is setting the GitHub project where you will store your resources. Scroll down to the **Pipeline** section and choose the **Pipeline** script from **SCM**, shown as follows:

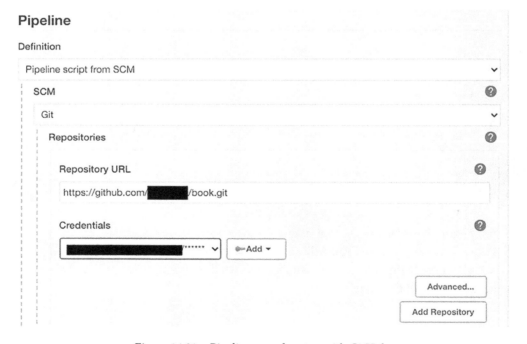

Figure 14.21 – Pipeline as code setup with GitHub

As shown in *Figure 14.21*, you will have to choose **Git** from the **SCM** dropdown and provide your **Repository URL** and **Credentials** details. **Repository URL** follows the format of `https://github.com/[your userid]/[project].git`.

4. Click **Apply** and then **Save**. You are now ready to learn about configuring your Jenkinsfile to execute the requisite stages for your DevOps pipeline.

You should take some time to map out the steps you will be needing to build your pipeline. To sum up, here is what you need to consider:

- Plan on creating in your Jenkins workspace a folder to check out your Git resources using the Git `clone` command.

- Determine which catalog you will publish to as your Dev environment.

- Determine the `apic` CLI commands to issue to perform the step within your Jenkins stage.

- Capture the test URL that you will use to perform the unit test.

- Plan on what action you will take in the event of a failure.

These steps were considered when the Jenkinsfile was created so you will be able to observe how this is done by reviewing the Jenkinsfile provided. Let's see that next.

Review of the Jenkinsfile

In this section, you'll be walked through the key steps within the Jenkinsfile. You can download the Jenkinsfile from the book's repository on GitHub and follow the steps during this walk-through. The key areas are as follows:

1. `GitCheckout`: You will be executing the following shell command that will run `git clone` in your workspace:

```
def GitCheckout(String workspace, String url, String
branch, String credentialsId) {
withCredentials([[$class:
'UsernamePasswordMultiBinding', credentialsId:
'github-credentials-repo-book', passwordVariable:
'password', usernameVariable: 'user']]) {
    int urlDelim = url.indexOf("://")
    String gitUrlString = url.substring(0, urlDelim + 3) +
    "\"${user}:${password}\"@" + url.substring(urlDelim
    +3);
```

```
//Fetch Jenkins workspace
  directoryName = sh (
  script:""" basename ${workspace} """,
          returnStdout: true
  )


//Clean up old workspace so you can populate with
updated GitHub resources
  sh """
  cd ${workspace}
  cd ..
  rm -fr ${directoryName}
  mkdir ${directoryName}
  cd ${directoryName}
  git clone -b ${branch} ${gitUrlString} ${workspace}
  """
  }
}
```

Once the workspace is loaded, you will need to prompt the user for the target environment. That is accomplished by bringing up a prompt:

```
def environmentChoices =
['Dev','Test','Stage'].join('\n')
def environment = null
environment = input(message: "Choose the publishing
target environment ?",parameters: [choice(choices:
environmentChoices, name: 'Environment')
])
```

The input function allows the user to select among the three defined environments where they wish to run the build and deploy functions. As shown in the preceding code, the catalogs in API Connect are titled Dev, Test, and Stage.

2. The next key step is the logon to API Connect. You need to log on to API Connect in order to execute the apic CLI commands:

```
def Login(String server, String creds, String realm){
  def usernameVar, passwordVar
  withCredentials([[$class:
```

```
'UsernamePasswordMultiBinding', credentialsId:
"${creds}", usernameVariable: 'USERNAME',
passwordVariable: 'PASSWORD']])
  {
  usernameVar = env.USERNAME
  passwordVar = env.PASSWORD
  sh "apic login --server ${server} --username
${usernameVar} --password ${passwordVar} --realm
${realm}"
  }
}
```

You can see the Login command being issued through a shell. The credentials are pulled from the Jenkins credential vault so they are not visible. server and realm are pre-loaded from a properties file that is downloaded from GitHub.

3. If the logon is successful, and assuming we chose Dev as the target environment, the next key step is the publish stage:

```
def Publish(String product, String catalog, String
  org, String server, String space = ""){
  echo "Publishing product ${product}"
  if (!space.trim()) {
  def status = sh script: "apic products:publish
    ${product} --catalog ${catalog} --org ${org} --
      server ${server}", returnStatus: true
    if (status == 0) {
    return status
    }
  }
  else {
    def status = sh script: "apic products:publish
    --scope space ${product} --space ${space}
    --catalog ${catalog} --org ${org} -server
    ${server}", returnStatus: true
    if (status == 0) {
      return status
      }
```

```
      }
  }
```

As can be seen from the preceding code, you can publish to a `Catalog` or a `Space`. Again, the parameters are pre-set from earlier loads of property files. `returnStatus:true` returns irrespective of whether the script was successful.

4. The final key stage is running the unit tests that were created in *Chapter 13, Using Test and Monitor for Unit Testing*:

```
def Runtest(String apikey, String apisecret, String
testurl) {
  echo "Publishing product on ${testurl}"
  def status = sh(script: 'curl -k -X POST \
    -H X-API-Key:706724d1-de96-43a6-9854-928a8ad17b2f \
    -H X-API-
Secret:86062657054832da7c24420167f095fb7fd6123db6fb022
cada76d262a2e5cb8 -H Content-Type:application/json
    -d " { "options": {"allAssertions":
    true,"JUnitFormat": true},"variables": { string:
    string, }}" https://hub.apicisoa.com/app/api/rest/
    v1/68d05760-98be-4ee6-b9a4-d4d188d31ee3867/tests/
    run',
returnStatus: true)
  if (status == 0) {
    return status
  }
```

As shown in the preceding code block, the `Runtest` function executes the unit test that was created. The values for `X-API-Key` and `X-API-Secret`, as well as the test URL, are shown for presentation purposes only. Normally, they would be shown as the passed-in parameters `apikey`, `apisecret`, and `testurl`.

That completes the Jenkinsfile walk-through. You were able to review a working Jenkinsfile that executes the build, test, and deploy stages and the APIC CLI commands used for each. There are plenty of other stages and steps you can add to the pipeline, such as implementing life cycle management and publishing to production, but they can be added in the future.

Summary

Building a DevOps pipeline is a key component in driving digital transformation. In this chapter, you learned about a number of products that help drive your DevOps initiative. You learned about Git and how it manages version control and participates in `Pipeline` as `Code`. You also learned that you can use a tool such as Jenkins to drive your DevOps. Jenkins provides many features and multiple plugins that help make the effort easier. A critical component of the API Connect DevOps process involves using the API Connect CLIs. You learned that those commands need to be installed on all Jenkins nodes and that a tool such as Ansible Automation could help manage the updates to the CLI.

From a publishing standpoint, you also learned that you can use the Platform APIs that come with API Connect. What you learned about the Platform APIs is an alternative way to run an APIC configuration. Whenever you see a feature within the API Manager GUI, you now know that there is a supporting Platform API that can be used for your own purposes. A sample was shown so you know how to execute them.

Finally, you learned how to build a Jenkins pipeline and incorporate the CLIs to deploy your products to multiple catalogs. In addition, you learned how to incorporate API Test hooks that were generated using API Connects Test and Monitor. With a working Jenkinsfile model, you are now ready to embark on your own DevOps pipeline for your company.

Coming up next is how you can utilize and customize the analytics generated in API Connect. Equally as important as analytics is customizing the Developers Portal. You will learn about these two topics in the next chapter.

15
API Analytics and the Developer Portal

Toward the end of *Chapter 2, Introducing API Connect*, you were briefly introduced to API Connect components and how they can assist your adoption of digital transformation and modernization. Now is a good time to learn in deeper detail two features of API Connect specifically related to how it impacts your digital transformation and/or modernization effort. Those two key features are the Developer Portal and analytics. In this chapter, you will see API Connect has several valuable components and having a better understanding of them will be helpful in your understanding of how best to take advantage of their capabilities.

Here, we're going to cover the following main topics:

- Customizing the Developer Portal
- Introducing analytics
- Creating visualizations and dashboards

Some of the new skills you will learn from this chapter are the following:

- Understand how to apply simple Portal changes.
- Learn about the basic configuration of the Developer Portal.

- Understand how to change themes to match your corporate branding.

- Identify where analytics are presented to improve your company's insights.

- Learn how to create new visualizations and dashboards to meet your needs.

Creating new user experiences is always fun and you'll understand how easy it is to do by the end of this chapter. You won't need any other tools to install so let's get started.

Customizing the Developer Portal

As you learned in *Chapter 1*, *Digital Transformation and Modernization with API Connect*, social interaction in digital transformation can bring valuable marketing value and product awareness but on the flip side, this social interaction can also bring negative attention. That is why customizing your Developer Portal is as important as developing APIs. In this section, you will learn about the beneficial features provided by API Connect and its Drupal implementation of the Developer Portal. You'll begin by learning about what comes out of the box with Drupal 9 and how API Connect has hooks to make the experience a simple and natural experience for the developer. The out-of-the-box theme of the API Connect Drupal Developer Portal is shown here:

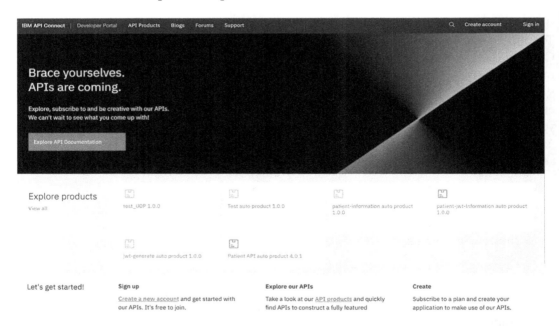

Figure 15.1 – API Connect Developer Portal default view

It is rather bland, but does show the API components that your consumers will be looking to discover and subscribe to. Wouldn't it be nice to make it match your company branding? We will start on learning these capabilities in the next section.

Reviewing Drupal 9 capabilities

Drupal 9 is an open source content management system. IBM has improved the open source implementation to provide an enterprise version that is more secure and has API Connect modules that extend the developer's experience.

> **Important Note**
> It should be noted that the Portal implementation does not support PHP framework code.

Drupal supports many additional modules that can be added. Some of the out-of-the-box capabilities are as follows:

- **Content creation**: Build new content to welcome the potential consumers and provide additional documentation to the developers about your APIs.

- **Blog capabilities**: Allow a provider to blog on the ongoing development activities at your company, providing valuable information for app developers.

- **Forums**: Provide forums to allow developers to learn and discuss your APIs and much more.

- **Social media links**: Links to social apps including LinkedIn, Twitter, and Facebook.

- **Security add-ons**: This helps the organization secure and lock down the portal from malicious activities.

There are many more features that you can find in the Portal documentation at `https://www.ibm.com/docs/en/api-connect/10.0.x?topic=apis-configuring-developer-portal-site`.

To provide the best user experience with the Developer Portal, having a resource that is skilled in Drupal is required, but for simple extensions, a combination of administrator capabilities and some web design skills such as Cascading Stylesheets (CSS or Sass). To understand how to customize Drupal, you need to be familiar with Drupal 9 concepts. By understanding these concepts, it will be easier to create new content. The **node** is the first concept you should learn.

Node types can be thought of as content objects. Items such as pages, blog posts, or forum topics are nodes. When modifying how you will navigate around the portal, you work with menus and blocks. Various types of **fields** can be added to add more metadata to the node. **Regions** are the areas to which the content just discussed can be added. The content within regions is contained in **blocks**. Drupal operates using a layered approach. *Figure 15.2* shows those layers.

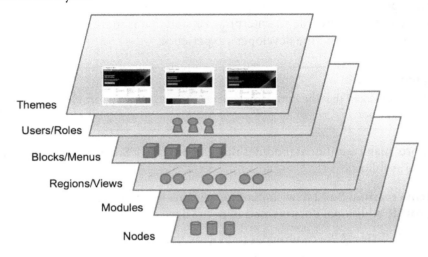

Figure 15.2 – Overview of Developer Portal components

The layers allow you to apply these features to build your portal site to your specification. You define what and where you want to place using content blocks and menus and apply these to one of the packaged themes provided with the API Connect installation. By having layers, you can make updates to layers independently without impacting the existing view. Then whenever you wish to apply new visual components, you simply apply the changes where you want them.

To further your understanding, *Figure 15.3* shows a visual layout of how you can build your user experience:

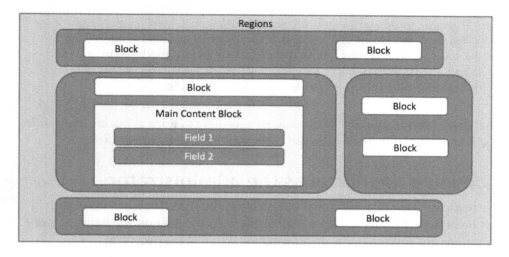

Figure 15.3 – Drupal layout

All of these regions and blocks form the theme of the portal. A Drupal theme is divided into regions where you can place blocks or create your own custom blocks. You have the ability to create and extend themes either by creating them from scratch or using an existing theme created by IBM for API Connect or the Drupal community.

> **Tip**
>
> It is a recommended practice to utilize the existing theme provided by IBM and create a sub-theme for your customization.

The IBM Documentation website provides a considerable list of tutorials on the Developer Portal. You can visit the site at `https://www.ibm.com/docs/en/ api-connect/10.0.1.x?topic=apis-developer-portal-tutorials`.

There are many tutorials on how to customize Portal that will help your organization provide the user experience you desire. With that in mind, what you will learn in this section is what you can accomplish as the Drupal administrator. These are simpler cases that may not require web developer skills.

> **Tip**
> When developing the portal look and feel on Drupal 9, you should understand the concepts of blocks and Drupal theming. There are many websites to learn these capabilities, but as a starting point please see the official Drupal documentation at `https://www.drupal.org/docs`.

With that short introduction out of the way, it's now time to learn about customization. Let's begin with some of the activities the Portal administrator can perform.

Customizing the Portal as the administrator

Since there are so many tutorials already available in the IBM Documentation on API Connect on how to customize the portal user interface, this section will focus on some customizations not covered in the documentation, but that are beneficial.

Those customizations are as follows:

- Removing blogs and forums if not needed
- Adding backup administrators to allow customization using the administration role
- Adding new themes envisioned by your marketing teams
- Updating the Welcome Banner

We start with learning how to remove blogs and forums.

Removing blogs and forums

Not every organization desires to offer blogs or forums to its consumers – for instance, if you are using API Connect for on-premise utilization between the internal departments of an organization. So, if you want to disable the links to the blogs and forums you can do that with just a few quick clicks. *Figure 15.4* shows the admin user logged on to the portal.

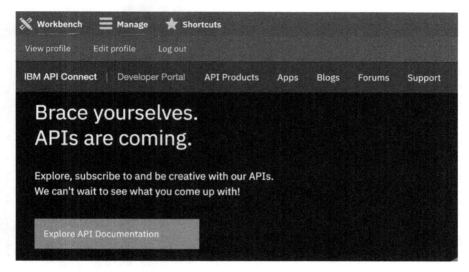

Figure 15.4 – The administrator view

You can see that there are some additional menu links for **Workbench** and **Manage**. You will be using the **Manage** menu to begin your customization.

> **Reminder**
>
> The admin user is established when the provider organization creates a Catalog and assigns a new portal instance. When that occurs, an email is sent to establish the password for the admin account. You use this account to perform customization on the portal.

In *Figure 15.5*, you can see the main menu for the Developer Portal. You will be removing the menu options for **Blogs** and **Forums**.

Figure 15.5 – Main menu showing Blogs and Forums

You'll begin by logging on as the admin for the particular Catalog instance of the Developer Portal. Once logged on follow these steps:

1. Navigate to **Manage | Structure | Menus | Main navigation**. *Figure 15.6* shows the displayed screen:

Figure 15.6 – Removing Blogs and Forums

As you can see, it shows the menu items displayed in the menu. You'll want to disable **Blogs** and **Forums**.

2. To disable the inclusion of **Blogs** and **Forums** simply uncheck those items as shown in *Figure 15.6*. This will remove them from the menu but not remove them from the system.

3. Once done, click the **Save** button and click **Back to Site** in the upper left corner.

When the screen returns, you will notice that **Blogs** and **Forums** are no longer displayed.

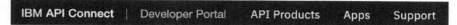

Figure 15.7 – An updated menu without Blogs and Forums

Having a backup administrator account is important because you don't always want to use the default admin user to perform customizations. You'll learn how to do that next.

Adding another user with the administrative role

It's easy to make changes as an admin user but once those changes occur it may be a while before you need to make additional changes. It's not unheard of that the admin password is forgotten. This is why you want a backup user with administration permissions. This user should be someone from your corporation and not an external consumer. Here is how you accomplish that:

1. With the admin user logged on, navigate to **Manage | People**. You will see a list of users who have joined the portal. *Figure 15.8* shows a sample of the list:

Figure 15.8 – Always add a backup administrator

As you can see there are three usernames. There is also an **Action** called **Add the Administrator role to the selected user(s)**. Ensure that is the action listed.

2. Select a username that you wish to add to the administrator role and click **Apply to selected items**.

    You will now see that under the **ROLES** column the user has been given the administrator role.

You have now added a backup administrator, which you can use as the primary portal web designer.

Next, you need to know the steps for how to modify the default look and feel of the Developer Portal. You'll do that next.

Creating a sub-theme from the API Connect default theme

Everybody loves choices. When it comes to user experience and brand awareness your marketing team will definitely be looking for color changes, logo updates, and default content updates. In this section, you'll learn the basics of taking the existing default API Connect theme and making some minor modifications so you can learn how to do updates. Let's begin with the following steps:

1. Log on as a user with the administrator role and navigate to **Manage | Appearance | Generate sub-theme**.

2. Provide a sub-theme name and sub-theme type CSS.

3. Under **Template**, choose one of the six provided themes. See *Figure 15.9*. In our example, we chose **Ruby Red**.

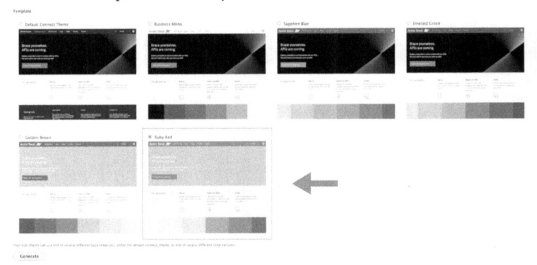

Figure 15.9 – Option themes to create sub-themes

4. Click **Generate** to create the sub-theme.

5. When your sub-theme is created, download the sub-theme to your workstation. The file is zipped so you need to unzip it.

6. Locate the CSS folder and edit the `overrides.css` file. This is where you or your web designers can make color changes. If you do not want to make color changes you can skip this step.

7. Save your changes and re-zip the download. Now go back to your portal and navigate to **Appearance | Install new theme**.

8. Upload your updated ZIP file using the **Browse** button. Click **Install**.

9. You'll see a success message and **Next steps**. Select **Enable newly added themes** as seen in the following figure:

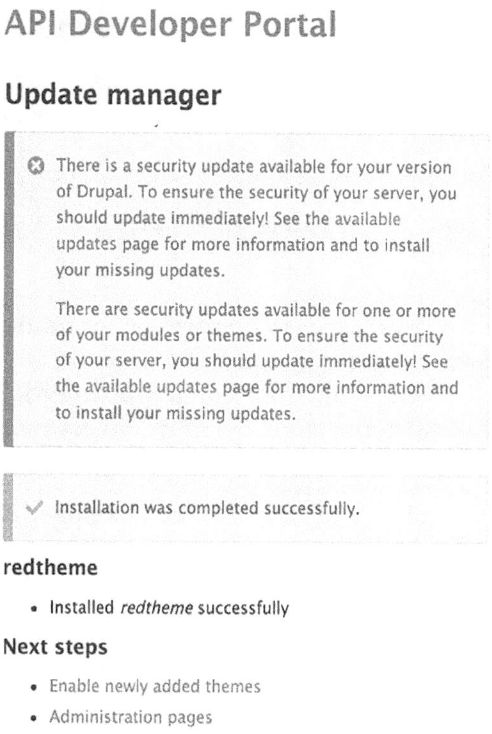

API Developer Portal

Update manager

There is a security update available for your version of Drupal. To ensure the security of your server, you should update immediately! See the available updates page for more information and to install your missing updates.

There are security updates available for one or more of your modules or themes. To ensure the security of your server, you should update immediately! See the available updates page for more information and to install your missing updates.

✓ Installation was completed successfully.

redtheme

- Installed *redtheme* successfully

Next steps

- Enable newly added themes
- Administration pages

Figure 15.10 – A new theme has been installed

10. You have now created a new sub-theme. Click **Back to Site** to review the differences.

Your final change will be to update the Welcome Banner from **Brace yourself. APIs are coming** to **Digital Transformation and Modernization with API Connect. Building APIs that transform your enterprise**.

11. While on the home page move your mouse to the section where **Brace yourself** is shown. A pencil should appear in the right corner of that section if you click anywhere in that area. Click the pencil to bring up the editor and choose **Edit**.

12. Change the text in the **Body** field from `Brace yourself. APIs are coming` to `Digital Transformation and Modernization with API Connect. Building APIs that transform your enterprise`

See *Figure 15.11* as a guide:

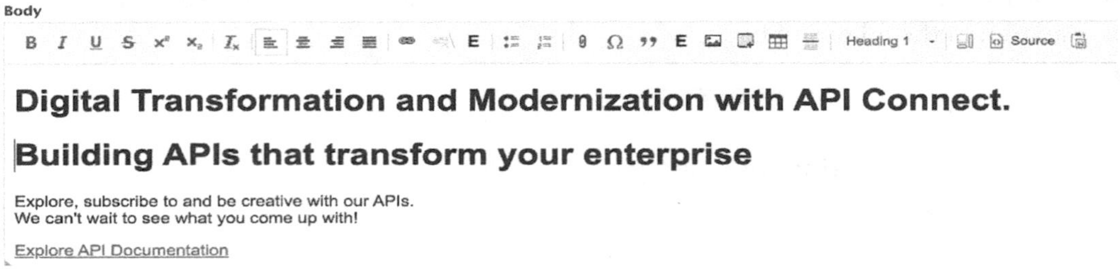

Figure 15.11 – Updating the Welcome Banner

13.   Click **Save** and you are now done. Admire your work.

You have completed some simple activities to prepare you for future updates. You learned how to navigate the portal site using the administrator account. I'm certain that this will spur your marketing team to ask more questions on how to improve the site. With the tutorials provided in the IBM Documentation on API Connect you should be able to continue to customize your portal to your satisfaction.

Another component that is equally as important as the Developer Portal is the analytics service. Analytics can provide you and your organization important information about the use and performance of your APIs. Let's explore the analytics capability next.

Introducing Analytics

This section will introduce the analytics that comes with API Connect and how to begin customizing it for your needs. The analytics that come with API Connect is implemented with open source from the **Elastic Stack** (**ELK** – Elasticsearch, **Log Stash, and Kibana**). In version 5 of API Connect, this component was installed on the management server. With version v10, analytics is separated from the management service and provides greater flexibility and better performance. Analytics servers are associated with Gateways. The administrator configures this relationship in the **Topology** view in the Cloud Manager. The Gateway collects the metric data for each API executed and sends the data to the associated analytics server.

Architect Tip

While this book is primarily for architects and developers, understanding some basic planning and configurations for analytics can be valuable knowledge. For instance, you should consider using MQ queueing to ensure your Gateway analytics aren't lost due to the loss of network storage. This will provide extra reliability in maintaining your analytics. You should also be aware that you can have multiple Gateways associated with a single Analytics service, but keep in mind that each Gateway can be associated with only one Analytics service. There are other planning considerations an architect should be aware of so you should review the IBM documentation on planning the Analytics deployment: `https://www.ibm.com/docs/en/api-connect/10.0.x?topic=deployment-planning-your-analytics`.

Understanding the analytics initial setup

Although the deployment of the Analytics server is outside the scope of this book there are a few contextual bits of information that will help you better understand the scope of the analytics provided by API Connect. Having this background will help you understand how analytics are captured and certain organizational traits that will make your analytics more aligned to your deployment. Next, you will learn, at a high level, how the analytics server is set up in the Cloud Manager topology.

Associating the Analytics server in the Cloud Manager topology

After your API Connect components (Cloud Manager/API Manager, Gateways, Portal, and Analytics server) are installed the cloud administrator will set up within the default availability zone the API Connect components. You can set up components in different availability zones but that is a topic beyond the scope of this book.

In *Figure 15.12*, you see the **Topology** page from the Cloud Manager user interface. What you are seeing are the components that are implemented in the **Default Availability Zone**. The Cloud Manager is not shown because it exists at a higher level and is the manager of the overall topology across the availability zone.

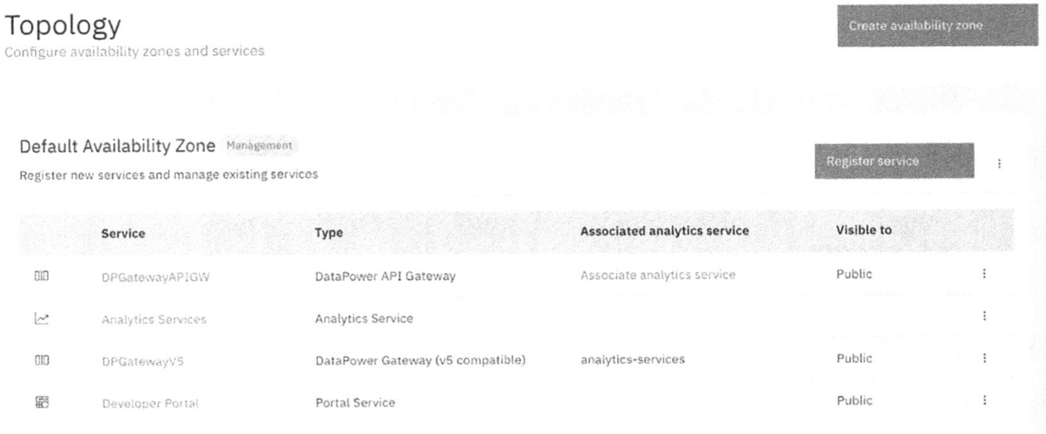

Topology
Configure availability zones and services

Create availability zone

Default Availability Zone Management
Register new services and manage existing services

Register service

	Service	Type	Associated analytics service	Visible to	
	DPGatewayAPIGW	DataPower API Gateway	Associate analytics service	Public	⋮
	Analytics Services	Analytics Service			⋮
	DPGatewayV5	DataPower Gateway (v5 compatible)	analytics-services	Public	⋮
	Developer Portal	Portal Service		Public	⋮

Figure 15.12 – Topology showing associating Analytics with a Gateway

Looking closer at *Figure 15.12* you will see two references to DataPower Gateways. **DPGatewayV5** is for the version 5 compatible Gateway and **DPGatewayAPIGW** is for the new higher performance Gateway simply known as the API Gateway. You also see a single **Analytics Service** that was deployed. In this simple case, there is only one analytics server. That is not a limitation because you can have more analytics servers deployed in different availability zones or the same availability zone as the other services. There is also a Portal server, which you learned about earlier in this chapter.

When the administrator is configuring the topology, one task is to associate a Gateway to an Analytics server. In *Figure 15.12*, you see that the **DPGatewayV5** service is associated with the **analytics-services** service. So that means all analytics generated by the **DPGatewayV5** Gateway will be sent to that specific Analytics Service. You'll also notice that **DPGatewayAPIGW** is not associated yet because it shows the **Associate analytics service** link.

If the administrator wants to associate that Gateway with the same analytics service, they simply click the link and will be presented with a popup to perform the association. *Figure 15.13* shows that popup:

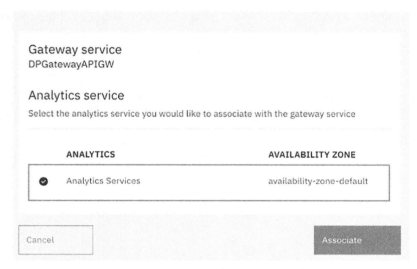

Figure 15.13 – Associate an Analytics server with a Gateway

Once this association is completed the Cloud Manager interacts with the Gateway to inform the Gateway to send its analytics to the associated analytics service. Once the Gateway is associated with an analytics service it cannot be associated with another analytics service.

You have been briefly introduced to how the topology setup is accomplished and should have a good idea of how Gateways send analytics to a designated analytics service. There are more sophisticated scenarios that can be configured. If you are interested in diving even deeper into this subject, you can review https://www.ibm.com/docs/en/ api-connect/10.0.x?topic=topology-creating-availability-zone to learn more about the flexibility of availability zones.

Now it's appropriate to begin learning how to use the Analytics user interface within API Connect. You'll be learning where you can find the Analytics user interfaces and how to create or modify existing dashboards and visualizations.

Viewing analytics

There are several locations where you can view analytics using the API Connect management and portal interfaces. What if your company utilizes a different tool to report analytics, such as Splunk? API Connect allows you to download the analytics information to your own servers to use alternative tools to render them. We won't cover that in this chapter, but you should be aware that the capability exists.

The first place you can review analytics is within a **Catalog**. You log on to the API Manager and select **Manage** to see your Catalogs. Once you choose a Catalog, you then click on the **Analytics** tab as shown in *Figure 15.14*. If your **Analytics** tab is not enabled, it is because the associated Gateway for the Catalog has not been associated with an Analytics service.

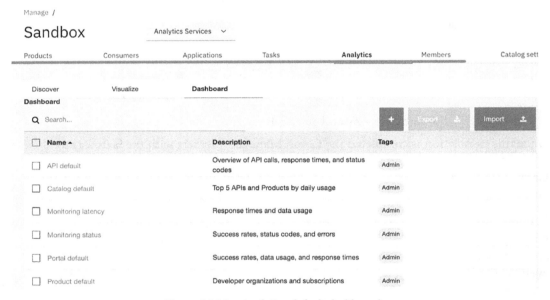

Figure 15.14 – Analytics default dashboards

The dashboards you see in *Figure 15.14* are the default dashboards provided by API Connect. A dashboard is a view of a comprised list of visualizations. Visualizations are the metrics rendered in various graphs and charts that you define. You will learn this in the *Creating visualization and dashboard* section later in the chapter. Dashboards can also be customized to contain relevant visualization.

You will also find **Analytics** under **Spaces** if you have created Spaces within a Catalog. The view is identical to *Figure 15.14*. You learned about **Spaces** in *Chapter 3, Setting Up and Getting Organized*.

The other place you can review **analytics** is within the Developer Portal. You may notice in *Figure 15.14* that there is a Portal default dashboard. This default out-of-the-box dashboard contains relevant information for the API consumers. As a consumer, you find those analytics under the **Applications** link on the portal. *Figure 15.15* shows the default dashboard for the portal:

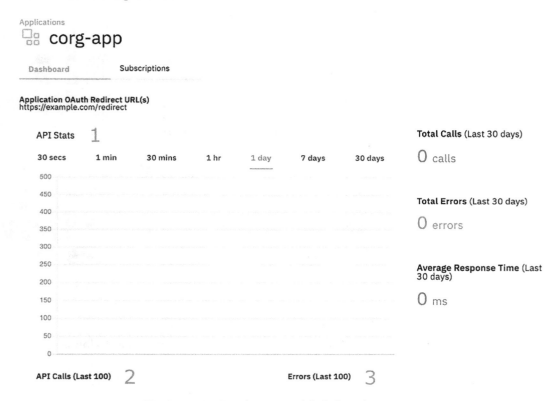

Figure 15.15 – Developer portal default analytics view

A consumer who has created an application with your APIs will be interested, for example, in various API statistics organized by time periods, total **API Stats**, **Total Errors**, and **Average Response Time**. You can see them in *Figure 15.15* as numbered items 1 through **3**.

It should be noted that the consumer cannot create their own dashboards but they can request that the portal dashboard be customized by someone within the provider organization.

You now know where to find the **Analytics** user interfaces in API Connect. It's time to understand how **analytics** are organized. That organization is done with dashboards and you'll learn that next.

Understanding dashboards

In the last section, you heard about dashboards, but what are they really? A **dashboard** is where you aggregate various visualizations to provide a custom view. There are several out-of-the-box dashboards that you can choose from, or you can create your own customized dashboards. When you navigate to any Catalog and click on the **Analytics** tab, you are by default shown the **Dashboard** view as shown in *Figure 15.16*:

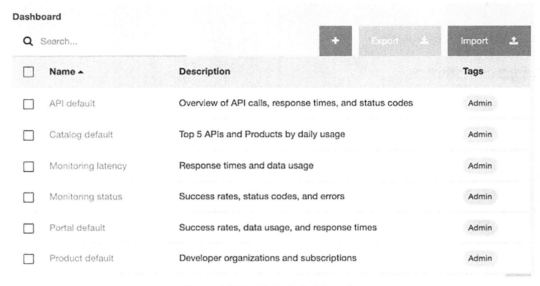

Figure 15.16 – Default dashboards

As you can see there are six default dashboards provided by API Connect as follows:

- **API default**
- **Catalog default**
- **Monitoring latency**
- **Monitoring status**
- **Portal default**
- **Product default**

If you were to click on the **API default** you would see various visualizations added to the API default dashboard for you to review. This is shown in the following figure:

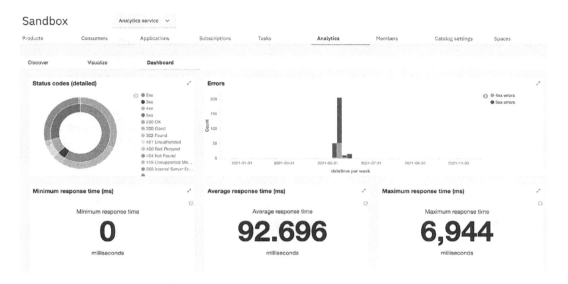

Figure 15.17 – Default API dashboard visualization

Figure 15.17 is just one example of the provided dashboard. You should peruse the others to see how they are constructed.

Understanding the analytics hierarchy

As you may recall, analytics are associated with a Gateway. In API Connect, Catalogs have associated Gateways. So, dashboards are related to the Gateway associated with an analytics service. Since portals are also associated with a Catalog it makes sense that the Developer Portal dashboard is also associated with the same analytics server. You see this relationship in *Figure 15.18*:

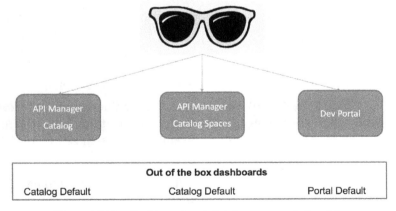

Figure 15.18 – Dashboard and Catalog/Space relationship

When you view the dashboards, the out-of-the-box implementation supports dashboards at the Catalog and Space levels as well as for the Developer Portal. A hierarchy exists between Catalogs and Spaces. Catalog analytics visualization and dashboards are propagated over to a **Space**. If you want to modify the search criteria, the dashboard, or a visualization seen in a **Space** of a **Catalog**, you must clone the Catalog dashboard first before modification. See `https://www.ibm.com/docs/en/ api-connect/10.0.1.x?topic=page-cloning-dashboard` for more details. For clarity's sake, it doesn't work in reverse. You cannot clone a Space dashboard to update a Catalog dashboard.

How does API Connect identify which analytics object was created within this hierarchy? It utilizes tags. **Tags** are used to designate the originator of the dashboard, visualization, and search. *Figure 15.19* shows the various tags and where they are related:

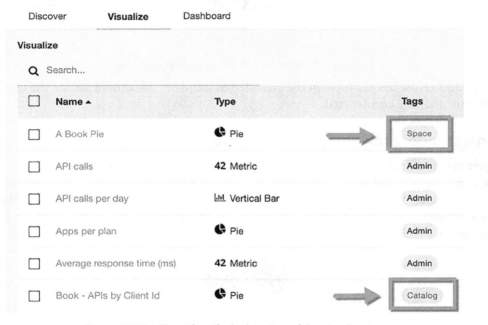

Figure 15.19 – Tags identify the location of the visualization

As you can see in *Figure 15.19*, **A Book Pie** has a tag that signifies that it was created for **Space** whereas **Book – APIs by Client Id** was created at the **Catalog** level. The Admin tag represents visualizations that are part of the default set of visualization provided by API Connect.

So now that you know where to find the analytics it's time to learn how to navigate within the analytics to find metrics. Let's learn about searching next.

Searching within analytics

Collecting data for metrics is necessary and with that data, you need to be able to categorize it and display it in multiple ways. When you launch the analytics dashboard user interface you are presented the default time range of **Last 15 minutes**. Clicking on that link allows you to choose a different time range. That view is shown in the following figure:

Figure 15.20 – Narrowing collection time

Figure 15.20 shows you the **Quick** time range that has a list of predetermined time ranges. If you want to be more specific you can choose either **Relative** or **Absolute** to refine your search even deeper. **Relative** gives you the ability to specify a start time relative to the current time. So, if you were looking for events that happened 20 minutes ago or 3 days ago, you could set those values to retrieve that information. **Absolute** provides you the capability to set a specific date range and retrieve information only for those days. For instance, if you were to specify May 3 to May 13, 2021, you would be able to see the information for those 10 days.

You can also use the **Discover** tab to customize searches. When you do a search, the default response shows you all elements that were captured and you may not need to see all the elements. You can select individual display elements by clicking on the **Available Fields** items on the left side of the page. *Figure 15.21* provides an example of all the available elements displayed:

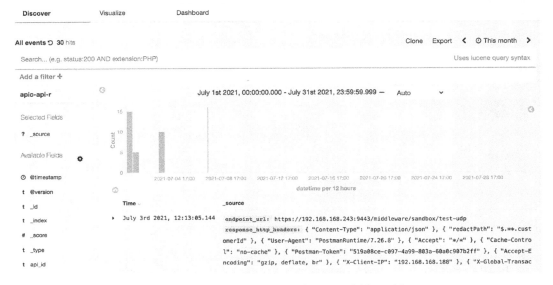

Figure 15.21 – Select elements from Available Fields

So, if you were to select **api_name**, **catalog_name**, **product_name**, and **transaction_id**, your new search results screen would look like this:

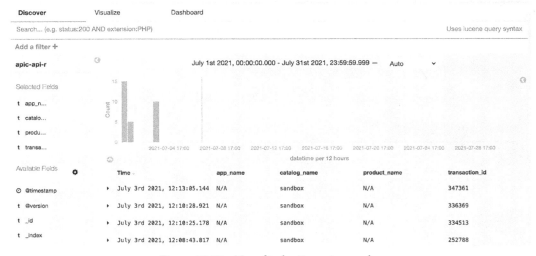

Figure 15.22 – New display items in search

As you can see in *Figure 15.22*, the presentation is much cleaner and more specific to what may be relevant. If you like this type of presentation, you can then save it with a search name and it will be available for future searches.

Narrowing your search criteria

Now that you know how to organize the data that is presented back to you, you can learn how to apply search criteria to narrow the number of returned results. To apply a query, you simply provide the name of the field and the value to search on, separated by a colon.

As you recall you have a list of the **Available Fields** that are located on the left side of the search panel. You can scroll through the list and select a field to run your query. For instance, if you wanted to find all APIs with the *api_name* of *member-api*, you can specify that in your query as api_name:member-api. *Figure 15.23* shows how that would render:

Figure 15.23 – Entering a query can narrow your search

So, your next question might be, *how do I use a compound query such as api_name and product_name?* It is rather simple. You enter `api_name:member-api AND product_name:member`. *Figure 15.24* shows you those results:

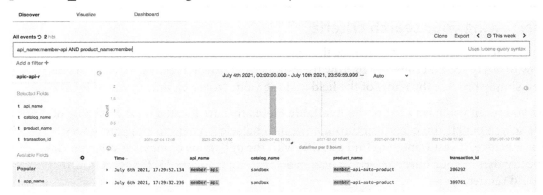

Figure 15.24 – Compound statements are allowed

There are other Boolean operators you can use, such as **OR** and **NOT**. In addition, if you want to search on ranges, you can enclose the range within braces (inclusive) or brackets (exclusive) with the word `TO` that separates the range values – for example, `transaction_id:[3532049 TO 3532225]`.

At this point, you have learned a lot about how analytics is organized and displayed. You are aware of the default dashboards that come with API Connect and most likely will use some of them for initial implementation. You now know how to rearrange your search views and create new searches. Given that background, it's a good time to learn how to create your own visualizations to add to the new dashboards you desire. You'll explore that in the next section.

Creating visualizations and dashboards

While the default dashboards address a number of common scenarios, you'll most likely have some additional ways you want to review your analytics. In this section, you'll learn how to create new visualizations and add them to the dashboards. You'll discover how to customize the data arrangements, but for more in-depth learning you should visit `https://www.elastic.co/training/free`.

> **Information**
> If you want to update one of the default dashboards that come with API Connect, you'll need to clone the dashboard first to create a new custom dashboard.

Since the dashboard is the primary view of your analytics, we'll begin by creating dashboards first.

Creating a new dashboard

When you click on the **Analytics** tab in any of the Catalogs, the first thing you see is the default dashboard. If you want to create a completely new dashboard, you simply click the + button. If you want to update one of the default dashboards, you need to click the dashboard name and then clone it. In this section, we will be creating a new dashboard. Here are the steps:

1. Navigate to the Catalog where you wish to create a new dashboard. When the dashboard page displays, click the + button, as shown in *Figure 15.25*:

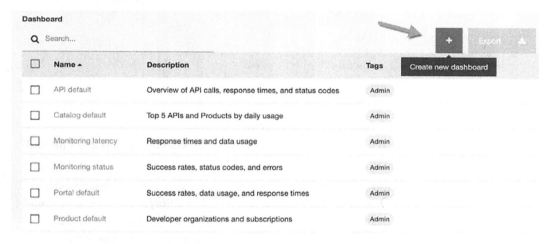

Figure 15.25 – Creating a new dashboard

2. You'll be presented with a new page that allows you to add visualizations. Since your dashboard is empty you need to click the **Add** link to add visualizations.

3. You will now be presented with existing visualizations. Scroll through the list of visualizations and select the visualizations you want on your new dashboard. This is shown in *Figure 15.26*:

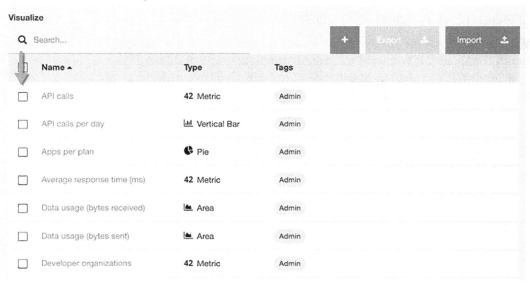

Figure 15.26 – Choosing visualizations

If you want to review the visualization first, click on the visualization name link and review what its capabilities are.

4. When you have selected all the visualization you desire, save the dashboard with a new name. Review your new dashboard and make any adjustments that you would like to see. Your new dashboard is also added to the list of dashboards. It will also be propagated to any Spaces you have created in the Catalog. You will also notice that the dashboard list will show a tag of Catalog that displays at what hierarchy level it was created.

> **Note**
>
> Since analytics is Catalog-specific, you cannot create a dashboard that includes data from multiple Catalogs. You should also remember that any new dashboard created is scoped to the Catalog or Space in which the API exists.

You have created your first new dashboard with existing visualizations. What if you didn't find a visualization that properly represents the analytics the way you want? This is when you need to create your own visualization with the data received from your APIs. You'll perform those steps in the next section.

Creating a new visualization

When you create a new visualization, you are taking metrics and creating viewable graphics that provide insight. The following table presents the various visualization types available to choose from:

Area chart	Markdown widget
Bar chart	Metric
Coordinate map	Pie chart
Data table	Region map
Gauge	Tag cloud
Goal	Tile map
Heat map	Vertical bar chart
Line chart	

Table 15.1 – Available types of visualizations

As you can see in *Table 15.1* there is a wide variety of graph types. Creating a visualization is essentially a two-step process. You first choose the graph/visualization type (bar chart, line chart, and so on) and then define the content and layout.

When you define the content and layout, you will use the **Data** tab to configure the metric and bucket objects. If you want to add a legend, you can use the **Options** tab.

For this introduction to creating new visualizations, you'll choose one graphic that fits some of the API transactions you have run in the book. We'll choose a **Pie chart** so we can choose the buckets to display segmentation of the total amount of API calls. A bucket is a collection of organized data that has been filtered (similar to a SQL GROUP BY statement).

Let's start creating a new visualization:

1. Navigate to the **Analytics** tab within a Catalog and then choose **Visualize**. Then click the + button to create a new visualization, as shown here:

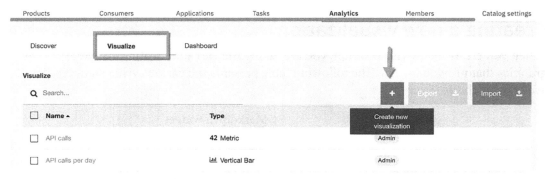

Figure 15.27 – Creating a new visualization

After clicking the + button, you can select the type of visualization to create. *Figure 15.28* shows a partial list of the available visualization types:

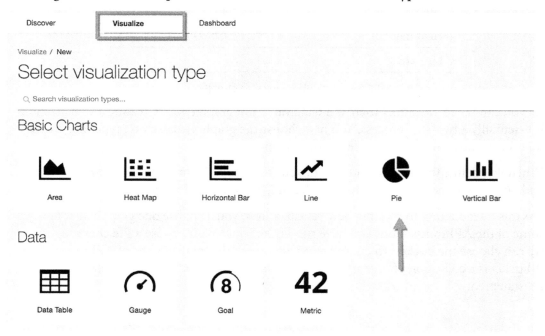

Figure 15.28 – Choosing the Pie visualization type

2. Choose the **Pie** visualization type and you will be presented with a panel to choose the type of data. This is shown in *Figure 15.29*:

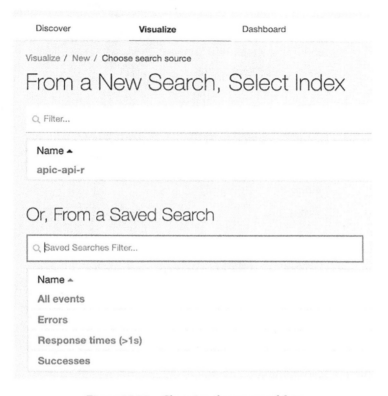

Figure 15.29 – Choosing the source of data

In *Figure 15.29*, you see some predefined searches. These were created by API Connect. **All Events** are all the events that have been captured on the Gateway. You will choose **All Events** since it provides many event fields to choose from.

3. On the left side, under **buckets**, choose **Split Slices** as shown in *Figure 15.30*:

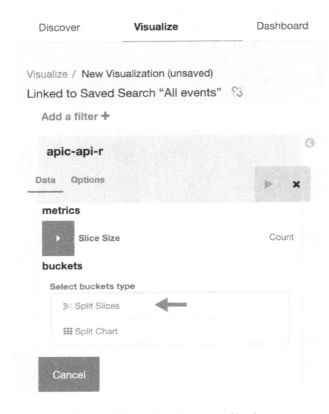

Figure 15.30 – Choosing type of bucket

4. When you chose **Split Slices** the panel changes and you will see an aggregation dropdown. Under the aggregation dropdown, choose **Terms**.

5. The dynamic nature of the panel will then re-draw the panel with a new set of fields. In the **Field** dropdown, you choose a field that appropriately represents your needs. Scroll down and choose **client_ip**.

6. Click the *play* button next to the **Data** and **Options** tabs to see the results.

    Figure 15.31 shows the results from our limited sample size. Your results will be different. You may also have to change the time range to ensure you have the appropriate metrics.

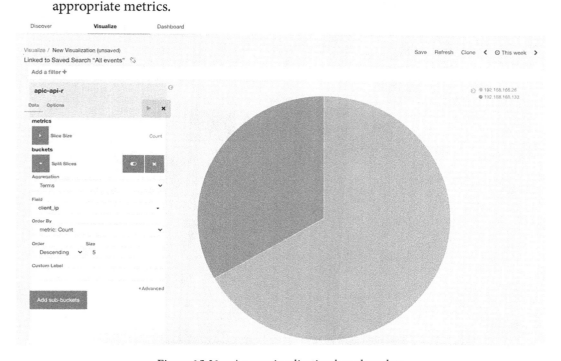

Figure 15.31 – A new visualization based on data

The new visualization shows which client IP made API calls through the configured Gateway.

7. Let's further enhance the same visualization by adding a sub-aggregation. Locate the **Add sub-buckets** button and click on it. This is shown in *Figure 15.32*:

Figure 15.32 – Add sub-buckets to do sub-aggregation

When you click **Add sub-buckets**, another entry field setup screen will appear so you can split the slice even more. You will need to click on **Split Slices**. The panel will refresh with more fields visible.

8. Click the **Sub Aggregation** dropdown and select **Terms**. Then choose **app_id** under the **Field** dropdown and click the *play* button again. See *Figure 15.33*:

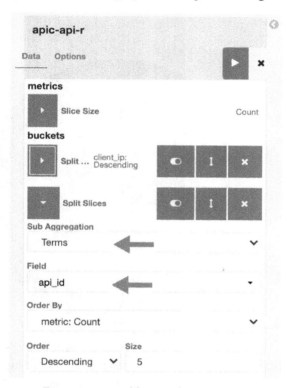

Figure 15.33 – Adding a sub aggregation

9.　*Figure 15.34* now show your new sub-aggregation pie chart:

Figure 15.34 – Sub-aggregation example

This shows visually (by client IP) the various applications calling the same API. You have now completed creating a visualization. You can add it to a dashboard of your choice.

You can also save your new visualization. *Figure 15.35* shows how this is done:

Figure 15.35 – Saving your visualization

You have now finished creating a custom visualization and dashboard. There will be many more opportunities to create and update your analytics.

> **Note**
>
> Metric aggregations give you the ability to calculate metrics based on values extracted from indexed fields. The Analytics service indexes the field received from the Gateway.
>
> You can think of metrics as the y axis and buckets as the x axis.

Most likely in the real world, you will be making your visualization and dashboard changes in a lower environment, such as Developer or User Acceptance Testing. After all your hard work, you'll want to ensure the same views are available in Production. The way you ensure the dashboards and visualization are available in those environments is by exporting and importing them. You'll do that in the next section.

Deploying analytics to new environments

In many cases, API Connect is deployed in multiple environments, whether on-premises or on the cloud. Now that you've created and tested your dashboards and visualizations, you are ready to promote them to upper environments. This is simple, and can be accomplished by exporting and importing the resource. The process is the same for both visualizations and dashboards. Let's begin by showing you how to do this with a visualization and then you can perform the same steps for dashboards on your own:

1. If you are not on the **Analytics** tab, please navigate to the appropriate Catalog and select **Analytics**. Select the visualization and a new set of buttons appear:

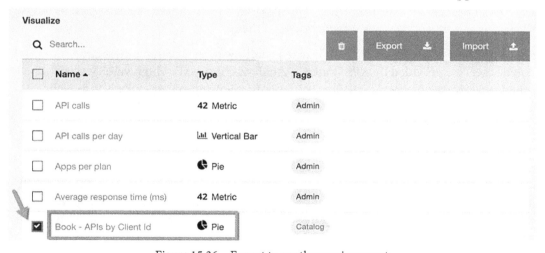

Figure 15.36 – Export to another environment

2. Select **Export**. It creates and downloads an export.json file. You can rename the .json file to represent your new visualization and later import it into other Catalogs. *Figure 15.36* shows where the **Import** button is located. In your target environment, just follow the same process to get to the **Visualization** tab and then import your visualization.

That completes the section on customizing analytics. You learned a number of new skills that will help you create meaningful analytics that can be shared with your organization. The skills you learned for accessing the analytics tab and creating new visualizations and dashboards will be invaluable to your organization.

Summary

This chapter was all about customizing API Connect's user interfaces. We covered the customization of the Developer Portal and Analytics service.

In the Developer Portal sections, you learned how to make visual changes to the portal using the administrator role. As you know, the social aspects of your digital transformation are extremely important and having the ability to apply your branding and social interaction with your users is vital.

The important portal adjustments you learned included ensuring you have backup administrators, as well as learning how to develop a new sub-theme where you can change colors simply by adjusting the CSS. You also learned how to update the Welcome Banner. Since not all portals are external facing, you explored how to disable features such as blogs and forums as some organizations may not need them.

Another topic in this chapter that you learned was how to make simple changes to the Analytics service provided by API Connect. Analytics have a very high value in your digital transformation effort. Providing analytics to your internal organization as well as your consumers shows how well thought out your API implementation was.

In this chapter, you found out how Analytics is organized and how there exists a hierarchy between Catalogs and Spaces. You also learned how the Analytics service is associated with your Gateways and that you can have multiple Gateways output metrics to an analytics service. You learned how to locate where **Analytics** data is viewed within the API Connect environment and how it is displayed on the **Applications** page on the Developer Portal. Finally, you learned how to modify and create new dashboards and visualizations. You also learned how to export your changes and import them into other environments of your choosing.

Having these new skills provides you with the ability to craft new and exciting graphical views, which will add value to your organization and consumer experience.

Your journey of learning about API Connect is nearly complete. Who knows what the future holds? You'll learn about upcoming technologies in the next chapter, so prepare yourself to learn even more.

16
What's Next in Digital Transformation Post-COVID?

Alas, all good things must come to an end, but what about Digital Transformation? As you learned in *Chapter 1*, *Digital Transformation and Modernization with API Connect*, business changes are constantly occurring, and what business change has had more impact than COVID-19? Here's a quick review of the key business changes that we are coming across:

- New start-ups are going digital at inception.

- Customers/consumers are changing and so are their expectations. Customer loyalty is waning.

- Information and customer feedback affect your products.

- Innovations are changing rapidly. You need to keep up or be left behind.

- The delivery of goods has drastically improved and that affects pricing.

- Similar products have made innovation and pricing differences indifferent. *Best-of-breed* is no longer a consideration.

As you look at each item, I'm sure you can already sense the impact that COVID still has on businesses. Companies had to turn on a dime to remain viable. Others needed to increase productivity and expand in ways never imagined. Remote work forced many organizations to rethink the need for office space and proceed using video conferencing applications to maintain continuity within operations. The automation of operations became even more necessary as employees were displaced by remote work and the challenges of getting remote workers online increased. These changes forced even existing products to improve security and expand capabilities. An example would be Zoom. Its explosion during COVID reached new users but exposed security holes that required patching. The companies who already could deliver goods and services did extremely well in these difficult times. It is now common to see packages delivered to the door of households that never took advantage of that capability before.

Digital transformation continues for the same reasons it began. Businesses change and improve and those who did not have either closed or have suffered irreparable damage. So, as digital transformation moves beyond the COVID-19 pandemic, there are still more capabilities and changes on the horizon. In this chapter, we are going to cover new topics and revisit some topics that will continue to drive businesses post COVID. The areas you will learn about are the following:

- Understanding API Connect late-breaking changes

- Understanding API Connect and hybrid cloud – CP4I, OpenShift

- Understanding AIOps

- Exploring 5G Edge computing

There will probably be more innovations beyond these but to dovetail effectively with your API Connect experience these are extremely important since they are extending digital transformation to an even greater extent. When you finish with the chapter, you will have new knowledge and skills on how to continue with your transformation journey. Those skills are as follows:

- Where to check for updates and fixes

- How hybrid cloud is utilized to bridge the gap during transformation

- How automation is assisting with Modernization and Transformation

- Where 5G technology is opening new Digital opportunities.

Let's learn about these new capabilities and envision how they may drive innovation and capabilities in your organization.

Technical requirements

In this chapter, there isn't any code walk-through so there are no technical requirements that need to be downloaded from GitHub. This being the final chapter, you should ensure you have downloaded all previous chapters in case there have been updated resources due to any late-breaking changes while you have been reading.

Understanding API Connect late-breaking changes

In the course of authoring this book, over a period of 5 months, IBM has provided several updates to API Connect for a variety of reasons. We discussed the versioning of API Connect releases in *Chapter 2*, *Introducing API Connect*.

A number of them address security and prevent cybercriminals from disrupting your business, but the majority of these updates provide improvements and bug fixes. During the 5 months, we have seen a modification release, two fix pack releases, and one -ifix release. Here is a list of those updates. This book started with version 10.0.1.1:

- Version 10.0.1.1
- Version 10.0.1.2
- Version 10.0.1.2-ifix1
- Version 10.0.2.0 (Continuous Deployment)
- Version 10.0.3.0 (Continuous Deployment)
- Version 10.0.1.5 (rollup of v10.0.2.0 and v10.0.3.0)

Each one of the releases addresses bug fixes and additional product features. Care should be taken when updating to certain releases. In some cases, the releases must be implemented in sequence, and in other cases, they contain cumulative updates. You should refer to the IBM Documentation page, https://www.ibm.com/docs/en/api-connect, anytime a new release is announced to determine whether there are special considerations.

> **Note**
>
> IBM refers to supported versions of API Connect as **Long-Term Support (LTS)**. As new features are developed, IBM releases a **Continuous Deployment (CD)** version so customers can try out the new features. These CD releases are not supported. v10.0.2.0 and v10.0.3.0 were CD releases but are now included in v10.0.1.5.

You should also pay attention to what's new. In some cases, there could be updates that improve how configurations are applied. An example is how custom user policies are applied. In version 10.0.1.5, you can now apply the custom policies with the user interface. In previous releases, you needed to install them using the CLI. Also, keep in mind that with any release there are supporting components such as the CLI that require download and re-distribution. This applies to customers using DevOps to build and deploy artifacts using the CLI commands. In version 10.0.1.5, IBM made performance enhancements, as well as a number of other enhancements to improve the developer's experience.

It is important to have a good understanding of how IBM provides fixes and upgrades. It provides you with the knowledge on how often fixes are released and the choices you have to apply them. You can refer to those upgrades by visiting the *What's new in the latest release* section of the IBM documentation: `https://www.ibm.com/docs/en/api-connect/10.0.1.x?topic=overview-whats-new-in-latest-release-version-10015`.

> **Note: v10.0.1.5 Enhancements**
>
> There is a considerable amount of changes released in v10.0.1.5. A few noteworthy updates are vertical and horizontal pod gateway scaling, GraphQL enhancements, and performance enhancements, to name a few.

Given that digital transformation and modernization are taking advantage of cloud capabilities you should be aware of how API Connect is continuing advances towards hybrid cloud and multi-cloud implementations. Having this knowledge will help your organization plan for future strategies for cloud implementations and the approaches that are provided. Having a good understanding of how API Connect merges into a hybrid cloud will be important. You will learn that next.

Understanding API Connect and Hybrid Cloud

In *Chapter 2, Introducing API Connect*, we discussed the deployment models of API Connect. IBM has placed considerable emphasis on the hybrid cloud and that can't be any more apparent than its acquisition of Red Hat for $34 billion. IBM's hybrid cloud strategy and its Watson AI technology are becoming a differentiator in the corporate hybrid architectural strategy and Digital Transformation journey.

To be successful in hybrid cloud, customers will be looking for a common platform and the ability to continue integration with new cloud applications and on-premise systems of record.

Let's begin with the OpenShift common platform from Red Hat.

OpenShift

Is there a notion of an open hybrid cloud architecture? With containerization and open source Kubernetes, you may believe that there is, but Kubernetes in each cloud provider has its challenges. There are nuances in every implementation and Kubernetes is a huge paradigm shift. Similar to Object-Oriented Programming, there is an associated learning curve that comes with Kubernetes that makes customers cautious.

Consider the challenges your company would face to undertake a **Do-it-yourself (DIY)** Kubernetes implementation. Here is a short list:

- Building a full DIY Kubernetes stack
- Testing your Kubernetes stack
- Staying abreast of Kubernetes changes
- Patching security CVEs (weekly)
- Troubleshooting Kubernetes stack problems
- Repairing Kubernetes problems

The big question is do you have the expertise in-house? If not, do you increase staff to find qualified resources? Consider the complexities of implementing Kubernetes shown in *Figure 16.1*:

Setup	Deployment	Security	Operation
• Operating System Setup • Validating Node setup • Setup common parameters (replicas, labels, configmaps, image names and other settings	• App Monitoring and Alerts mechanism • Choosing Storage and Persistence • Configuring Ingress and Egress • Picking Host Registry for Images • Establish Build and Deploy Models • Determining Identity and Security Access	• Cluster and Pod Monitoring and Alerts • Resource monitoring and Metrics pipelines • Kubernetes Security Hardening and risk mitigation • Securing Images • Security Certification • Network Ingress and Egress Policy • High availability and Disaster Recovery • Network and Container Segmentation	• Operating System Upgrades • Manage Kubernetes Versions • Container Image Upgrades • App Upgrade and Fixes • Security CVE Patches • Continuous Security Scanning • Multi/Hybrid Cloud Rollout • Maintaining Enterprise Container Registry • Managing Cluster and App Elasticity • Monitor, Alert, Remediate • Log Consolidation

Figure 16.1 – Complexity of Kubernetes implementation

All of these unplanned complexities warrant looking for something simpler and supporting critical functionality such as security, monitoring, and easy upgrades. This is where Red Hat OpenShift makes sense.

OpenShift is deployable both on cloud and on-premises. Utilizing the OpenShift platform addresses many of the issues identified in this section. *Figure 16.2* highlights the difference between what you gain by standardizing on OpenShift versus doing it on your own with Kubernetes:

	OpenShift		Kubernetes	
Container orchestration	✓	Kubernetes	✓	Kubernetes
Container image	✓	OCI-compliant/docker	✓	OCI-compliant choices
Container runtime	✓	CRIO/docker	✓	OCI-compliant engine choices
Container build	✓	RHCC/S2I/dockerfile	✗	
Container registry	✓	Quay/OSS docker registry	✗	
Container scanner	✓	OSCAP/Clair Secure monitor	✗	
CI/CD automation	✓	Jenkins 2	✗	
IDE	✓	Eclipse Che Code Ready	✗	
Web UX	✓	Web console	✓	Web console
CLI UX	✓	oc/odo/kubectl	✓	kubectl only
Service catalog	✓	Operators/Open Service Broker	✓	Operators
Secrets management	✓	Kubernetes Secrets	✓	Kubernetes Secrets
Supported/preferred runtime	✓	EAP, Spring boot, vert.x, node.js	✗	Kubernetes

Figure 16.2 – OpenShift and Kubernetes comparison

As you can see in *Figure 16.2*, not only is it easier to use Kubernetes under OpenShift but you also get additional critical components for your digital transformation journey such as a web user experience (**Web UX**) for your administrators and developer tools such as Spring Boot, a built-in capability for DevOps, and container build capabilities such as source to image (**S2I**).

Wherever you choose to deploy, whether in the cloud, on-premises, or hybrid, having a common platform has its benefits. *Figure 16.3* provides a good picture of those choices:

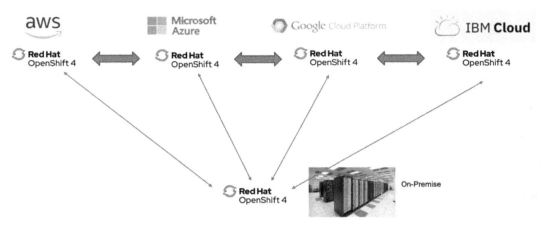

Figure 16.3 – OpenShift on multiple clouds and on-premise

The strategic decision is two-fold. Having a choice of clouds provides your organization the flexibility to change to another cloud because OpenShift is portable in a *write once, run anywhere* open hybrid cloud platform architecture.

> **Note**
> Be careful about cloud platform lock-in. There are cloud platform services that are unique to the provider and are tightly coupled. If you decide to change providers and want to move data and applications to another cloud, you may find the cost of re-coding or vendor-required advanced notice prohibitive.
> In a DIY implementation, you can easily lock yourself into that cloud platform by adding cloud-specific implementations and/or additional components.

Once you have decided on how you want to approach digital transformation and modernization (whether you decide on OpenShift or not), your next challenge will be integrating all of the new and existing services in a cloud-friendly manner. This is where Cloud Pak for Integration fits nicely, especially if you have on-premise integrations you wish to move up to the cloud. You'll learn about that in more detail next.

Cloud Pak for Integration (CP4I)

Integration continues to be an important business objective for all enterprises. In fact, since the introduction of cloud platforms, there have been even more integration points than ever before. The complexity of hybrid and multi-clouds has increased with security and performance requirements. The ability to securely implement event processing and APIs is paramount. The transfer of data and transformation and routing are extremely important as the systems of records and applications are now located in multiple data centers.

As you already know from *Chapter 2, Introducing API Connect*, API Connect was re-architected to run in containers and be cloud-ready. Integration tooling such as **App Connect Enterprise** (**ACE**) has also changed to support microservices and faster delivery times. Couple those with support for high-speed data transfer, event processing using Confluent/Kafka, and MQ messaging and you have assembled a cloud integration platform.

> **Note: v10.0.1.5**
> API Connect on Cloud Pak for Integration provides the capability for API Connect to provide runtime security for Kafka topics using the Event Gateway Service.

This platform can modernize your on-premise applications so you can now develop on-premise or in the cloud of your choice. Those are the capabilities of Cloud Pak for Integration. But then, what are the motivators to utilize Cloud Pak for Integration? We will review those next.

Hybrid Cloud modernization motivation

Change is perhaps one of the biggest motivators and the pandemic certainly introduced change. Whether it was working from home, changes in logistics, closed borders, supply chain disruptions, or all of the above, businesses needed to make difficult decisions to survive.

Change can also bring opportunity and many organizations took this opportunity to focus their attention on factors they desired and made them directives. Whether they were new or existing executives, the following topics ranked high on their lists:

- **Agility**: The business wanted to enable more rapid innovation that was guided by needs (especially during the pandemic).

- **Resilience**: The business required continuous availability that required less hands-on and more capacity. You might relate this to increased virus testing capabilities.

- **Scalability**: This relates to the need for scaling the business upon demand with limitless capacity. Think of home deliveries or increased online ordering capacity.

- **Cost**: With fewer people in the office or data center, taking advantage of cloud platforms.

- **Portability**: Can't be locked down. Need to be able to move between clouds if capabilities or costs dictate better options.

When business conditions like the ones mentioned are the motivating factors, a platform like CP4I makes total sense. *Figure 16.4* shows the various integration styles that comprise CP4I:

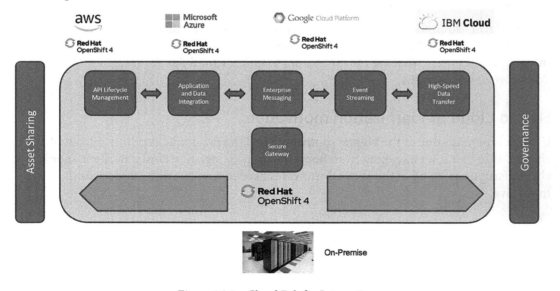

Figure 16.4 – Cloud Pak for Integration

As you can see, CP4I runs on multiple clouds as well as on-premise. This is accomplished by utilizing Red Hat OpenShift, which you learned about in the preceding section. CP4I is built using containers so agility, as well as scalability and resilience, are omnipresent. Portability is not an issue with container technology and OpenShift, so given the executive wish list, all the boxes are checked.

Without going into too much detail, the following are some of the highlights of CP4I:

- CP4I has a **Platform Navigator** component. It has centralized management and control over all the CP4I components.

- CP4I has an **Asset Repository** capability. It improves development agility with the reuse of integration assets.

- It offers **Common Services**. We can get logging, metering, monitoring, and many other key **foundational services** with IBM Common Services.

- All the functionality you have learned about API Connect is included here.

- It integrates all your business data and applications more quickly and is portable across any cloud.

- With a minimal ramp-up time, your existing on-premise App Connect developers can build reliable integrations using the skills they learned with App Connect Enterprise on CP4I.

- Send large files (SFTP) and datasets virtually anywhere, reliably, and at maximum speed.

- Utilize MQ to create persistent, security-rich connections between your on-premises and cloud environments.

> **Note**
> The added value of using CP4I beyond just API Connect is the ability for the enterprise to choose between a set of solution capabilities to build a complete solution and let API Connect be the method to connect the implementations.

From an API Connect standpoint, you just revisited OpenShift and Cloud Pak for Integration to round out your complete understanding of what motivates your organization and what it may be looking for in future modernization efforts.

Now that we have a better understanding of platforms such as OpenShift and Cloud Pak for Integration, transformation doesn't end here. Many people know that **Artificial Intelligence (AI)** has been a major force in cloud platforms. Let's now introduce you to another game-changer that meets one of the executive goals mentioned in this section—improved resilience and agility using artificial intelligence.

Understanding Artificial Intelligence for IT Operation (AIOps)

One executive goal that started to pick up steam before the pandemic was automation. Automation goes beyond DevOps. It encompasses efforts to reduce manual tasks and improve resiliency. Prior to the pandemic, many companies were performing **Proof of Concepts (PoCs)** using multiple tools to determine where automation could occur and which toolsets provided adequate coverage. You learned about Ansible Automation in *Chapter 14, Building Pipelines for API Connect.*

Pre-pandemic many companies were getting organized and gathering feedback on automation experience. While not fully implemented across the enterprise, some silos were showing the benefits of automation, but automation maturity has not been achieved. Various reasons for not reaching velocity are the need for skilled resources, the presence of undefined processes, cultural barriers, and the limitation of only looking at operational activities. The pandemic changed the focus.

During the pandemic, the initial efforts were forced. People needed to work from home, and that immediately put an emphasis on how quickly organizations were able to mobilize. Laptops needed to be procured, VPN and video conferencing became overwhelmed, and collaboration tools became predominately the means to communicate and connect remote resources. Those that had some automation benefited from those pre-pandemic efforts, but the need to have it done more quickly became apparent.

The reality of the pandemic and the associated changes that the organization made to keep their people safe and maintain business as usual were real eye-openers. Resiliency was redefined with greater clarity. Investing in automation throughout the organization became clearer and people and cultures began to change. Not only did they want more automation but they wanted the automation to be based on potential business changes. This is where **AI** plays a key role.

The role of AIOps

AIOps is used to enhance IT operations. With the myriad of data from events, logs, and other sources means the opportunity to ingest the information and allow machine learning to enable IT operations is possible. Some of the benefits you can achieve with AIOps are as follows:

- Improved response and resolution because AIOps can decipher root causes and suggest remediation faster than previous human troubleshooting.

- With AIOps machine learning, operations can be more predictive and proactive and will decipher alerts to identify critical events from lower-level events.

- With AIOps, operators can receive notifications based on thresholds and with relevant information to make a diagnosis and take corrective action faster.

From a Digital Transformation viewpoint, you may get a myriad of digital capabilities, so utilizing AIOps can help move past the complexities and relieve the burden on your IT teams. By taking this approach early on, you can avoid IT touchpoints and provide more agility and resilience as a part of the business goal.

While automation has been somewhat limited to on-premise data centers due to a hesitancy to perform automation on the complexity of cloud and multi-cloud resources, the utilization of AIOps can help alleviate the operational risks and concerns.

In *Chapter 14*, *Building Pipelines for API Connect*, we discussed automation for on-premise implementations but AIOps can certainly play a role.

Although the pandemic is continuing, the new normal has provided organizations clarity on how automation is a clear requirement. Extending that automation drumbeat into AIOps in the post-COVID era, you will see more adoption and capabilities going forward.

IBM, to bolster its capabilities with AIOps, has purchased Turbonomic. Turbonomic's platform uses AI to monitor and manage containers, virtual machines, databases, servers, and storage.

AI and Digital Transformation

The future of AI-Driven Cross-Cloud application operations using Turbonomics, an **Application Resource Management (ARM)**, and the use of Instana for **Application Performance Management (APM)** for automating the monitoring and management of their applications, Digital Transformation is now capitalizing on AI power to drive more innovation and efficiency.

So far, the discussion has been around Digital transformation and how it can be a focal point during the pandemic. As you know, Digital transformation is a focus on enabling better products, services, experience, and business cultural change. Moving applications and data to the cloud is but one method being portrayed. When it comes to data, new technology is leading to a shift of data and processing of data closer to the user or the *edge*. Edge computing will be discussed next.

Exploring 5G Edge computing

When discussing Digital Transformation, a majority of the descriptions describe moving toward cloud platforms to modernize and also take advantage of operational cost factors. When you consider cloud platforms, there are the predominant players that have huge capacities that correlate to economies of scale.

Visually, you might imagine the cloud as a hub and any integrations being the spokes. This is the classic hub and spoke model as shown in the following figure. So, if you are multi-cloud you may have multiple hubs. In the case of hybrid, your on-premise applications are part of the spokes if you are focusing on cloud deployments.

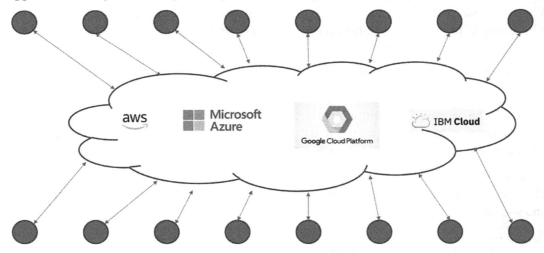

Figure 16.5 – Hub and spoke cloud

The pandemic accelerated the move to the cloud as a means to support working from home and expand the business outside the wall of private data centers. Now, if you were to look closely at the cloud models, you would find that although the cloud platforms have geographically disbursed cloud locations, there are consumers that are still considerable distances away from any of the hubs. Therefore, APIs that are connected to the hub are dependent on the closeness of hubs. If those APIs are based on mobile technology, then the need for speed is dependent on wireless performance and bandwidth. This is where 5G and its increased bandwidth is solving the problem.

The cloud models have been successful for many years and continue to be successful and are improving, but is there another extension to cloud computing on the horizon? Yes, there is. With the 5G rollout continuing, it's now called 5G Edge computing.

To provide a little more context on why this is important you should understand the advantage of Edge computing. You'll learn about that next.

Advantage of 5G Edge computing

You can categorize Edge computing as the practice of capturing, storing, processing, and analyzing data closer to the consumer, where the application data is created. With cloud platforms, this extension removes the cloud as the primary resource for all data processing and allows collection and processing at the Edge.

Some of the advantages that Edge computing provides are as follows:

- Data can be captured, analyzed, and processed near the client without relying on cloud provider integration.

- Compute power is now available on the edge thereby reducing compute costs in the cloud.

- Various edge type models such as **Internet of Things (IoT)**, manufacturing, financial, automotive, and Augmented Reality/Virtual Reality are greatly enhanced with 5G to support wide-area coverage, low latency, and cost-efficiency.

As you know, 5G is now being rolled out and with that, these new business models are being added to the digital transformation. 5G will improve performance and many Telcos are preparing to move more processing power to the edge. *Figure 16.6* provides a graphic on how Edge computing enables the capturing, storing, processing, and analyzing of data near the client, where the data is generated, instead of in a centralized data-processing warehouse:

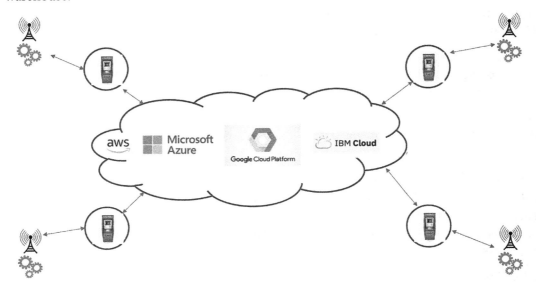

Figure 16.6 – Edge Computing

In *Figure 16.6*, the circled areas represent APIs running on various implementations that could be phones, kiosks, and other consumers. By building out an edge messaging integration, businesses can have a re-imagined infrastructure where they can take advantage of the speed of 5G, represented by the cell towers, with extremely low latency.

IBM is preparing for this journey too. With the purchase of Turbonomic, it will use Turbonomic network performance management tools for enterprise 5G deployments.

While Edge computing is still in its infancy, you can imagine the types of capabilities that can be accomplished using 5G and the edge. Just think, events are triggered from data sources and received by devices immediately, whether it's database updates, messages/ events, or secure streaming. All this can be possible as you move towards edge computing with 5G.

Not only did you learn about how AI can assist with digital transformation automation but also about the next extension of cloud technology, Edge computing. With Edge computing, you learned why it provides value above and beyond just using a cloud platform. It allows data to be closer to the edge where it impacts the user the most. As Telcos continue to upgrade their networks with 5G, more business models will appear that support improved AR/VR, IoT, financial and automotive industries, and others.

Summary

In this chapter, we discussed the impact of the COVID-19 pandemic and how it rewarded those who have achieved some digital transformation and motivated others to accelerate their efforts. With all change there is opportunity. You learned that as you move towards a hybrid cloud implementation, some considerations must be made about the platform and toolset you utilize to ensure you maintain the flexibility necessary to stay agile and cost-aware. You were introduced to OpenShift as a platform to address learning curves of existing resources as well as to provide a common platform that can support multi-cloud, on-premise, and portability. You also learned that OpenShift is more adaptable and provides more out-of-the-box features than DIY Kubernetes. You learned that DIY requires your organization to do the following:

- Build a full DIY Kubernetes stack

- Test your Kubernetes stack

- Stay abreast of Kubernetes changes

- Patch security CVEs (weekly)

- Troubleshoot Kubernetes stack problems

- Repair Kubernetes problems

Utilizing OpenShift helps organizations move toward modernization more comfortably.

You also learned about IBM's hybrid cloud enabling using Cloud Pak for Integration (CP4I). Integration is still key whether you are running on-premise or in one or many clouds. Digital transformation may be a mix of various technologies and SaaS applications and integrating that mix will take the proper set of tooling.

The combination of OpenShift and CP4I is proven technology that improves the success factors of digital transformation and modernization.

You also learned about how AI is taking IT operations to a new level of efficiency. With AIOps the vast amount of data collected in events and logs can be utilized to allow machine learning to improve operations. Challenges that hindered people-centric operations such as time to troubleshoot and the remediation of outages have been reduced and are more proactive than reactive.

Finally, you learned about late-breaking changes in API Connect and how the delivery of fixes and enhancements are managed.

You have obtained so much knowledge of API Connect over these chapters. Hopefully, you will take this knowledge and apply it to your current organization to make it a rich implementation on your digital transformation journey. Best of luck with your implementation and continued success.

Packt.com

Subscribe to our online digital library for full access to over 7,000 books and videos, as well as industry leading tools to help you plan your personal development and advance your career. For more information, please visit our website.

Why subscribe?

- Spend less time learning and more time coding with practical eBooks and Videos from over 4,000 industry professionals

- Improve your learning with Skill Plans built especially for you

- Get a free eBook or video every month

- Fully searchable for easy access to vital information

- Copy and paste, print, and bookmark content

Did you know that Packt offers eBook versions of every book published, with PDF and ePub files available? You can upgrade to the eBook version at packt.com and as a print book customer, you are entitled to a discount on the eBook copy. Get in touch with us at customercare@packtpub.com for more details.

At www.packt.com, you can also read a collection of free technical articles, sign up for a range of free newsletters, and receive exclusive discounts and offers on Packt books and eBooks.

Other Books You May Enjoy

If you enjoyed this book, you may be interested in these other books by Packt:

API Testing and Development with Postman

Dave Westerveld

ISBN: 9781800569201

- Find out what is involved in effective API testing
- Use data-driven testing in Postman to create scalable API tests
- Understand what a well-designed API looks like
- Become well-versed with API terminology, including the different types of APIs
- Get to grips with performing functional and non-functional testing of an API
- Discover how to use industry standards such as OpenAPI and mocking in Postman

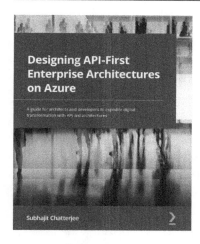

Designing API-First Enterprise Architectures on Azure

Subhajit Chatterjee

ISBN: 9781801813914

- Explore the benefits of API-led architecture in an enterprise
- Build highly reliable and resilient, cloud-based, API-centric solutions
- Plan technical initiatives based on Well-Architected Framework principles
- Get to grips with the productization and management of your API assets for value creation
- Design high-scale enterprise integration platforms on the Azure cloud
- Study the important principles and practices that apply to cloud-based API architectures

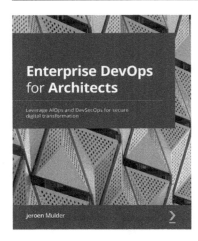

Enterprise DevOps for Architects

Jeroen Mulder

ISBN: 9781801812153

- Create DevOps architecture and integrate it with the enterprise architecture
- Discover how DevOps can add value to the quality of IT delivery
- Explore strategies to scale DevOps for an enterprise
- Architect SRE for an enterprise as next-level DevOps
- Understand AIOps and what value it can bring to an enterprise
- Create your AIOps architecture and integrate it into DevOps
- Create your DevSecOps architecture and integrate it with the existing DevOps setup
- Apply zero-trust principles and industry security frameworks to DevOps

Packt is searching for authors like you

If you're interested in becoming an author for Packt, please visit authors. packtpub.com and apply today. We have worked with thousands of developers and tech professionals, just like you, to help them share their insight with the global tech community. You can make a general application, apply for a specific hot topic that we are recruiting an author for, or submit your own idea.

Share Your Thoughts

Now you've finished *Digital Transformation and Modernization with IBM API Connect*, we'd love to hear your thoughts! Scan the QR code below to go straight to the Amazon review page for this book and share your feedback or leave a review on the site that you purchased it from.

https://packt.link/r/1801070792

Your review is important to us and the tech community and will help us make sure we're delivering excellent quality content.

Index

Symbols

5G Edge computing
  advantage 545
  exploring 544, 545
{n + 1} request problem 305

A

access code 247
access token 230
additional security measures
  adding 269, 270
Administrator 225
advanced transformations
  implementing, with
    GatewayScript 292, 297-301
  implementing, with XSLT 292-296
Agile integration
  references 9
analytics
  about 504
  associating, in Cloud Manager
    topology 505-507
  dashboards 510
  default dashboards 508

deploying, to environments 527
initial setup 505
viewing 508, 509
Analytics subsystem
  about 45
  associating, to gateway 46
Analytics tab, Catalog
  about 46
  time range, changing 47, 48
Ansible
  installing 468
  suggestions, for utilization 470
Ansible automation 467
Ansible playbook
  creating 468
  running 469
  successful execution 469
API and product lifecycle 374
API Blueprint
  URL 105
API Catalog
  about 74
  configuring 75-79
API Connect
  about 17, 28, 29, 154
  cloud topology 33

components 30-32
deploying options 29
digital transformation, enabling 20
goals of digital transformation,
    aligning 21, 22
high-level organization 69
implementation choices,
    modernizing 23, 24
late-breaking changes 533, 534
out-of-the-box security capabilities 223
pre-built transformation
    policies 272, 273
reference link, for deployment
    white paper 30
reference link, for installment
    requirements 29
sample pipeline 483
SOAP capabilities 156-158
system requirements 30
API Connect components
  Analytics 45-48
  API Manager 34, 35
  Cloud Manager 32-34
  Developer Portal service 38, 39
  gateway 41-45
API Connect Developer Portal 22
API Connect Gateway (apigw)
    service 44, 55
API Connect, IBM Documentation page
  reference link 533
API Connect toolkit
  download link 56
APIC security implementation
  Authentication URL user registry,
      configuring 225-227
  LDAP user registry,
      configuring 227, 228

OAuth providers, configuring 230-233
OpenID Connect (user
    registry) 228, 229
preparing for 224
user registry, creating 225
API development 107-110
API environment
  discoverability 406, 407
  naming standards 407, 408
  standards, applying to 406
API key 234
API-led architectural approach 15
API-led digital transformation 4
API lifecycle
  approvals 383-386
API Manager
  about 34, 35, 95
  Administrator 391
  API Administrator 391
  APIs and products, developing 36, 37
  Catalogs 38
  Community Manager 391
  Developer 391
  Member 391
  Owner 391
  roles 35, 389-392
  Viewer 391
API Manager user interface
  reference link 191
API operations, in Products
  Rate Limits, defining for 344, 345
API properties 117-119
API Proxy
  about 106
  creating 107
  prerequisites 107
  testing 114

APIs
about 106
creating 104
responsibilities 18
testing 112, 113
types 15, 16
API security
Basic authentication,
implementing 234-238
Client ID (API key),
implementing 234-238
enabling 111, 112
APIs, within Products
Rate Limits, defining for 340, 342
App Connect Enterprise (ACE) 539
Application Performance
Management (APM) 543
Application Resource Management
(ARM) 543
Artificial Intelligence for IT
Operation (AIOps)
about 542
benefits 543
assembly
Rate Limits, defining 346, 347
Assembly Function object 423
Authentication URL user registry
configuring 225-227
reference link 226
authorization code 230
authorization code grant type 231
Authorization Server 230
Authorization URL 233
Availability Zones (AZs)
setting up 33

B

backend for frontend (BFF) APIs 14
Basic authentication
implementing, in API security 234-238
Built-in Policies
reference link 120
business transformation 9

C

Catalog
about 38, 225
API, publishing to 359-361
configuring 356, 357
settings 357, 358
Catalog by Name option
configuring 350-355
catalog scoped user-defined policies
about 412
creating 413
Catch extension 145
CLI commands
about 56
using, for development process 58
CLI commands and parameters
reference link 351
Client 230
Client ID
about 223
implementing, in API security 234-238
Client Secret 223
CLIs
using 462, 463
cloud implementations
about 50, 51
IBM Cloud (reserved instance) 52
IBM CP4I 51, 52

Cloud Manager
    about 21, 32, 33
    roles 387-389
Cloud Pak for Integration (CP4I)
    about 24, 25, 29, 539
    API management 52
    application integration 52
    Asset Repository capability 541
    Common Services 541
    end-to-end security 52
    enterprise messaging 52
    event streaming 52
    features 541
    high-speed data transfer 52
    Platform Navigator component 541
    reference link 25
cloud topology
    in API Connect 33
command syntax, for count
        and burst limits
    reference link 356
completed .yaml file 418, 419
consumer interaction 362-371
consumer organizations
    configuring, at Catalog level 362-367
context variables
    about 115
    api 115
    client 115
    message 115
    reference link 115
    request 116
    system 116
Continuous Deployment (CD) 29, 534

D

dashboard
    creating 516-518
DataPower gateway
    about 41
    clustering 43-45
    Docker 42
    Linux 42
    Physical 42
    reference link 43
    Virtual 42
default dashboards, analytics
    about 510
    analytics hierarchy 511, 512
    search criteria, narrowing 515, 516
    searching, within analytics 513, 514
    visualization 511
deployment models
    about 48
    cloud implementations 50, 51
    HA 54
    Hybrid cloud 53, 54
    on-premises implementation 49
Designer
    about 95
    connecting, to LTE 102-104
    installing 96
Developer Portal
    about 85
    Administrator 393
    configuring 85-89
    Developer 393
    Owner 393
    roles 392
    Viewer 393

Developer Portal service
  about 38
  administrator view 499
  blogs and forums, removing 498-501
  customizing 494
  customizing, as administrator 498
  default view 495
  digital transformation portal
      API Advertising 39-41
  Drupal 9 capabilities, reviewing 495-498
  home page 39
  sub-theme, creating from API
      Connect default theme 502, 503
  user, adding with administrator role 501
  welcome banner, updating 504
development tools 95
digital framework
  business cases 13
  considerations 11
  customers 12
  performance 12
  processes 12
digital modernization
  about 4, 18
  approaches 19
digital transformation
  about 4, 18
  failure, avoiding 10
Drupal 7 38

E

Elastic Stack 504
Enterprise Service Buses (ESBs) 8
environments
  logical segregation 400-403
  physical segregation 404
  segregating 400
  working with 451-453
error conditions
  handling, on failed invoke 121-125
error handling 145-150
Evidence API
  updating, with operation switch 208-219
Extensible Stylesheet Transformation
    Language (XSLT)
  advanced transformations,
      implementing 292-296

F

failed invoke
  error conditions, handling on 121-125
Fast Healthcare Interoperability
    Resources (FHIR)
  about 53, 58
  API Manager, logging into 61
  draft FHIR API, creating 63, 64
  draft FHIR API, uploading 63, 64
  example target API Connect
      environment 59
  government-mandated FHIR
      interfaces 180
  IDPs, determining 60
  Product, creating with APIs 65
  Product, publishing to Catalog 65
  realms, determining 60
  resources 177
  target Catalog and Provider
      organization, setting for
      deployments 61, 62
  working with 58
FHIR API
  logic policies, applying 193, 194

FHIR architecture
  broker adapter for FHIR 178
  proprietary mixed APIs 178
  vendor-neutral clinical repository 178
FHIR resources
  about 177-179
  elements 177
  metadata 177
FHIR server 179, 180
fields
  refactoring 283, 284
  removing, from GraphQL 327-330

G

gateway, API Connect 41
Gateway by Name option
  configuring 355
GatewayScript
  advanced transformations,
    implementing 297-301
General-Purpose (GP)/Backend for
    Frontend (BFF) APIs 16
Generate Test button 444
Git
  capabilities 478, 479
  using, as SCM 477
global scoped user-defined policies
  about 412, 422
  packaging 430-433
  planning 422-426
  publishing 430-433
government-mandated FHIR interfaces
  about 180
  Patient Access 180
  Provider Directory 180

GraphQL
  considerations, on
    performance 322, 323
  costs, setting 321
  fields, removing from 327-330
  need for 304, 305
  URL 305
  warnings, addressing 319-321
  weights, refining 324-326
  weights, setting 321
GraphQL anatomy
  about 306
  mutation 308
  query 308
  resolver 308
  types 306-308
GraphQL API
  creating 311
GraphQL Express server
  installing 309, 310
GraphQL proxy
  adding 312-319
GraphQL server 311
GraphQL user interface 311
Groovy 481

H

HA
  about 54
  achieving, in API Connect 54
  setting up 33
HAPI FHIR Test server
  reference link 180
hard limit 339
Header Control property 127

HL7 organization 176
HL7 FHIR
  URL 178
HL7v2 176
hooks
  monitoring 471-477
  testing 471-477
HTTP Client 439
HTTP method
  modifying, in Invoke policy 126
HTTP request
  body 440
  headers 440
  HTTP method 440
  params 440
  Request url 440
  response 440
  send 440
Hybrid cloud
  about 53, 535
  modernization motivation 539, 540
  reference architecture 53
hybrid reference architecture 14

I

IBM Cloud Pak for Integration
  reference link 52
IBM Cloud (reserved instance) 52
IBM CP4I
  about 51
  capabilities 52
Identity Provider (IdP) 228
ID token 251
If and Switch logic policy
  applying 194-200
iFix 28

Infrastructure as a Service (IaaS) 15, 50
Inline Referencing technique 130
Input message type 161
interoperability 176
Invoke policy
  about 121
  HTTP method, modifying in 126
  properties, accessing 130
  variables, accessing 130

J

Jenkins
  about 480
  housekeeping tasks 483-485
Jenkinsfile
  about 481
  review 487-490
Jenkins pipeline
  constructing 479
JSON to XML policy
  applying 287-289
JSON Web Key (JWK)
  about 252, 253
  versus Crypto Object 254
JSON Web Token (JWT)
  about 257, 258
  generating 259-264
  policies, using 258
  verifying 264-269

K

Kubernetes
  versus OpenShift 537

L

LDAP user registry
  about 227
  configuring 228
Lightweight Directory Access
        Protocol (LDAP) 33
Local Testing Environment (LTE)
  about 31
  Designer, connecting to 102-104
  installation, verifying 101
  installing 97-100
  prerequisites 97
logic policies 119
logic policies, for FHIR API
  applying 193, 194
  If and Switch logic policy 194-199
  operation switch logic policy 200-208
Long-Term Support (LTS) 29, 534

M

Map policy
  using 273-282
marketing 6
Member API
  POST request 442
  test request 440
Message Transmission Optimization
        Mechanism (MTOM) 153
methods, for creating Jenkinsfile
  declarative method 482
  scripted method 481, 482
mutation 308

N

Node.js
  reference link 16

O

OAS 3.0 105
OAuth2 224
OAUTH 2.0
  about 239
  API, enabling with security
        definition 239-241
  applying 239
  client, creating 242-245
OAuth flow
  prerequisites, for testing 246
  testing 245
  three-legged flow, testing 247-250
OAuth flow changes
  about 255
  effect 251
  Fetch Access Token 256, 257
  Fetch Authorization Code 255, 256
OAuth provider changes 252-255
OAuth provider configuration changes
  effect 251
OAuth providers
  configuring 230-233
OAuth security
  implementation steps 239
on-premises implementation
  about 49, 50
  bare-metal Kubernetes 49
  OpenShift Container
        Platform (OCP) 49
  VMware ESX 49
OOTB security features 269

OpenAPI
  design 104
OpenAPI document 105
OpenAPI Specification (OAS) 104
OpenID Connect
  about 228
  creating 229
  implementing 251
  OAuth flow changes 255-257
  OAuth provider changes 252-255
OpenShift
  about 535, 536
  on multiple clouds 538
  on-premise 538
  reference link 50
  versus Kubernetes 537
OpenShift Container Platform (OCP) 23
operation switch logic policy
  applying 200-208
  Evidence API, updating with 208-219
Organization Owner 225
Out-of-the-Box (OOTB) Policies 104
out-of-the-box security capabilities, APIC
  API key 223
  basic authentication 223
  OAuth 223
Output message type 161
over-fetching 305

P

Parameter Control property 127
parameters
  header type 130
  path type 130
  query type 130
passion 8

Payment Card Industry (PCI)
  security compliance 33
Payment Card Industry (PCI)
  standards 283
people 8
Pipeline item
  building 485-487
pipelines
  about 461
  stages 461, 462
place 7
Plan by name option
  configuring 348, 349
Plans
  configuring 334-338
  Rate Limits, defining 340
  working with 332, 333
Platform-API interface 58
Platform APIs
  execution 464, 465
  using 463
Platform as a Service (PaaS) 15, 50
playbooks 467
policies
  about 115, 119
  logic constructs 120
portal API Advertising
  using, for digital transformation 39-41
POST Operation 161
pre-built transformation policies,
    API Connect 272, 273
prescription benefit management
    (PBM) 80
price 6
process 8
Process/Interactive APIs 16
product 6

product lifecycle
  deleted 376
  deprecated 376
  draft 376
  example 376-382
  published 376
  retired 376
  staged 376
  stages 375
Products
  configuring 334-338
  subscribing 368-371
  working with 332, 333
promotion 7
Proof of Concepts (PoCs) 542
Provider Organization
  about 70
  configuring 70-74
publish Platform API
  calling 466, 467

Q

quorum 43

R

Rate Limits
  creating 339
  defining, for APIs within
    Product 340-342
  defining, for select APIs
    within Product 343
  defining, for specific API operations
    in Product 344, 345
  defining, in assembly 346, 347
  defining, in Plan 340
  setting 321, 326, 327

realms 60
redaction policy 286
replication fetching 305
resolver 308
Resource Owner 230
Resources 230
REST API proxy
  example 131-134
RESTFul FHIR API
  creating 181
  working with 182-193
RESTful Service Description
    Language (RAML)
  URL 105
REST proxy
  configuration, reviewing 168-171
  creating 166, 167
  testing 171-173
roles 386
roles, API Manager
  administrator 35
  API administrator 35
  community manager 35
  developer 36
  member 36
  organization owner 35
  reference link 36
  viewer 36

S

Schema Definition Language (SDL) 306
SCM
  GIT, using as 477
Security Definition 223
Security Enforcement 105, 224
security policies 119
Security Scheme 105

select APIs, within Products
  Rate Limits, defining for 343
service chaining
  with double Invokes 140-144
Service-Level Agreements (SLAs) 43, 121
Service Oriented Architect (SOA) 8
Simple Mail Transport Protocol
      (SMTP) server 33
Simple Object Access Protocol (SOAP)
  about 153
  capabilities, in APIC 156-158
slaves 480
SOAP API Proxy
  testing 164-166
SOAP proxy
  reviewing 161-163
  testing 164-166
SOAP service
  testing 164
SoapUI 155
soft limit 339
Software as a Service (SaaS) 15
Software Development Lifecycle
      (SDLC) 461
source to image (S2I) 537
spaces
  about 68
  configuring 81-84
  utilizing 80, 81
split-brain scenario 54
stages, pipelines
  build 461
  deploy 462
  pull 461
  release 462
  unit test 461

Switch policy
  building 135-139
System APIs 16

T

Tekton 479
test cases
  monitoring 454-456
Test Suite
  creating 443, 444
three-legged flow
  about 245, 247
  testing 247-250
Throw Policy 145
Token URL 233
toolkit 95
transforms policies 119
Transport Layer Security (TLS) 269

U

UDP configuration file
  creating 426-429
UDP .yaml file
  assembly section 416
  attach section 415
  building 414
  gateways section 415
  information section 414
  properties section 415
  specification version 414
under-fetching 140, 305
Uniform Resource Locator (URL) 33
unit tests
  configuring 438-450

Universal Description, Discovery
    and Integration (UDDI) 153
user-defined policies
    about 119
    catalog scoped user-defined policies 412
    global scoped user-defined policies 412
    implementing 421, 422
    installing 419, 420
UX APIs 16

V

variables
    about 115
    context variables 115, 116
    using 115
Version 5 Compatibility (v5c) gateway 44
version numbering scheme 395, 396
versions
    managing 395
version scheme
    for healthcare organization 397-399
visualization
    creating 519-526
    saving 526
VRMF - Version, Release, Modification,
    Fix pack strategy 28

W

Web Services APIs 8
web user experience (Web UX) 537
WS-Attachments 153
WS-Policy 153
WS-Security 153
WS-Trust 153

X

XML to JSON policy
    applying 287, 290-292